Lecture Notes in Artificial Intelligence   8399

Subseries of Lecture Notes in Computer Science

LNAI Series Editors

Randy Goebel
University of Alberta, Edmonton, Canada
Yuzuru Tanaka
Hokkaido University, Sapporo, Japan
Wolfgang Wahlster
DFKI and Saarland University, Saarbrücken, Germany

LNAI Founding Series Editor

Joerg Siekmann
DFKI and Saarland University, Saarbrücken, Germany

For further volumes:
http://www.springer.com/series/1244

Annalisa Appice · Michelangelo Ceci
Corrado Loglisci · Giuseppe Manco
Elio Masciari · Zbigniew W. Ras (Eds.)

New Frontiers in Mining Complex Patterns

Second International Workshop, NFMCP 2013
Held in Conjunction with ECML-PKDD 2013
Prague, Czech Republic, September 27, 2013
Revised Selected Papers

 Springer

Editors
Annalisa Appice
Michelangelo Ceci
Corrado Loglisci
Department of Computer Science
Università degli Studi di Bari Aldo Moro
Bari
Italy

Giuseppe Manco
Elio Masciari
ICAR
CNR
Rende
Italy

Zbigniew W. Ras
Department of Computer Science
University of North Carolina
Charlotte, NC
USA

ISSN 0302-9743   ISSN 1611-3349 (electronic)
ISBN 978-3-319-08406-0   ISBN 978-3-319-08407-7 (eBook)
DOI 10.1007/978-3-319-08407-7
Springer Cham Heidelberg New York Dordrecht London

Library of Congress Control Number: 2014942657

Printed on acid-free paper

Springer is part of Springer Science+Business Media (www.springer.com)

Preface

New Frontiers in Mining Complex Patterns (NFMCP 2013)

Modern automatic systems are able to collect huge volumes of data, often with a complex structure (e.g. multi-table data, XML data, web data, time series and sequences, graphs and trees). This fact poses new challenges for current information systems with respect to storing, managing, and mining these sets of complex data. The Second International Workshop on New Frontiers in Mining Complex Patterns (NFMCP 2013) was held in Prague in conjunction with the European Conference on Machine Learning and Principles and Practice of Knowledge Discovery in Databases (ECML-PKDD 2013) on September 27, 2013. It was aimed to bring together researchers and practitioners of data mining who are interested in the advances and latest developments in the area of extracting patterns from complex data sources like blogs, event or log data, medical data, spatio-temporal data, social networks, mobility data, sensor data and streams, and so on.

This book features a collection of revised and significantly extended versions of papers accepted for presentation at the workshop. These papers went through a rigorous review process to ensure their high quality and to have them comply with Springer-Verlag publication standards. The individual contributions of this book illustrate advanced data mining techniques which preserve the informative richness of complex data and allow efficient and effective identification of complex information units present in such data.

The book is composed of four parts and a total of sixteen chapters.

Part I gives a view of **Data Streams and Time Series Analysis** by illustrating some complex situations involving temporal and spatio-temporal data. It consists of five chapters. Chapter 1 describes data driven parameter estimation measures for mining flock patterns with a validation procedure to measure the quality of these extracted patterns. Chapter 2 presents a simple yet effective and parameter-free feature construction process for time series classification. Chapter 3 proposes a learning algorithm, combining conditional log-likelihood with Bayesian parameter estimation, designed for analyzing multivariate streaming data. Chapter 4 investigates the problem of mining frequent trajectories by resorting to frequent sequential pattern mining. Chapter 5 details a process mining approach that uses predictive clustering to equip an execution scenario with a prediction model.

Part II analyses issues posed by **Classification, Clustering, and Pattern Discovery** in presence of complex data. It consists of six chapters. Chapter 6 describes a novel methodology that combines principal component analysis and support vector machines in order to detect emotions from speech signals. Chapter 7 illustrates a sequential pattern mining algorithm that allows us to discover lengthy noise-tolerant sequential patterns over item-indexable databases. Chapter 8 studies the problem of efficiently mining frequent partite episodes that satisfy partwise constraints from an input event sequence. Chapter 9 proposes an estimation algorithm for the copula of a continuous multivariate distribution. Chapter 10 focuses on extending the ReliefF algorithm for regression, in order to address the task of hierarchical multi-label classification (HMC). Chapter 11 addresses the task of learning both global and local models for predicting structured outputs.

Part III presents technologies and applications where complex patterns are discovered from **Graphs, Networks, and Relational Data**. It contains three chapters. Chapter 12 describes a generative model for random graphs with discrete labels and weighted edges. Chapter 13 proposes a pre-processing strategy to simplify Semantic Similarity Networks based on a hybrid global-local thresholding approach based on spectral graph theory. Chapter 14 investigates the translation of the complex data represented by natural language text to complex (relational) patterns that represent the writing style of an author.

Finally, Part IV gives a general overview of **Machine Learning and Music data**. It contains two chapters. Chapter 15 explores a set of personalized emotion classifiers that are learned using feature data extracted from audio and tagged with a set of emotions by volunteers. Chapter 16 addresses the problem of using random forests, in order to identify multiple musical instruments in polyphonic recordings of classical music.

We would like to thank all the authors who submitted papers for publication in this book and all the workshop participants and speakers. We are also grateful to the members of the Program Committee and external referees for their excellent work in reviewing submitted and revised contributions with expertise and patience. We would like to thank João Gama for his invited talk on "Evolving Data, Evolving Models". Special thanks is due to both the ECML PKDD workshop chairs and to the members of ECML PKDD organizers who made the event possible. We would like to acknowledge the support of the European Commission through the project MAES-TRA - Learning from Massive, Incompletely annotated, and Structured Data (Grant number ICT-2013-612944). Last but not the least, we thank Alfred Hofmann of Springer for his continuous support.

January 2014

Annalisa Appice
Michelangelo Ceci
Corrado Loglisci
Giuseppe Manco
Elio Masciari
Zbigniew W. Ras

Organization

Program Chairs

Annalisa Appice   University of Bari, Italy
Michelangelo Ceci   University of Bari, Italy
Corrado Loglisci   University of Bari, Italy
Giuseppe Manco   ICAR-CNR, Italy
Elio Masciari   ICAR-CNR, Italy
Zbigniew W. Ras   University of North Carolina at Charlotte, USA
                             & Warsaw University of Technology, Poland

Program Committee

Nicola Barbieri   Yahoo Research, Spain
Petr Berka   University of Economics Prague, Czech Republic
Sašo Džeroski   Jozef Stefan Institute, Slovenia
Floriana Esposito   University of Bari, Italy
Dimitrios Gunopulos   University of Athens, Greece
Mohand-Saïd Hacid   University Claude Bernard Lyon 1, France
Dino Ienco   IRSTEA Montpellier, UMR TETIS, France
Kristian Kersting   Fraunhofer IAIS, Germany
Arno Knobbe   University of Leiden, The Netherlands
Stan Matwin   University of Ottawa, Canada
Dino Pedreschi   University of Pisa, Italy
Jean-Marc Petit   INSA-Lyon, LIRIS, France
Fabrizio Riguzzi   University of Ferrara, Italy
Henryk Rybiński   Warsaw University of Technology, Poland
Eirini Spyropoulou   University of Bristol, UK
Jerzy Stefanowski   Poznan University of Technology, Poland
Maguelonne Teisseire   IRSTEA Montpellier, UMR TETIS, France
Herna Viktor   University of Ottawa, Canada
Alicja Wieczorkowska   Polish-Japanese Institute of IT, Poland
Włodek Zadrożny   IBM Watson Research Center, USA
Djamel Zighed   Université Lumière Lyon 2, France

Additional Reviewers

Gianni Costa   Hai Phan Nhat
Stefano Ferilli   Matteo Riondato
Massimo Guarascio   Riccardo Ortale
Ayman Hajja

Evolving Data, Evolving Models
(Invited Talk)

João Gama

LIAAD-INESC TEC and Faculty of Economics, University of Porto, Porto, Portugal

Abstract. In recent years we witnessed an impressive advance in the social networks Field, which became a "hot" topic and a focus of considerable attention. The development of methods that focus on the analysis and understanding of the evolution of data are gaining momentum. The need for describing and understanding the behavior of a given phenomenon over time led to the emergence of new frameworks and methods focused in temporal evolution of data and models. In this talk we discuss the research opportunities opened in analysing evolving data and present examples from mining the evolution of clusters and communities in social networks.

Contents

Graphs, Networks and Relational Data

Machine Learning and Music Data

Data Streams
and Time Series Analysis

Parameter Estimation and Pattern Validation in Flock Mining

Rebecca Ong[1], Mirco Nanni[1], Chiara Renso[1(✉)], Monica Wachowicz[3], and Dino Pedreschi[2]

[1] KDDLab, ISTI-CNR, Pisa, Italy
chiara.renso@isti.chr.it
[2] KDDLab, University of Pisa, Pisa, Italy
[3] University of New Brunswick, Fredericton, Canada

Abstract. Due to the diffusion of location-aware devices and location-based services, it is now possible to analyse the digital trajectories of human mobility through the use of mining algorithms. However, in most cases, these algorithms come with little support for the analyst to actually use them in real world applications. In particular, means for understanding how to choose the proper parameters are missing. This work improves the state-of-the-art of mobility data analysis by providing an experimental study on the use of data-driven parameter estimation measures for mining flock patterns along with a validation procedure to measure the quality of these extracted patterns. Experiments were conducted on two real world datasets, one dealing with pedestrian movements in a recreational park and the other with car movements in a coastal area. The study has shown promising results for estimating suitable values for parameters for flock patterns as well as defining meaningful quantitative measures for assessing the quality of extracted flock patterns. It has also provided a sound basis to envisage a formal framework for parameter evaluation and pattern validation in the near future, since the advent of more complex pattern algorithms will require the use of a larger number of parameters.

1 Introduction

The increasing availability of data pertaining to the movements of people and vehicles, such as GPS trajectories and mobile phone call records, has fostered in recent years a large body of research on the analysis and mining of these data, to the purpose of discovering the patterns and models of human mobility. Examples along this line include [3,4], which highlight the broad diversity of mobility patterns. A few authors concentrated on the problem of characterising and detecting *flocks*, i.e., patterns describing a set of objects that stay closely together during an interval of time, either moving together along the same route (a *moving flock*), or staying together in a specific location (a *stationary flock*) [5,8,10].

A. Appice et al. (Eds.): NFMCP 2013, LNAI 8399, pp. 3–17, 2014.
DOI: 10.1007/978-3-319-08407-7_1, © Springer International Publishing Switzerland 2014

In this paper, we follow the definition where a flock is a group of at least k objects that, observed during a time interval ΔT with a sampling rate R, remain spatially close to each other within a distance ϵ. While this definition and other variations in literature are useful for detecting flocks, it is apparent that setting such parameters $(k, \epsilon, \Delta T, R)$ makes it complex for an analyst to use flock mining in different contexts. These parameters clearly depend on the data under analysis, and may vary greatly in different settings. We observed remarkable differences in datasets pertaining to pedestrian and car movements, which are the two trajectory datasets used in this paper. Such differences can be expected as well when observing other types of moving objects, e.g., bird trajectories. Despite this complexity, no prior work addressed the problem of parameter setting, which is a barrier especially for mobility experts who would like to use flock mining as a black box. To this aim, we address the *parameter estimation* problem of finding appropriate values for the parameters using a systematic data-driven method, based on the trajectory dataset that is being analysed. This paper provides an empirical evaluation of the effects of parameters in two different moving objects datasets. This is an initial step towards delineating a data-driven parameter estimation method for flock mining.

The structure of this paper is as follows: Sect. 2 presents some related approaches in the literature. Section 3 describes the experiments performed on two datasets in order to study the effect of the different parameters. Section 4 proposes and tests several measures for validating the set of flocks found by the flock extraction algorithm. Finally, Sect. 5 sums up the conclusions derived from the study.

2   Related Works

Although the problem of finding realistic parameter values in data mining is well recognized in literature, very few papers have addressed this problem. Paper [2] is a well known work that proposes a solution for parameter estimation, and has inspired our approach. In this work, the authors propose a heuristic technique for determining the parameter values of the density-based clustering algorithm DBSCAN: *Eps* (radius) and *MinCard* (minimum cardinality of clusters). A function called *k-distance* is defined to compute the distance between each object and its k-th nearest neighbor. These values are then plotted with objects ordered in descending order, in the so-called *sorted k-distance plot*. This plot is then used to find an estimation of the *Eps* parameter, basically corresponding to the point where a sudden increase in the k-distance occurs. The objects plotted to the left of the threshold will be considered as noise while other objects will be assigned to some cluster.

Another work on parameter estimation related to flocks is found in [7] where the authors propose a set of algorithms for detecting convoys in trajectory datasets. They proposed a guideline for determining the parameters δ and λ of the Douglas-Peucker (DP) trajectory simplification algorithm. The DP algorithm uses δ as a tolerance value for constraining the spatial distance between the original and the simplified points. The algorithm uses another additional parameter λ, which refers to the length of the time partitions. The determination of a good value for δ has the goal of finding a trade-off value giving a good simplification of original trajectories while maintaining a tight enough distance. For finding a good value for δ authors propose to run the DP algorithm with δ set to 0. They consider the actual tolerance values at each simplification step and find the values with the largest difference with their neighbour before averaging them to obtain the final parameter value. Meanwhile, a good λ is computed by taking the average probability of each object having an intermediate simplified point that is not found in other trajectories. However, the parameter estimation techniques were applied to the preprocessing step of the convoy algorithm rather than applying it directly to the parameters related to the flock or convoy definition.

3   Flock Algorithm and Parameter Estimation

Our study for parameter estimation has been designed with reference to the flock algorithm introduced in [10]. The algorithm finds moving flocks, each of which is a group of objects consisting of at least *min_points* members that are spatially close together while moving over a minimum time duration *min_time_slices*. The algorithm requires four user-defined parameters: *synchronisation_rate(R)* - refers to the rate, specified in seconds, at which observation points (e.g., GPS recordings) are sampled for each moving object; *min_time_slices(ΔT)* - is the minimum number of consecutive times slices for which the objects remain spatially close; *min_points(κ)* - is the minimum number of objects in a moving flock; *radius(ϵ)* - defines the spatial closeness of a group of moving objects at a specific time instance.

We start with a description of the datasets used for our empirical evaluations and an overview of some flock quality measures, since these are necessary to understand the impact of the parameters on the results. These are followed by a discussion of the effect of each parameter, before closing with an approach on finding a suitable *radius* value.

3.1   Context Awareness and Flock Cohesion Distance

We performed the study on two datasets that have two entirely different settings of two different types of moving objects. The first dataset, called DNP, contains 370 trajectories, one for each visitor and consisting of a total of 141,826 sample points. These were recorded using GPS devices given to the visitors of a natural park at the parking lots where they have started their visits. Due to the sparsity of this dataset, we have combined the data in different days into one day. The second dataset, called OctoPisa, contains the trajectories of $\approx$40,000 cars for a total of $\approx$1,500,000 travels covering a temporal span of 5 weeks in a coastal area of Tuscany around the city of Pisa. From this large dataset, we concentrated on a subset of trajectories occurring on June, 29, 2010 in order to be able to perform a more detailed study on a specific time period. This is one of the days with the highest number of moving cars. It contains 28,032 trajectories (corresponding to 557,201 observed GPS points) of 10,566 vehicles.

The initial set of parameter values that we used for the DNP dataset is as follows: $R = 5$ min., $\Delta T = 3$, $\kappa = 3$, and $\epsilon = 150$ m. Meanwhile, the initial set used for the OctoPisa dataset is as follows: $R = 1$ min., $\Delta T = 3$, $\kappa = 3$, and $\epsilon = 150$ m. For both datasets, we maximized R to a value that does not cause large distortion (from the domain expert's perspective) among the input trajectories. We selected a value of 3 for both ΔT and κ since using 2 is too small while 4 is quite large for finding a good number of flocks. Then, using the values for R, ΔT, κ, and the type of entity (i.e., pedestrian and car) in consideration, we derived a feasible and logical value for ϵ. In observing the effects of the individual parameters, we only modify the value of the parameter in consideration and retain the initial values for the rest.

A first step in parameter estimation is to understand how the parameters influence the results obtained by the flock extraction algorithm. In doing so, it is important to have an objective measure of this influence in order to understand whether decreasing or increasing the parameter values improves or worsens the quality of discovered flocks.

In our study, we used three measures, which are extensions of measures used for cluster evaluation. These measures include cohesion, separation and silhouette coefficient.

Flock cohesion distance is a measure of spatial closeness among members of a discovered flock. It can be computed using Eq. 1, which evaluates a specific flock F_i by computing the distance between each flock member m_j with the base $m_{i(b)}$ (the central object of the flock, as returned by the extraction algorithm [10]). $|F_i|$ is the number of F_i's members.

$$flock\ coh(F_i) = \frac{\sum_{\substack{m_j \in F_i \\ m_j \neq m_{i(b)}}} prox_{intra}(m_j, m_{i(b)})}{|F_i| - 1} \tag{1}$$

The $prox_{intra}$ between a flock member m_j and the flock base $m_{i(b)}$ can be computed by averaging the Euclidean distance among (x, y) points that were sampled simultaneously as described in Eq. 2. T refers to the flocking duration and it consists of a set of sampled time instances. x_j^t and y_j^t refer to the x, y components of member j at time instance t. $x_{i(b)}^t$ and $y_{i(b)}^t$ are the x and y components of flock i's base.

$$prox_{intra}(m_j, m_{i(b)}) = \frac{\sum_{t \in T} euclDist((x_j^t, y_j^t), (x_{i(b)}^t, y_{i(b)}^t))}{|T|} \tag{2}$$

The overall flock cohesion distance of a set F of flocks can be computed by averaging the flock cohesion distance scores for each flock in F, as shown in Eq. 3. Naturally, a flock with a low cohesion distance score is considered as a high quality flock.

$$overall\ flock\ cohesion\ distance(F) = \frac{\sum_{F_i \in F} flock\ coh(F_i)}{|F|} \tag{3}$$

Complementary, **flock separation** is a measure of spatial or spatio-temporal detachment of a flock from the rest and it can be computed using Eq. 4.

$$flock\ sep(F_i) = \sum_{\substack{F_j \in F \\ i \neq j}} prox_{inter}(m_{b(i)}, m_{b(j)}) \qquad (4)$$

While $prox_{intra}$ measures the distance among members of a flock, $prox_{inter}$ measures the distance among different flocks. The distance between a pair of flocks is computed by computing the distance between their respective bases. both spatial and temporal dimensions ($prox_{inter(XYT)}$) and one that considers only the spatial similarity, i.e. the route followed ($prox_{inter(routeSim)}$). Choosing between these two depends on the similarity level that the user is interested in.

$prox_{inter(XYT)}$ computes the spatial distance among the portion of the base trajectories that overlap in time as was done for $prox_{intra}$. The remaining portion that does not overlap incurs a penalty $pnlty$, which is the maximum possible distance obtained from the overlapping portion plus an arbitrary value. In the case that the bases being compared are disjoint, a maximum penalty score is incurred. More specifically, Eq. 5 describes how $prox_{inter(XYT)}$ is computed. $nonOverlapLTI$ refers to the number of un-matched time instances in the longer trajectory, $euclDistOvlp$ is the sum of Euclidean distances among pairs of points (from each base) that overlap in time, and $maxTD$ refers to the length of the longer base trajectory in terms of the number of time instances.

$$prox_{inter(XYT)} = \frac{pnlty * (nonOverLTI) + euclDistOvlp}{maxTD * (|F_i| - 1)} \qquad (5)$$

For $prox_{inter(routeSim)}$, we adapted an existing algorithm for computing the route similarity distance. The algorithm ignores the temporal component of the bases and computes the distance in terms of the spatial components by comparing the shape of the trajectories.

As with the overall flock cohesion of an obtained flock result, the overall flock separation can be computed by averaging individual flock separation scores.

Finally, the **flock silhouette** coefficient is a combination of the previously discussed measures as shown in Eq. 6. Note that computed scores can range from -1 (large intraflock distances and small interflock distances) to 1 (small intraflock distances and large interflock distances).

$$flock\ Sil(F_i) - \frac{flock\ sep(F_i) - flock\ coh(F_i)}{max\{flock\ sep(F_i), flock\ coh(F_i)\}} \qquad (6)$$

As with overall flock cohesion and separation, the overall silhouette coefficient of a flock result can be computed by the averaging silhouette score of each flock.

3.2 Observing the Effect of Varying the Parameters

This part discusses the observations derived from investigating the effect of different parameter values on the obtained flock results for the DNP and the OctoPisa datasets. The following subsections provides a discussion of the individual effect of each parameter.

Effect of *synchronisation_rate* **(R).** Out of the 4 parameters of the algorithm, the *synchronisation_rate* can affect the quality of the input dataset. More specifically, a very large value of R can distort the input trajectories whereas a very small value requires a longer processing time.

We have observed how dataset cohesion changes for different values of R. Dataset cohesion describes how each individual trajectory is cohesive with respect to the rest of trajectories in the dataset. To compute this value, we applied the flock separation measure and treated each trajectory as a base trajectory.

Experiments demonstrate that the synchronisation step can indeed modify the input and its cohesiveness but at the same time, the variation is small in the two datasets for smaller values of R. Using XYT cohesion, the largest difference between the smallest cohesion score compared to the other scores obtained using larger R is 371.67 m when $R = 11$ min. in the DNP dataset. Meanwhile, the largest difference is 1987.77 m using route similarity cohesion when $R = 15$ min. in the DNP dataset. For the Octopisa dataset, the largest difference is 480.28 m and 10789.69 m using XYT and route similarity cohesion, respectively. The route similarity cohesion score varied more compared to the XYT cohesion score. This plot thus suggests the use of smaller values of R.

We now present the effect of the *synchronisation_rate* on the discovered flocks themselves. Figure 1 illustrates how different values of R can affect the flock results by observing the change in the number of moving flocks discovered, the overall flock cohesion, the overall flock separation (based on the spatio-temporal coordinates XYT and route similarity), and the overall silhouette coefficient (XYT and route similarity based).

In general, fewer flocks are found as R increases. Furthermore, the overall flock cohesion varies slightly for different R values, while the overall flock separation tend to decrease with increasing R values. In the case of $R = 8$ min. in Octopisa and $R = 13$ min. in DNP, the discovered flocks becomes 0 and hence, the cohesion and separation scores are no longer applicable. Considering the plots for the XYT and route similarity separation scores, it is advisable to set R to a value less than 6 min. in DNP and a value less than 4 min. in Octopisa due to the sudden drop in the separation scores. A sudden drop occurs when no

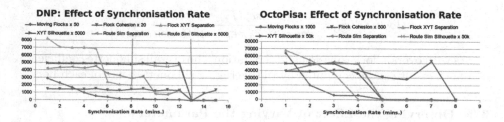

Fig. 1. Effect of the *synchronisation_rate* parameter for the two datasets.

flock or only a single flock is discovered, or when the distance among the discovered flocks is small. Very large XYT separation scores indicates that the flocks are temporally disjoint. On the other hand, very large route similarity scores indicates that different flocks are following different routes. Finally, the silhouette coefficients summarize the effect of R on both the cohesion and separation scores. The silhouette coefficients are generally close to 1 (i.e., ideal case), except for cases wherein the silhouette coefficient is 0. These cases refer to instances wherein only a single flock or no flock was found, making the silhouette coefficient inapplicable.

Effect of the *min_time_slices* (ΔT) Parameter. *min_time_slices* and *synchronisation_rate* are parameters that are both related to time. Since the plots and observations for these parameters are generally similar, we no longer present the plots for *min_time_slices*.

As observed with *synchronisation_rate*, an increasing value of ΔT results in fewer number of discovered moving flocks, and, generally, lowering the XYT and route similarity separation scores. The XYT- and route similarity-based silhouette coefficients are either close to 1 when more than a single flock is found, or 0, otherwise. Based on the experiments, a value of 2 or 3 time slices is ideal for the DNP dataset since there is a large drop in the XYT separation score when $\Delta T = 4$. Same is true for the OctoPisa dataset.

Effect of the *min_points* (κ) Parameter. *min_points* refers to the minimum objects that should consist a flock. Our experiments show that its selection is not critical, and it is also relatively easy to choose based on the objectives of the analysis: it is a tradeoff between having more flocks with fewer members (low threshold value), or having few flocks with more members.

Effect of the *radius* (ϵ) Parameter. Lastly, we have also observed the effect of the *radius* parameter, which defines the spatial closeness among flock members. As observed in Fig. 2, the number of flocks generally increases as ϵ increases. Meanwhile, the flock cohesion degrades (i.e., intra-distance increases) as ϵ increases. The XYT flock separation score tends to improve as ϵ increases when excluding the cases wherein the discovered flocks do not overlap in time (i.e., maximum XYT separation score is obtained) or no flocks were found. Meanwhile, the route similarity separation score generally improves as larger values of ϵ are used. As with previously observed parameters, the silhouette coefficients for varying ϵ remains close to 1.

The effect of *radius* as compared with the effect of the other parameters is as follows:

1. Out of all the scores used in assessing the effects of the parameters, the number of moving flocks has been the most sensitive. Generally, its value is directly proportional to the value of ϵ while it is inversely proportional to the other parameters. It is also worth noting that a higher number of moving flocks does not necessarily mean that the obtained flock results are better.
2. Compared to other flock validity measures, we consider flock cohesion as most important since it is in harmony with and explicit in the definition of a

flock (i.e., a flock consists of members that are spatially close together over a specific time duration). While the flock cohesion score linearly increases (i.e., flock cohesion degrades) as the ϵ increases, the cohesion score did not change as much with respect to changing values of the other parameters. Thus, we can conclude that the *radius* has a larger impact on the obtained flock results compared to the other parameters. Excluding the cases wherein no flocks are discovered (i.e., the flock separation score is irrelevant) and the cases wherein there is no overlap in time among the discovered flocks (i.e., the XYT separation score is set to the maximum), higher ϵ generally improves the separation scores whereas higher values for the other parameters generally degrades the separation scores.

3. As a final point, the silhouette coefficient scores obtained by varying different parameters for both datasets were consistently close to 1, except for cases where less than 2 flocks were found.

Based on these experiments, we conclude that (1) The selection of minimum number of points is relatively straightforward; (2) the most crucial parameter is the *radius*, since it exhibited a larger effect on the flock cohesion score compared to the other parameters; lastly, (3) while *radius* is the most crucial parameter, it is still important to choose good values for the other parameters since they still affect the quality of the discovered flocks.

3.3   Finding a Suitable *radius* Value

The results show that *radius* is a crucial parameter of the flock algorithm and a method for finding a good value for it can provide great support for mobility data analysts. In this section we propose the following technique, which is an extension of the technique introduced for the *Eps* parameter of DBSCAN. Since DBSCAN deals with single n-dimensional data points while the flock algorithm deals with 3D data points (spatial component plus time) that are connected through object IDs, adjustments to their technique are necessary to accommodate the points linked by the same object IDs. The general idea of the extended technique is to compute the k-th distance among objects that co-occur in the same time instant where k is $min\_points - 1$ and k-th distance refers to the distance of a point from its k-th nearest neighbour. Once the k-th distances have been computed for

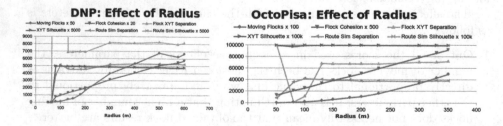

Fig. 2. Effect of the *radius* parameter on flock quality measures in the two datasets.

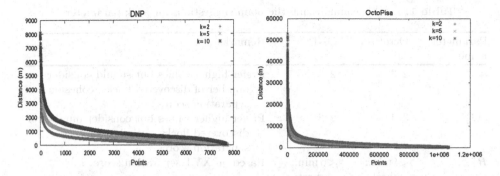

Fig. 3. Plot for k-th nearest neighbours for selected k's in the DNP (left) and the OctoPisa (right) datasets.

each point, they are sorted in non-ascending order and plotted as a line graph. The portion in which there is a sudden decrease in the k-th distance suggests an upper bound for the *radius* parameter of the algorithm. Figure 3 reports the plots obtained over the DNP (left) and the OctoPisa (right) datasets.

Using the left part of Fig. 3 for the DNP dataset, a suggested *radius* value should be below the 500 m–2000 m range for flocks with at least 3 members (i.e., $k = 2$). The obtained plot for the OctoPisa dataset is shown on the right part of Fig. 3. It suggests 3000 m–4000 m as an upper bound for the *radius*. This is reasonable since the OctoPisa dataset covers a wider spatial area (about 4600 Km^2 vs about 48 Km^2 of DNP).

It is also worth noting that the plots suggest different *radius* values for varying k's and yet, the division between the objects that would be included in some flock and those that are considered as noise is almost the same. Combining the suggested upper bound with contextual knowledge and the observation on the effect of *radius*, we recommend that a good range of values for *radius* is between 80 m to 300 m for DNP and between 50 m to 300 m for OctoPisa. Table 1 summarizes the main recommendations for a good range of parameter values for the two datasets.

4   Flock Pattern Validation

To assess the effectiveness of the algorithm and of the flocking results, validation must be performed. This provides users with confidence in using the algorithm and trusting its results. Furthermore, it also guides them on whether the obtained results using specific parameter values are accurate enough. In the case that the results are not accurate, the user may modify the initial parameter settings accordingly in order to obtain better results.

Flock validation is a challenging problem since the optimal flocking results are usually unknown, except for a few rare cases. The aim is to compute the goodness of a flocking algorithm and its flocking results in terms of a quantitative measure despite of not having a standard algorithm and/or result to compare it with.

Table 1. A table summarizing the main suggestions for flock parameters

Parameter name	OctoPisa	DNP	Remarks
κ	2–3	2–3	Prefer higher values but should consider number of discovered flocks, cohesion and separation scores
ΔT	3–4	2–3	Prefer higher values but consider number of discovered flocks, cohesion and separation scores
R	<4 min.; best: 1–2	<6 min.; best: 1 & 4	Based on XYT separation score
ϵ	50 m to 300 m	80 m to 300 m	the DBSCAN-based plot (gives the optimal result in terms of cluster assignment) and the plots on moving flocks, cohesion and separation scores

The next paragraphs provide more details about unsupervised and supervised measures.

The proposed approaches fall under two main categories, namely, supervised and unsupervised validation. Supervised approaches require validation of the flocking results with respect to some "ground truth" while unsupervised approaches assess the validity of the flocking results based on their intrinsic properties.

4.1 Supervised Validation

As mentioned previously, the optimal flocking result that can be obtained from a specific dataset is usually unknown. However, we can evaluate the quality of the results by comparing them with known flocks from controlled experiments or synthetic or semi-synthetic datasets, if they are available. That is, supervised validation could be performed by comparing the obtained flocking results with the expected flocking results. The flock similarity measure described in Subsect. 3.2 can be used to assess the similarity between these results.

In the next paragraphs, we provide a discussion of visual inspection of trajectories in discovered flocks, and use of controlled experiments for the purpose of flock validation.

Visual Inspection of Trajectories in Discovered Flocks Since flocking is often associated with an image of birds or other types of entities that remain close together over a specific time duration, the most natural way of checking for the occurrence of flocks is through the sense of sight. Hence, the trajectories of flock members over the duration of flocking can be plotted using existing visualization tools [1,9]. Figure 4 gives two plots, the left plot being a flock example while the other is not. The left plot illustrates how the entities moved closely together over

Fig. 4. The left plot is a flock example while the right plot is a non-example.

the flocking duration. On the contrary, the non-example shows that the entities were moving in different directions.

Since this technique involves inspection of plots by a knowledgeable observer, it is advisable to apply this technique for a sampled set of flocks when there is a large set of discovered flocks.

Use of Controlled Experiments Datasets In this part, we discuss the supervised validation of the flock algorithm using a dataset that describes the movement of a small group of pedestrians in the DNP park. The pedestrian movement in this dataset is controlled in the sense that the pedestrians were given instructions on how they should move. For this reason, we refer to this as the *controlled DNP dataset*. There were 6 teams of pedestrians who participated in the controlled experiment. Each team was given a GPS device for tracking their movement and a set of instructions on how they should move within a small portion of the park. At certain times, some teams were moving with two other teams while at other times, some teams were moving alone. We have formulated the instruction such that flocking will occur exactly twice. More specifically, Table 2 provides a description of the actual details of flocking based on the images taken by the participants and an animation of the flocks.

In order to validate the flocking algorithm, we run it on this dataset and the following parameters were used: $R = 1$ min., $\Delta T = 3$, $\kappa = 3$, and $\epsilon = 100$ m. Table 2 also provides a comparison of the actual and obtained results. Since the number of expected and discovered flocks are small, it is easy to verify that the obtained results are consistent with the expected results. There are only some minor differences in the flocking times. We have also computed the similarity

Table 2. Actual and discovered flocking times on day December 12, and flock members in the controlled DNP dataset.

Flock ID	Start time	End time	Flock members
Actual flocks in the controlled dataset			
0	13:41 2010	13:46 2010	6; 22; 32
1	14:02 2010	14:09 2010	13; 25; 27
Discovered flocks			
0	13:41 2010	13:48 2010	6; 22; 32
1	14:02 2010	14:10 2010	13; 25; 27

score using the flock similarity score described earlier and a high score of 0.95 was obtained.

4.2 Unsupervised Validation

An alternative way to validate the algorithm when ground truth results are not available is to perform an unsupervised validation. We use the random model validation, which is based on the null hypothesis principle [6], for concluding that the obtained flocks are inherent in the dataset and not by random chance. The main idea behind this validation technique is to initially assume that the extracted flocks are obtained by chance. In order to disprove this hypothesis, different versions of the original dataset are generated through some random simulation technique. Afterwards, the flock discovery algorithm is executed having each random version as input. The results obtained from applying the algorithm to the original dataset and to the random versions are then compared. Obtaining different flock results from different input datasets demonstrates that the algorithm produces results that are not obtained by mere chance but are based on the nature and the characteristics of the input dataset.

We have also introduced two random simulation techniques, which are radius-based distortion of spatial coordinates and simulation through Markov-chain random models.

Radius-Based Distortion of Spatial Coordinates. It is known that the collected observation points contain inaccuracies due to the limitation of current location technologies. Considering this uncertainty, we propose a generation technique that modifies the x-, and y-values in the dataset by using a radius distortion threshold as a measure of uncertainty. The research assumption here is that a larger threshold would produce a dataset that is very different from the original while a smaller threshold would produce a dataset that is comparable to the original.

Six random versions of the DNP dataset were generated by replacing (x, y) pairs in the original dataset with a new value bounded by the user-specified radius distortion threshold. Table 3 presents an extract of the discovered flocks from each of the simulated datasets using varying radius distortion values. The flock algorithm was run on all datasets, including the original and random versions, with the same parameters: $R = 5$ min., $\Delta T = 3$, $\kappa = 3$, and $\epsilon = 150$ m.

Likewise, we have performed the simulation procedures on the Octopisa dataset. Table 4 provides a summary of the discovered flocks from each simulated dataset. The flock algorithm was executed on all datasets with the same parameters: $R = 1$ min., $\Delta T = 3$, $\kappa = 3$, and $\epsilon = 150$ m.

The result of this investigation shows that different flocks are obtained when the dataset is simulated by a large enough radius distortion, such as 100 m in the DNP dataset. Therefore, the obtained flocks depend on the dataset and are not discovered by chance. Moreover, as the radius threshold of the simulation algorithm is decreased, the extracted flocks become more similar to those found in

Table 3. Flock results for different random versions of the DNP. As expected, we can notice a high similarity when the radius distortion is small. The similarity score increases by almost 50 %–60 % when the radius distortion is increased by 50 m.

	Similarity score	Start time	End time	Flock members
Original	1	12:15	12:30	15; 96; 288
		12:45	13:00	139; 140; 141
		12:50	13:00	129; 139;140
Version 1 (100 m)	0	12:05	12:20	85; 125;147;
		12:05	12:15	23; 85; 125
Version 2 (50 m)	0.63	12:15	12:25	15; 96; 288
		12:45	12:55	139; 140; 141
		12:35	12:50	217; 223; 264
Version 4 (30 m)	0.67	12:15	12:30	15; 96; 288
		12:45	13:00	139; 140; 141
Version 6 (10 m)	0.98	12:15	12:30	15; 96; 288
		12:45	12:55	139; 140; 141
		12:50	13:00	129; 139; 140

Table 4. Flock results for different random versions of OctoPisa obtained selecting one hour (12 to 13). As expected, the values reflect a behavior similar to the DNP dataset.

	Similarity score	Start time	End time	Flock members
Original	1	12:41	12:43	1; 2; 3
		12:26	12:29	4; 5; 6
		12:30	12:32	4; 5; 6
		12:51	12:55	7; 8; 9
Version 1 (100 m)	0.24	12:45	12:47	10; 11; 12
		12:27	12:29	4;5; 6
		12:11	12:19	13; 14; 15
		12:16	12:24	16; 17; 18
Version 2 (50 m)	0.25	12:41	12:43	1; 2; 3
Version 4 (30 m)	0.25	12:41	12:43	1; 2; 3
Version 6 (10 m)	0.75	12:41	12:43	1; 2; 3
		12:26	12:29	4; 5; 6
		12:30	12:32	4; 5; 6

the original dataset. This demonstrates the robustness of the flocking algorithm to uncertainties in the observation points of the input dataset.

Simulation Through Markov-chain Random Models. Another technique for generating a distorted version of the synchronised input dataset is to build a Markov chain based on the dataset distribution. This model assumes that only the current state affects the next state while past states and future states are irrelevant. The initial step involves building the model by computing the probabilities of transitions between the cells, which is a region of x-,y- coordinate values.

The initial step involves building the model by computing the probabilities of transitions between the x-,y- coordinate values. A transition exists from (x_i, y_i) to (x_{i+1}, y_{i+1}) if they belong to the same moving entity, and time instance $i + 1$ immediately follows time instance i. Since there are a large number of varying x- and y-values in the datasets, the x- and y-values are grouped into grids. Then, the probabilities of transitions between these grids is computed. Analogous to an existing transition between a pair of x- and y-values, a transition exists from $grid_i$ to $grid_{i+1}$ if there exists a transition from (x_i, y_i) to (x_{i+1}, y_{i+1}) such that (x_i, y_i) belongs to $grid_i$ and (x_{i+1}, y_{i+1}) belongs to $grid_{i+1}$ and time instance $i + 1$ immediately follows time instance i.

This simulation technique modifies the x and y values based on the described Markov chain while retaining the original entity ID and the original time values found in the synchronized dataset. The (x, y)-pair for the first time instance of an entity is a random value biased towards the most probable initial (x, y)-pairs found in the dataset. The succeeding (x, y)-pairs are determined based on the immediately preceding (x, y)-pair and the Markov chain. Once this grid is determined, the next (x, y)-pair can be computed as a random value limited by the bounds of this grid. In the case that there is no next probable grid, a new trajectory is started by randomly picking a most probable initial (x, y)-pair and continuing in a manner as described before.

We have generated random versions of the DNP and the OctoPisa dataset by building a Markov chain based on their underlying data distribution of spatial points. In running the flock discovery algorithm, κ and ΔT were both set to 3 in both datasets. Meanwhile, the ϵ parameter was set to 150 m for both datasets. Lastly, R was set to 300 s for DNP and 60 s for OctoPisa.

The simulation algorithm was ran several times for each dataset. In the case of DNP, there were 4 generated datasets that yielded some flocks. There was 1 flock obtained from 3 of the generated datasets, and there were 6 flocks extracted from the other generated dataset. Recall that 11 flocks were found in the original dataset. Aside from varying in the number of flocks discovered, the members and durations of the flocks themselves were also very different.

Ten simulated versions were generated for the Octopisa dataset. However, none of the generated datasets contain moving flock patterns. This means that there is no specific grid movement that has a high probability in the Markov chain and it was difficult to create arbitrary flocks in the generated datasets. In other words, the sequence of movement for each car is likely to vary compared to those of other cars.

5   Conclusions and Future Work

This paper provides an empirical evaluation of the effects of flock parameters in two diverse moving objects datasets, one for pedestrians and another for vehicles. Validation techniques for flock patterns were also proposed in this work. The aim is to delineate a data-driven parameter estimation method for flock mining.

Future work includes experiments over a wider set of trajectory datasets, for instance to consider animal movements or vessels, to further validate our results and to propose a formal framework for general flock mining parameters evaluation. Human intervention in the current approach can be minimised by automating the estimation techniques for *synchronisation_rate* and *radius* while allowing the analyst to adjust *min_points* and *min_time_slices*. Further study on recursively combining parameter estimation and pattern validation techniques is an interesting research direction for finding optimal parameter values.

References

1. Andrienko, G., Andrienko, N., Wrobel, S.: Visual analytics tools for analysis of movement data. SIGKDD Explor. Newsl. **9**(2), 38–46 (2007)
2. Ester, M., Kriegel, H.-P., Sander, J., Xu, X.: A density-based algorithm for discovering clusters in large spatial databases with noise. In: KDD, pp. 226–231 (1996)
3. Giannotti, F., Pedreschi, D. (eds.): Mobility, Data Mining and Privacy - Geographic Knowledge Discovery. Springer, New York (2008)
4. Gonzlez, M.C., Hidalgo, C.A., Barabsi, A.-L.: Understanding individual human mobility patterns. Nature **453**, 779–782 (2008)
5. Gudmundsson, J., van Kreveld, M.J.: Computing longest duration flocks in trajectory data. In: GIS (2006)
6. Hand, D.J., Smyth, P., Mannila, H.: Principles of Data Mining. MIT Press, Cambridge (2001)
7. Jeung, H., Yiu, M.L., Zhou, X., Jensen, C.S., Shen, H.T.: Discovery of convoys in trajectory databases. In: Proceedings of the VLDB Endow, pp. 1:1068–1:1080 (2008)
8. Laube, P., Imfeld, S., Weibcl, R.: Discovering relative motion patterns in groups of moving point objects. Int. J. Geogr. Inf. Sci. **19**(6), 639–668 (2005)
9. Trasarti, R., Rinzivillo, S., Pinelli, F., Nanni, M., Monreale, A., Renso, C., Pedreschi, D., Giannotti, F.: Exploring real mobility data with M-Atlas. In: Balcázar, J.L., Bonchi, F., Gionis, A., Sebag, M. (eds.) ECML PKDD 2010, Part III. LNCS, vol. 6323, pp. 624–627. Springer, Heidelberg (2010)
10. Wachowicz, M., Ong, R., Renso, Ch., Nanni, M.: Finding moving flock patterns among pedestrians through collective coherence. Int. J. Geogr. Inf. Sci. **25**(11), 1849–1864 (2011)

Feature Extraction over Multiple Representations for Time Series Classification

Dominique Gay$^{(\boxtimes)}$, Romain Guigourès, Marc Boullé$^{(\boxtimes)}$, and Fabrice Clérot

Orange Labs, 2, Avenue Pierre Marzin, 22307 Lannion Cedex, France
{dominique.gay,Romain.Guigoures,marc.boulle,Fabrice.Clerot}@orange.com

Abstract. We suggest a simple yet effective and parameter-free feature construction process for time series classification. Our process is decomposed in three steps: (i) we transform original data into several simple representations; (ii) on each representation, we apply a coclustering method; (iii) we use coclustering results to build new features for time series. It results in a new transactional (i.e. object-attribute oriented) data set, made of time series identifiers described by features related to the various generated representations. We show that a Selective Naive Bayes classifier on this new data set is highly competitive when compared with state-of-the-art times series classification methods while highlighting interpretable and class relevant patterns.

1 Introduction

Time series classification (TSC) has been intensively studied in the past years. The goal is to predict the class of an object (a time series or a curve) $\tau_i = \langle(t_1, x_1), (t_2, x_2), \ldots, (t_{m_i}, x_{m_i})\rangle$ (where $x_k, (k = 1..m_i)$ is the value of the series at time t_k), given a set of labeled training time series. TSC problems differ from traditional classification problems since there is a time dependence between the variables; in other terms, the order of the variables is crucial in learning an accurate predictive model. The increasing interest in TSC is certainly due to the wide variety of applications: from e.g., medical diagnosis (like classification of patient electrocardiograms) to the maintenance of industrial machinery. Other domains, where data might be time series, are also concerned: finance, meteorology, signal processing, computer network traffic, ... The diversity of applications has given rise to numerous approaches (see Sect. 2 for detailed related work). However, most efforts of the community have been devoted to the following three-step learning process: (i) choosing a new data representation, (ii) choosing a similarity measure (or a distance) to compare two time series and (iii) using the Nearest Neighbor (NN) algorithm as classifier on the chosen representation, using the chosen measure. Ding et al. [10] offer a survey of the various data representations and distances found in the literature and an extensive experimental study using the NN classifier. They conclude that NN classifier coupled with Euclidean distance (ED) or Dynamic Time Warping (DTW) show the highest predictive performance for TSC problems using the original

A. Appice et al. (Eds.): NFMCP 2013, LNAI 8399, pp. 18–34, 2014.
DOI: 10.1007/978-3-319-08407-7_2, © Springer International Publishing Switzerland 2014

time domain. Recently, Bagnall et al. [1] experimentally show that the performance of classifiers significantly increases when changing data representation (compared with original temporal domain); thus, for a given classifier, there is a high variance of performance depending on the data transformation at use. To alleviate this problem, an ensemble method TSC-ENSEMBLE [1] based on three data representations (plus the original data) and NN algorithm is suggested. The experimental results demonstrate the importance of representations in TSC problems and show that a simple ensemble method based on several data representations provides highly competitive predictive performance. However, with the good performance of NN-based approaches also come the drawbacks of lazy learners: i.e., there is no proper training phase, therefore the training set has to be entirely stored and all the computation time is postponed until deployment phase; these weaknesses do not meet the requirements of deployment in resource-limited and/or real-time applications. Another weakness of the NN approaches is the lack of interpretability; indeed NN only indicates the nearest series w.r.t. the used similarity measure.

The method we suggest takes the pros and leaves the cons of the methods listed above: we come back to the *eager*[1] paradigm, benefit from the combination of multiple representations, build and select valuable features from multiple representations. More precisely, in this paper, we suggest a parameter-free process for constructing valuable features over multiple representations for TSC problems. Our contribution is thus essentially methodological. The next section discuss further related work to give a wider view of existing solutions for TSC problems. Section 3 motivates and describes the three steps of our unsupervised process: (i) transformation of original data into several new data representations; (ii) coclustering on various data representations; (iii) the exploitation of coclustering results for the construction of new features for the data. The output of the process is then a traditional data set (i.e. labeled objects described by attributes) ready for supervised feature selection and classification. We report the experimental validation of our approach in Sect. 4 before concluding.

2   Related Work

In TSC problems, DTW-NN is recognized by the community as a hard-to-beat baseline and it is confirmed by our experiments (see Sect. 4). However, there exist alternative approaches: besides the numerous similarity measures coupled with Nearest Neighbor algorithm [10], very recent novel metric has been proposed [25] as well as fusion of distance measures [7] and NN ensembles over multiple representations [1]. Concerning the intra-class variance, to deal with the lack of objects that cover sub-class pattern, Grabocka et al. [13] suggest to create virtual transformed objects for the training set; it results in a significant improvement of SVM predictive performance.

---

[1] In contrast to lazy learning, eager learning has an explicit training phase and generally deploys faster.

For the sake of interpretability, an extension of decision tree has been proposed recently [15]. On the other hand, feature-based approaches have also been intensively studied. Feature-based approaches for TSC aim at extracting class-relevant characteristics of series so that a conventional classifier can used. A wide variety of features has been studied: e.g., global, trends [16], symbolic, intervals, distance-based [9], features coming from spectral transforms [20,32] or a combination of several types of features [11].

Shapelet-based approaches [12,30], a subtopic of feature-based approaches, have drawn much attention in recent years. Shapelets are time series subsequences that are representative of a class. First approaches have embedded extracted shapelets in a decision tree [12,30,31], others in a simple rule-based classifier [29], while very recently, Lines et al. [19] have designed a shapelet-based transform.

Our approach generates similarity-based features and histogram features over multiple representations (see next section); the former allows us to reach predictive performance comparable to the best similarity-based NN classifiers, with the latter we gain some insight in the data. The closest works are from the inspiring works of Bagnall et al. [1] who establish competitive predictive performance by combining multiple representations in an ensemble classifier.

3   Feature Construction Process

Notations. In TSC problems, we define a time serie as a pair (τ_i, y_i) where τ_i is a set of ordered observations $\tau_i = \langle (t_1, x_1), (t_2, x_2), \ldots, (t_{m_i}, x_{m_i}) \rangle$ of length m_i and y_i a class value. A time series data set is defined as a set of pairs $D = \{(\tau_1, y_1), \ldots, (\tau_n, y_n)\}$, where each time series may have a different number of observations (i.e. with different length). Notice that the time series of a data set may also have different values for $t_k, (k = 1..m_i)$. The goal is to learn a classifier from D to predict the class of new incoming time series $\tau_{n+1}, \tau_{n+2}, \ldots$ To achieve this goal, we suggest the feature construction process summarized as follows:

1. We transform original data into multiple data representations.
2. We process a coclustering technique on each representation.
3. We build a set of features from each coclustering result and obtain a new data set gathering the various sets of features.

The new data set is thus object-attribute oriented and ready for supervised classification phase. Since our main contribution is methodological, we will take some time to motivate each step of the process, and when necessary, to make the paper self-contained, we will recall the main principles of the tools used in each step.

3.1   Transformations and Representations

Numerous data transformation methods for time series has been suggested in the literature: e.g., polynomial, symbolic, spectral or wavelet transformations, (see [10,18] for well-structured surveys on experienced data representations). The underlying idea of using data transformation is that transformed data might contain class-characteristic pattern that are easily detectable (i.e. patterns unreachable in the original time domain). The following example illustrates and confirms the relevance of using representations, and highlights simple interpretable features that might arise from data representations.

Motivating example. Graphs from Fig. 1 confirm the relevance of changing data representation: indeed, from original data (a) it is challenging to separate the two classes (blue/red). It has been shown [1] that the accuracy of NN-based classifiers on the original data is about 60 %. On the other hand a simple transformation, like double derivate (b) facilitate class discrimination. For example, after computing the double derivate transformation, we see that curves with some values above 6 (or below −6) are red whereas curves with most of its values between −6 and 6 are blue. On this data example, a simple transformation and two interpretable features are enough to characterize the two classes of curves.

To illustrate and instantiate our process, we use the original representation and we pick six representations among the numerous ones existing in the literature.

Derivatives: DV et DDV. We use derivatives and double derivatives of original time series (computed between time t et $t-1$). These transformations allow us to represent the local evolution (i.e., increasing/decreasing, acceleration/deceleration) of the series.

Cumulative integrals: IV et IIV. We also use simple and double cumulative integrals of the series, computed using the trapeze method. These transformations allow us to represent the global (cumulated) evolution of the series.

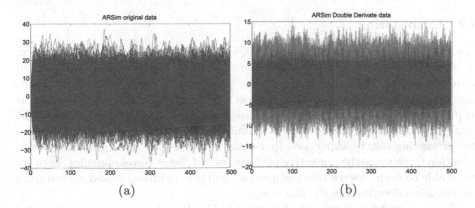

Fig. 1. ARSim 2-class data: original data versus double derivate transformation (Color figure online).

Power Spectrum: PS. A time series can be decomposed in a linear combination of sines and cosines with various amplitudes and frequencies. This decomposition is known as the Fourier transform. And, the Power Spectrum is $PS(\tau_i) = \langle (f_1, a_1), \ldots, (f_{m_i}, a_{m_i}) \rangle$, where f_k represent the frequency domain and a_k the power of the signal (i.e. the sum of the Fourier coefficients squared). This transformation is commonly used in signal processing and plunges the original series into the frequency domain.

Auto-correlation function: ACF. The transformation by auto-correlation (ACF) is: $\tau_{i\rho} = \langle (t_1, \rho_1), \ldots (t_{m_i}, \rho_{m_i}) \rangle$ where

$$\rho_k = \frac{\sum_{j=1}^{j=m_i-k} (x_j - \bar{x}) \cdot (x_{j+k} - \bar{x})}{m \cdot s^2}$$

and where $\bar{x}$ and s^2 are the mean and variance of the original series. ACF transformation describes the correlation between values of the signal at different times and thus allow us to represent auto-correlation structures like repeating patterns in the time series.

We do not pretend that the chosen representations are suitable for all TSC problems; there are many other transformation techniques in the literature. Depending on the application, the domain expert remains the best to select potentially suitable representations for the problem at hand. Let us just recall that the time complexity for computing the chosen representations is at most sub-quadratic w.r.t. the number of points.

Thus, for a given time series data set D_{orig}, we build six new data representations: D_{DV}, D_{DDV}, D_{IV}, D_{IIV}, D_{PS} and D_{ACF} depending on the transformation used. In the following, for the sake of generality, an object from one of these representations will be called "curve" instead of time series since D_{PS} does not use the time domain.

3.2   Coclustering

In classification problems (also in TSC), there might exist intra-class variance, i.e. the variations between objects of the same class might be numerous and of various aspects. Using clustering as a pre-processing step to supervised classification is not new and is a solution to deal with intra-class variance. The idea is to pre-process the data set by grouping together similar objects and to highlight local patterns that might be class-discriminant: e.g., Vilalta et al. [26] suggest a pre-processing step by supervised (per-class) clustering using Expectation Maximization to enhance the predictive performance of Naive Bayes classifier. In order to be able to derive interesting features, we will use an unsupervised coclustering technique as described in the following.

A curve can be seen as a set of points (X, Y), described by their abscissa and ordinate values. A set of curves is then also a set of points (C_{id}, X, Y) where C_{id} is the curve identifier. This tridimensional representation (one categorical variable and two numerical variables) of a curve data set is needed to apply coclustering methods. Indeed, the goal is to partition the categorical variable and to discretize the numerical variables in order to obtain clusters of curves and intervals for X and Y. The result is a tridimensional grid whose cells are defined by a group of curves, an interval for X and an interval for Y.

For that purpose, we use the coclustering method KHC [6] (Khiops Coclustering). Originally designed for clustering functional data [23], it is also suitable for the particular case of curve data as defined above and it is directly applicable for our pre-processing step. KHC method is based on a piecewise constant non-parametric density estimation and instantiates the generic MODL approach [4] (Minimum Optimized Description Length) – which is similar to a Bayesian Maximum A Posteriori (MAP) approach. The optimal model M, i.e. the optimal grid, is obtained by optimization of a Bayesian criterion, called *cost*. The *cost* criterion bets on a trade-off between the accuracy and the robustness of the model and is defined as follows:

$$cost(M) = -\log(\underbrace{p(M \mid D)}_{\text{posterior}}) = -\log(\underbrace{p(M)}_{\text{prior}} \times \underbrace{p(D \mid M)}_{\text{likelihood}})$$

Using a hierarchical prior (on the parameters of a data grid model) that is uniform at each stage of the hierarchy, we obtain an analytic expression for the *cost* criterion:

$$cost(M) = \log n + 2 \log N + \log B(n, k_C) \tag{1}$$

$$+ \log \binom{N + k - 1}{k - 1} + \sum_{i_C=1}^{k_C} \log \binom{N_{i_C} + n_{i_C} - 1}{n_{i_C} - 1} \tag{2}$$

$$+ \log N! - \sum_{i_C=1}^{k_C} \sum_{j_X=1}^{k_X} \sum_{j_Y=1}^{k_Y} \log N_{i_C j_X j_Y}! \tag{3}$$

$$+ \sum_{i_C=1}^{k_C} \log N_{i_C}! - \sum_{i=1}^{n} \log N_i! + \sum_{j_X=1}^{k_X} \log N_{j_X}! + \sum_{j_Y=1}^{k_Y} \log N_{j_Y}! \tag{4}$$

where n is the number of curves, N the number of points, k_C (resp. k_X, k_Y) is the number of clusters of curves (resp. the number of intervals for X and Y), k the number of cells of the data grid, n_{i_C} the number of curves in cluster i_C, N_i the number of points for curve i and N_{i_C} (resp. N_{j_X}, N_{j_Y}, $N_{i_C j_X j_Y}$) is the cumulated number of points for curves of cluster i_C (resp. for interval j_X of X, interval j_Y of Y, for cell (i_C, j_X, j_Y) of the data grid. Notice that $B(n, k_C)$ is the number of divisions of n elements into k subsets. The two first lines stand for the prior and the two last lines relates to the likelihood of the model. Intuitively, low *cost* means high probability $(p(M \mid D))$ that the model M arises from the data D. From an information theory point of view, according to [24], the negative logarithms of probabilities may be interpreted as code length. Thus, the *cost* criterion may also be interpreted as the code length of the grid model plus the code length of data D given the model M, according to the Minimum Description Length principle (MDL [14]). Here, low *cost* means high compression of the data using the model M.

The *cost* criterion is optimized using a greedy bottom-up strategy, (i) starting with the finest grained model, (ii) considering all merges between adjacent clusters or intervals, for the curve and dimension variables, and (iii) performs the best merge if the criterion decreases after the merge. The process loops until no further merge improves the criterion. The obtained grid constitutes a non-parametric estimator of the joint density of the curves and the dimensions of points.

KHC is parameter-free, robust (avoids over-fitting), handles large curve data sets with several millions of data points and its time complexity is $\Theta(N\sqrt{N}\log N)$ (sub-quadratic) where N is the number of data points: thus, KHC meets our problem needs (for full details, see [6]).

An example of visualization of coclustering results. The Fig. 2 shows an example of visualization of two clusters of curves of the optimal grid obtained on ARSim data set (in DDV representation). Figure (a) (resp. (b)) shows a cluster whose curves are essentially from class c_1 (blue in Fig. 1 of the motivation example), (resp. c_2, red in the same example). The optimal grid obtained with KHC is made up of 43 clusters of curves, 13 intervals for X and 12 intervals for Y (i.e. DDV values). The joint density estimation (i.e. the optimal grid) is much finer than needed by the classification problem. Indeed, the ARSim

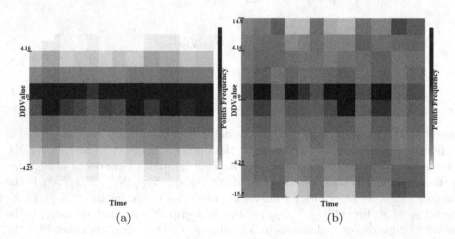

Fig. 2. Representation of the frequency of the cells for two clusters of curves obtained with KHC on ARSim data set (in DDV representation, DDV on y-axis and time on x-axis): (a) a cluster whose curves are mostly of class c_1; (b) a cluster whose curves are mostly of class c_2. For a given cell, stronger color indicates high point frequency.

data set is a 2-class classification problem and we found 43 groups of curves. This finer granularity gives us the potential for finer class characterization if the data representation is relevant for the task.

3.3 Feature Construction

Feature construction for TSC problems [21] aims at capturing class-relevant properties for describing time series. The generated features goes from simple ones like minimum, maximum, mean, standard deviation of time series to more complex ones like e.g., coefficients of spectral decompositions [20,32] or local pattern extracted from temporal abstractions of the series [2]. The main advantage of feature-based approaches is the final transactional (or vector) representation of the data which is suitable for conventional classifiers like Naive Bayes or decision trees. In our process, we generate features from coclustering results as follows.

For each coclustering result obtained with KHC on a data representation $(D_{orig}, D_{DV}, D_{DDV}, D_{IV}, D_{IIV}, D_{PS}, D_{ACF})$, we create a set of new features: $\mathcal{F}_{orig}, \mathcal{F}_{DV}, \mathcal{F}_{DDV}, \mathcal{F}_{IV}, \mathcal{F}_{IIV}, \mathcal{F}_{PS}, \mathcal{F}_{ACF}$. The new features are the descriptive attributes of the new data set whose objects are curves.

Let D_{rep} be one of the seven representations described above. Let $M_{rep} = KHC(D_{rep})$ be the tridimensional optimal grid obtained by coclustering with KHC on D_{rep}. We denote k_C the number of clusters of M_{rep} and k_Y the number of intervals of M_{rep} for dimension Y. We then create similarity-based features and histogram features.

Similarity-Based Features
Considering the good performance of (dis)similarity-based approaches (e.g., ED-NN and DTW-NN), we define a dissimilarity index based on the *cost* criterion.

Definition 1 (Dissimilarity index). *The dissimilarity between a curve τ_i and a cluster c_j of the optimal grid M_{rep} is defined as:*

$$d(\tau_i, c_j) = cost(M_{rep|\tau_i \cup c_j}) - cost(M_{rep})$$

i.e., the difference of cost between the optimal model M_{rep} and the model $M_{rep|\tau_i \cup c_j}$ (the optimal grid in which we add the curve τ_i to the cluster of curves c_j).

Intuitively, d measures the perturbation brought by the integration of a curve into a cluster of curves of the optimal grid (i.e. according to the *cost* criterion used for grid optimization). In terms of code length, if a curve τ_i is similar to the curves of cluster c_j, the total code length of the data is not much different from the total code length of the data plus τ_i. Thus, small values of $d(\tau_i, c_j)$ indicate that τ_i is similar to the curves of c_j whereas high values of d ($d(\tau_i, c_j) \gg 0$) mean that τ_i does not look like the curves of c_j.

According to the dissimilarity index d, we generate the following features:

- k_C numerical features (one for each cluster c_j of curves of M_{rep}). The value for a curve τ_i is the difference $d(\tau_i, c_j)$. Thus, for a given curve τ_i, these features tell how τ_i is similar to the clusters of curves of the optimal grid (according to d).
- One categorical feature indicating the index j of the cluster of curves that is the closest to a curve τ_i according to the dissimilarity d defined above (i.e., $\arg\min_j d(\tau_i, c_j)$).

Histogram Features
Taking up the idea of interpretable features (see motivating example and Fig. 1), we also generate the following features:

- k_Y numerical features (one for each interval i_Y of Y from M_{rep}) whose value for a curve τ_i is the number of points of τ_i in interval i_Y.

These histogram features quantify the presence of a curve in intervals of Y obtained in the coclustering step.

For a given curve τ_i, we now have the following informations provided by the new features (for each representation): (i) the dissimilarity values between τ_i and all the clusters of curves, (ii) the index of the closest cluster of curves and (iii) the number of points of τ_i in each interval of Y.

3.4 Supervised Classification Algorithm

We saw that our feature construction process may generate hundreds of new features for each representation. The whole set of features $\mathcal{F}_{tot}$ for our new data set may contain thousands of attributes. Therefore, the classifier at the end of our process has to be capable of handling a large number of attributes but also selecting the relevant attributes for the classification task. At this stage, we could use conventional classifiers like decision trees or SVM. However, we choose the Selective Naive Bayes classifier (SNB) that is parameter-free, performs efficient feature selection and outperforms classical Naive Bayes [5]. Notice that SNB exploits pre-processing techniques that discretize numerical variables, group values of categorical variables, weight and select features w.r.t. class-relevance by using robust conditional density estimators and following the MODL approach (see [3,4]). Thus, the generated features benefit from these pre-processing techniques and preserve a potential of interpretability; we lead specific experiments in the next section to support this claim. Moreover, SNB is parameter-free, so is the whole feature construction process. Its time complexity is $\Theta(KM \log(KM))$, where M is the number of objects and K the number of features.

4   Experimental Validation

The implementation of the classification process is based on existing tools (KHC for coclustering and SNB for supervised classification[2]). Connections between the

---

[2] KHC and SNB are both available at http://www.khiops.com.

tools are scripted using MATLAB. The whole process is named MODL-TSC. The experiments are led to discuss the following questions:

Q_1 Is MODL-TSC comparable with competitive contenders of the state-of-the-art in terms of accuracy?
Q_2 MODL-TSC employs and combines several representations. Are they all useful? Do they all bring the same impact?
Q_3 What kind of data insight do we gain using the coclustering-based features?

4.1 Protocol

We experiment our process on 51 time series data sets: 42 data sets are from UCR [17] and 9 new data sets introduced in [1]. A brief description of the data is given in Table 1. The benchmark data sets offer a wide variety in terms of application domains, number of series, length of series and number of class values. We lead experiments in a predefined train-test setting for each data set (see [17]). We compare the predictive performance of our process, called MODL-TSC, with a baseline, two of the most effective alternative approaches and a recently introduced interpretable classifier:

- ED-NN: the Nearest Neighbor classifier using the Euclidean distance. This approach is considered as a baseline.
- DTW-NN: the Nearest Neighbor classifier using the elastic distance Dynamic Time Warping, considered as hard to beat in the literature (see [28])
- TSC-ENSEMBLE [1] exploits multiple representations via an ensemble method and the NN algorithm. Its performance is comparable to DTW-NN
- FAST-SHAPELETS [22] mines shapelets (i.e., class relevant time series subsequences) that might be embedded in e.g., a decision tree

4.2 Results

For fair comparisons, we have rerun the experiments using implementations of ED-NN, DTW-NN, FAST-SHAPELETS and TSC-ENSEMBLE provided by E. Keogh, A. Bagnall and their teams. Performance results in terms of accuracy are reported in Table 1. The best result for each data set is written in bold. The last column indicates how many features per representation MODL-TSC has generated.

Comparisons with state-of-the-art. Firstly, global results (mean accuracy, number of wins and mean rank) show that MODL-TSC is very competitive compared to state-of-the-art methods. Proceeding the Friedman test [8] (at significance level $\alpha = 0.05$) and the Nemenyi post-hoc test lead us to the critical difference diagram in Fig. 3. Two groups of approaches emerge and we observe that MODL-TSC, TSC-ENS and DTW-NN perform significantly better than ED-NN and FAST-SHAPELETS; and there is no significant difference of performance

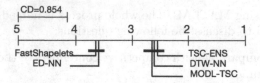

Fig. 3. Representation of difference of performance by critical difference diagram between MODL-TSC, ED-NN, DTW-NN, TSC-ENS and FAST-SHAPELETS.

inside each group. We also run Wilcoxon's sign rank test for pairwise comparisons (also with $\alpha = 0.05$) which confirms this result. This global view of performance results confirms that MODL-TSC is very competitive compared to two of the most effective contenders of the state-of-the-art and performs better than the baseline ED-NN and the very recent FAST-SHAPELETS approach.

Secondly, we observe the remarkable performance of MODL-TSC on ARSim, ElectricDevices, FordA and OSULeaf data. On these data, we outperform DTW-NN and TSC-ENSEMBLE: the difference of test accuracy is at least 10. Here, the added-value of the data representations (i.e., the new features) is at work. TSC-ENSEMBLE (exploiting only three representations) and DTW-NN (working in time domain) obtain only poor accuracy results. Conversely the performance of MODL-TSC is very low on Coffee, DiatomSizeReduction, ECGFiveDays and OliveOil data. The difference of test accuracy (about 10 compared with DTW-NN and TSC-ENSEMBLE) is now to our disadvantage. We think that this poor performance might due to one reason: the training set size of these data sets is very small (less than 30 of curves) and it could be insufficient for either learning relevant coclusters or learning a predictive model without over-fitting. Indeed, e.g., for OliveOil data, there are only 30 training curves, no cluster of curves is found by KHC whatever the representation and most generated features from intervals of the Y-axis are considered irrelevant by the pre-processing step of SNB classifier.

Added-value of the representations. In Table 1, we also report accuracy results of SINGLE-MODL-TSC using only one representation. We observe that using a single representation provide poor average accuracy results. Almost always, MODL-TSC using several representations outperforms SINGLE-MODL-TSC on $\mathcal{F}_{orig}$ (resp. $\mathcal{F}_{DV}, \mathcal{F}_{DDV}, \mathcal{F}_{IV}, \mathcal{F}_{IIV}, \mathcal{F}_{PS}, \mathcal{F}_{ACF}$). In some cases (e.g., Italy-PowerDemand, MALLAT or MedicalImages), the good performance of MODL-TSC can be attributed to the combination of several representations. Indeed, the gap of test accuracy between any SINGLE-MODL-TSC and MODL-TSC is about 10; thus the combination of features coming from different views of the data improves accuracy results. In other cases (e.g., ARSim or wafer), the good performance seems to be due to only one (or at most two) representation while the other representations are ignored. As an example, for ARSim data, the DDV representation is the most relevant. KHC obtains 43 clusters of curves and 12 intervals for Y_{DDV}. Most of the clusters are almost pure (only one class of curves per cluster). Moreover, as we saw in Fig. 2(a) and (b), the number of points in

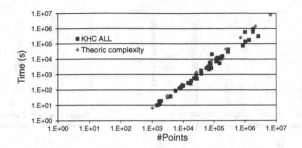

Fig. 4. Cumulative running time results for KHC on the seven studied representations and comparison with the announced theoretical complexity.

intervals generated by KHC above 4.16 and below −4.25 are class-discriminant since curves of class 1 almost never have points in these regions.

These experiments recall the very importance of representations in TSC problems and particularly in our feature construction process. Even the simple representations we chose to illustrate our process show good predictive performance. Depending on the application, we may still hope some improvement in performance if we could rely on expert domain knowledge to select relevant representations to use in our generic process.

4.3   Running Time Results

Among the three steps of our process, the coclustering step is the most demanding in terms of computational time. Moreover, for the largest data sets of our benchmark, other steps (building representations and features and learning the SNB classifier) are negligible compared with the coclustering step. For a better understanding of how much time KHC costs, we report in Fig. 4 the cumulative running time of KHC on the seven representations for each data set w.r.t. the number of points N in the training data set. We also draw the theoretical complexity announced for KHC in previous section: $\alpha N \sqrt{N} \log N$, where $\alpha = 3.10^{-5}$. We observe that for the most difficult data sets (ARSim, FordA and FordB, from 1 million to 1.8 million data points), KHC runs during one day for each representation to reach the optimal grid.

4.4   Interpretation: An Example

If we consider the cumulative integral (IV) representation of TwoPatterns data, the optimal grid obtained by KHC is made of 224 clusters of curves, 11 intervals for X and 9 intervals for Y_{IV}. According to the MODL pre-processing techniques, among all the attributes generated from all representations, the two most relevant attributes are from the IV representation:

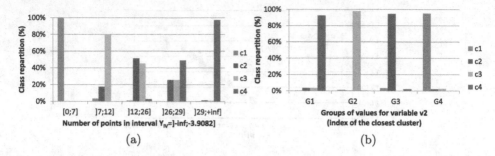

Fig. 5. Histogram representation of class repartition for discretization of variable v_1 and value grouping of variable v_2.

1. v_1, the number of points in interval $I_{Y_{IV}} =] - \infty; -3.9082]$
2. v_2, the index of the closest cluster

In the supervised learning step, the discretization for v_1 and the value grouping for v_2 provide the following contingency tables represented as histograms (see Fig. 5(a) and (b)): We observe (Fig. 5(a)) that the number of points p of a curve in interval $I_{Y_{IV}}$ (i.e. the number of points with value less than -3.9082) is class relevant. Indeed, in the learning phase, curves such that $p \le 7$ are of class c_1; when $p > 29$ (about 23 % of the points of the curve), curves are mostly of class c_4 and when $7 < p \le 12$ they are mostly of class c_3. This type of feature is similar to the ones in the motivating example and Fig. 1: for a given representation, some regions of y-axis (delimited by intervals) will be class-discriminant and the number of points of an incoming curve in this interval will also be class-discriminant.

In Fig. 5(b), we firstly see that, for variable v_2 ("index of the closest cluster"), MODL pre-processing by supervised value grouping provide 4 groups: G_1, (resp. G_2, G_3 et G_4) made of 56, (resp. 53, 53 et 62) indexes of clusters that are mostly of class c_4 (resp. c_3, c_2, c_1). The attribute v_2 is then class-relevant. Indeed, for example, if j is the index of cluster, that is the closest to a curve τ_i, and belongs to G_2 (i.e. $j \in G_2$), then τ_i is considered very similar to curves of class c_3. Moreover, the variable "index of the closest cluster" is an indicator of the relevance of the representation in our process for the current TSC problem. In this example, attribute v_2 alone, is enough to characterize 95 % of the data, therefore, IV data representation is very relevant for characterizing the classes of TwoPatterns data. Conversely, for the original representation (D_V), the optimal grid obtained with KHC is made of 255 clusters of curves but MODL pre-processing indicates that the variable "index of the closest cluster" is not relevant to characterize the classes of TwoPatterns; as a consequence, SINGLE-MODL-TSC on $\mathcal{F}_{orig}$ shows bad test accuracy results.

Table 1. Description and characteristics of time series benchmark data sets. Comparisons of accuracy results for MODL-TSC, FAST-SHAPELETS, DTW-NN, ED-NN, and TSC-ENSEMBLE. Accuracy results for SINGLE-MODL-TSC using each single representation and number of features generated per representation (last column).

Data	#Train	#Test	Length	#Classes	FASTSHAPELETS	DTW-NN	ED-NN	TSC-ENS	MODL-TSC	F_{orig}	F_{DV}	F_{DDV}	F_{IIV}	F_{IIV}	F_{PS}	F_{ACF}	#features
50words	450	455	270	50	44.29	69.01	63.08	68.57	60.22	59.78	59.12	47.03	34.51	28.35	51.65		645/23/56/646/203/30/830 27/30/29/32/34/31/63
Adiac	390	391	176	37	48.59	59.65	51.10	62.15	49.10	48.34	34.53	19.95	17.90	34.27	02.05		25/14/8/29/34/18/25 15/3/18/22/7/15
ArSim	2000	2000	500	5	91.40	60.36	61.13	67.85	61.05	30.00	99.95	50.00	99.95	98.55	50.00		22/24/22/38/52/20/28
Beef	30	30	470	5	55.33	50.00	53.33	60.00	53.33	46.67	30.00	20.00	36.67	33.33	40.00	50.00	15/3/18/22/7/15
CBF	30	900	128	3	94.71	66.67	85.22	82.44	96.56	98.00	33.11	33.11	66.56	78.44	33.11	53.11	56/72/104/66/82/33/64
ChlorineConcentration	467	3840	166	3	58.31	64.84	65.00	67.76	58.15	56.56	57.79	57.11	53.26	53.26	55.00	53.26	15/11/13/21/22/15/22
CinC_ECG_torso	40	1380	1639	4	82.64	82.14	89.71	94.57	87.61	48.04	78.91	53.57	29.86	33.33	67.25	43.48	113/74/81/102/76/25/85
Coffee	28	28	286	2	93.21	77.69	57.44	82.14	64.62	49.23	53.57	53.57	53.57	53.57	34.87	34.87	105/26/25/113/80/26/72
Cricket_X	390	390	300	12	47.32	77.69	57.44	60.26	64.62	49.23	36.41	34.10	33.33	27.69	37.18	34.87	12/15/20/15/19/25/19
Cricket_Y	390	390	300	12	49.52	77.23	64.36	65.50	74.87	63.33	30.26	29.74	44.10	35.13	31.54	35.23	17/14/14/21/28/15/22
Cricket_Z	390	390	300	12	45.27	79.23	62.05	62.56	65.64	56.67	25.13	31.03	31.54	30.26	41.54	33.08	16/9/7/23/23/14/16
DiatomSizeReduction	16	306	345	4	88.30	96.73	93.46	94.44	90.39	71.94	59.15	76.14	74.82	74.82	74.82	74.82	544/43/50/195/151/6/27
Earthquakes	322	139	512	2	73.38	70.50	69.78	73.38	71.94	79.00	74.82	74.82	74.82	74.82	74.82	73.64	124/61/80/163/109/126/137
ECG200	100	100	96	2	77.30	77.00	88.00	89.00	79.00	79.00	64.62	66.14	71.24	70.59	73.00	73.00	33/29/21/42/65/24/24
ECGFiveDays	23	861	136	2	99.59	76.77	79.67	98.72	64.46	63.61	49.71	49.71	63.76	53.89	49.71	59.61	16/12/29/22/34/11/18
ElectricDevices	8953	7745	96	7	58.93	82.95	78.41	71.60	70.47	66.29	58.58	35.86	45.74	25.50	31.78	50.91	21/17/18/28/46/21/21
FaceAll	560	1690	131	14	91.02	90.49	76.93	83.71	74.88	66.03	48.54	35.02	42.44	33.66	34.24	72.73	25/36/26/31/44/48/47
FaceFour	24	88	350	4	67.17	72.42	68.64	84.85	98.64	62.84	90.68	65.14	73.26	62.50	35.43	48.63	185/87/49/233/259/156/78
FacesUCR	200	2050	131	14	80.28	65.93	59.63	73.46	67.28	82.64	90.68	66.05	53.28	54.32	72.72	83.79	111/67/64/236/274/147/70
Fish	175	1320	463	7	84.77	90.67	91.33	94.67	98.00	84.00	94.00	85.33	77.33	71.33	79.33	63.83	76/365/308/94/118/126/83
FordA	3571	810	500	2	72.59	88.14	86.25	87.57	87.30	85.95	69.73	69.73	50.37	68.92	78.38	86.67	57/30/37/49/62/62/50
FordB	3601	810	500	2	93.93	37.56	37.01	40.26	42.86	41.56	41.56	31.82	32.47	37.01	79.33	78.38	119/24/27/101/94/45/99
GunPoint	50	150	150	2	87.33	88.36	34.18	31.82	100.00	84.00	41.56	28.91	23.45	22.91	32.18	37.01	12/7/8/13/14/7/13
HandOutlines	1000	300	2709	2	38.44	95.04	95.53	94.95	81.05	24.91	49.85	49.85	70.26	72.89	49.85	22.73	74/21/20/49/45/24/51
Haptics	155	308	1092	5	25.91	70.49	75.41	77.05	81.05	68.80	68.85	53.42	67.12	52.05	67.21	70.75	56/21/21/38/36/22/39
InlineSkate	100	550	1882	7	90.51	59.73	57.53	69.80	73.97	65.57	53.42	50.68	65.57	63.93	45.21	62.30	22/24/23/33/44/29/27
ItalyPowerDemand	67	1029	24	2	70.49	93.39	91.43	86.35	92.92	82.49	43.96	12.32	76.42	65.20	53.42	53.42	47/29/29/36/38/42/41
Lightning2	60	61	637	2	59.73	73.68	68.42	70.00	67.24	51.32	53.16	51.58	53.16	57.63	54.34	53.16	20/14/6/16/18/15/12
Lightning7	70	73	319	7	96.72	83.47	87.86	86.34	90.89	81.02	92.01	53.91	76.20	68.69	53.91	74.36	64/40/34/101/147/83/79
MALLAT	55	2345	1024	8	56.70	79.03	82.90	80.61	87.58	75.78	64.58	41.02	56.69	41.53	61.32	73.00	104/107/30/105/102/31/64
MedicalImages	381	760	99	10	78.28	86.43	87.99	87.99	89.72	79.80	52.77	52.77	48.09	68.04	68.04	78.37	45/29/16/40/56/26/46
MoteStrain	20	1252	84	2	75.43	94.97	86.67	86.67	40.00	40.00	40.00	40.00	40.00	40.00	40.00	40.00	18/59/62/12/17/7/11
NIFECG_Thorax1	1800	1965	750	42	78.94	86.43	51.65	57.85	74.38	49.59	60.33	72.73	38.43	35.12	42.98	48.35	140/76/82/11/18/11/11
NIFECG_Thorax2	1800	1965	750	42	78.67	72.55	69.55	74.21	68.05	60.40	72.21	81.53	55.07	53.24	42.93	42.93	148/515/839/103/122/142/79
OliveOil	30	30	570	4	64.09	33.11	85.94	86.15	87.51	25.70	76.50	93.37	89.78	73.14	86.25	61.70	26/29/32/33/46/19/30
OSULeaf	200	242	427	6	68.55	90.66	84.88	84.80	95.84	95.35	95.65	93.37	89.78	73.14	78.91	75.24	24/137/27/22/36/27/24
SonyRobotSurface	20	601	70	2	78.52	93.68	87.99	86.35	90.56	80.64	79.52	69.12	56.64	88.08	78.44	73.00	19/14/14/25/36/13/22
SonyRobotSurface2	27	953	65	2	93.68	93.24	89.95	92.46	86.83	86.63	75.08	64.62	70.35	56.08	56.08	51.73	32/13/5/35/39/22/35
StarLightCurves	1000	8236	1024	3	73.07	94.97	88.00	80.00	40.00	40.00	40.00	40.00	40.00	40.00	40.00	42.24	12/10/135/14/18/13/14
SwedishLeaf	500	625	128	15	93.24	99.33	95.72	91.33	67.67	92.67	54.00	54.00	90.00	80.33	49.67	52.01	765/42/40/234/156/17/38
Symbols	25	995	398	6	91.90	100.00	88.00	81.00	100.00	97.84	100.00	100.00	89.00	63.00	85.00	64.00	136/98/64/90/64/48/91
SyntheticControl	300	300	60	6	99.80	100.00	76.00	89.03	84.81	74.36	64.35	84.81	63.48	62.77	49.96	75.24	312/686/47/155/122/72/81
Trace	100	100	275	4	90.97	90.43	74.71	89.03	84.81	25.70	64.35	25.88	63.48	83.98	49.96	51.73	276/542/53/120/102/66/74
TwoLeadECG	23	1139	82	2	88.65	83.11	85.94	86.15	87.51	73.06	55.83	46.34	87.59	60.25	35.68	51.73	298/251/54/137/114/74/87
TwoPatterns	1000	4000	128	4	70.68	90.66	84.88	84.80	80.85	64.10	45.83	46.01	60.66	53.43	34.79	42.24	149/183/178/52/74/23/47
uWaveGestureLibrary_X	896	3582	315	8	60.83	63.40	66.16	67.65	99.96	64.10	55.83	46.01	60.66	53.43	42.24	52.01	194/179/77/118/89/55/129
uWaveGestureLibrary_Y	896	3582	315	8	63.56	65.83	64.96	66.36	74.06	67.59	59.32	52.40	63.12	53.88	36.24	52.01	50/54/72/84/38/41
uWaveGestureLibrary_Z	896	3582	315	8	99.64	97.99	99.55	99.72	100.00	92.67	54.00	100.00	63.12	53.88	49.67	64.00	
wafer	6174	6174	152	2	99.80	100.00	76.00	81.00	100.00	97.84	100.00	100.00	89.00	96.24	96.71	97.92	
WordsSynonyms	267	638	270	25	40.61	88.58	81.76	62.54	61.44	50.94	53.76	49.37	39.97	30.72	25.39	39.81	
yoga	300	3000	426	2	73.07	83.63	83.03	83.67	73.13	66.40	70.20	72.00	68.20	65.03	61.33	61.00	
Mean Acc					73.23	78.06	73.89	77.45	77.44	65.98	61.20	55.30	59.26	53.48	53.27	57.54	
#wins	5	18	2	8	19												
MODL-TSC wins vs.	35	25	34	29			49	47	49	50	50	49					
Average rank	3.6667	2.6078	3.5784	2.5294	2.6176												

5   Conclusion and Perspectives

We have suggested MODL-TSC, a simple yet effective and generic feature construction process for time series classification problems (TSC). Our process is
parameter-free, easy to use and the generated features offer a high potential of
interpretation. The three main steps of the process are: (i) transforming data
for generating multiple data representations; (ii) coclustering on each representation; (iii) constructing new features from coclustering results. The new data
set is made of objects (time series identifiers) and descriptive attributes from
the various representations. To predict the class of new incoming time series,
we use the Selective Naive Bayes classifier (SNB). The time complexity of our
process is sub-quadratic, thus time-efficient. Experimental results show that the
performance of MODL-TSC is highly competitive and comparable with two of
the most accurate approaches of the state-of-the-art (namely, DTW-NN and TSC-
ENS). In addition, MODL-TSC embraces the eager paradigm and unlike the lazy
approaches (ED-NN, DTW-NN and TSC-ENS), our approach has a proper learning phase and can deploy fast enough for real-world applications. Moreover, a
qualitative study has shown that the generated features give some insight in the
data representations: we are able to qualify the adequacy of a representation for
solving the TSC problem at hand and to identify class-discriminating regions of
values from data representations embedded in our process.

The results of this work are promising and also confirm the importance of
representations in TSC problems. Indeed, depending on the application domain,
a particular transformation will facilitate the discovery of class relevant patterns.
Moreover, the combination of multiple representations with MODL-TSC leads
to highly competitive predictive performance. We have used only a few simple
representations in the time, frequency and correlation domains to demonstrate
that our feature construction approach is well-founded. The literature offers
plenty of relevant data representations (see [27] for a wide view). Notice also
that designing new representations is still a hot topic (see e.g., [19]). It gives a
large potential of improvement for MODL-TSC on data sets and applications where
we are less performant than DTW-NN and TSC-ENSEMBLE since our methodology
allows us to use a large spectrum of representations.

Acknowledgments. We wish to thank Anthony Bagnall and his team from University
of East-Anglia for providing TSC-ENSEMBLE prototype, Eamonn Keogh and his team
from University of California Riverside for providing prototypes of DTW-NN and FAST-
SHAPELETS.

References

1. Bagnall, A., Davis, L.M., Hills, J., Lines, J.: Transformation based ensembles for
   time series classification. In: SDM'12, pp. 307–318 (2012)
2. Batal, I., Sacchi, L., Bellazzi, R., Hauskrecht, M.: Multivariate time series classification with temporal abstractions. In: FLAIRS'09 (2009)

3. Boullé, M.: A bayes optimal approach for partitioning the values of categorical attributes. J. Mach. Learn. Res. **6**, 1431–1452 (2005)
4. Boullé, M.: MODL: a bayes optimal discretization method for continuous attributes. Mach. Learn. **65**(1), 131–165 (2006)
5. Boullé, M.: Compression-based averaging of selective naive Bayes classifiers. J. Mach. Learn. Res. **8**, 1659–1685 (2007)
6. Boullé, M.: Functional data clustering via piecewise constant nonparametric density estimation. Pattern Recogn. **45**(12), 4389–4401 (2012)
7. Buza, K.A.: Fusion methods for time-series classification. Ph.D. thesis, University of Hildesheim (2011)
8. Demsar, J.: Statistical comparisons of classifiers over multiple data sets. J. Mach. Learn. Res. **7**, 1–30 (2006)
9. Rodríguez, J.J., Alonso, C.J., Boström, H.: Learning first order logic time series classifiers: rules and boosting. In: Zighed, D.A., Komorowski, J., Żytkow, J.M. (eds.) PKDD 2000. LNCS (LNAI), vol. 1910, pp. 299–308. Springer, Heidelberg (2000)
10. Ding, H., Trajcevski, G., Scheuermann, P., Wang, X., Keogh, E.J.: Querying and mining of time series data: experimental comparison of representations and distance measures. PVLDB **1**(2), 1542–1552 (2008)
11. Eruhimov, V., Martyanov, V., Tuv, E.: Constructing high dimensional feature space for time series classification. In: Kok, J.N., Koronacki, J., Lopez de Mantaras, R., Matwin, S., Mladenič, D., Skowron, A. (eds.) PKDD 2007. LNCS (LNAI), vol. 4702, pp. 414–421. Springer, Heidelberg (2007)
12. Geurts, P.: Pattern extraction for time series classification. In: Siebes, A., De Raedt, L. (eds.) PKDD 2001. LNCS (LNAI), vol. 2168, p. 115. Springer, Heidelberg (2001)
13. Grabocka, J., Nanopoulos, A., Schmidt-Thieme, L.: Invariant time-series classification. In: Flach, P.A., De Bie, T., Cristianini, N. (eds.) ECML PKDD 2012, Part II. LNCS, vol. 7524, pp. 725–740. Springer, Heidelberg (2012)
14. Grünwald, P.: The Minimum Description Length Principle. MIT Press, Cambridge (2007)
15. Hidasi, B., Gáspár-Papanek, C.: ShiftTree: an interpretable model-based approach for time series classification. In: Gunopulos, D., Hofmann, T., Malerba, D., Vazirgiannis, M. (eds.) ECML PKDD 2011, Part II. LNCS, vol. 6912, pp. 48–64. Springer, Heidelberg (2011)
16. Kadous, M.W., Sammut, C.: Classification of multivariate time series and structured data using constructive induction. Mach. Learn. **58**(2–3), 179–216 (2005)
17. Keogh, E., Zhu, Q., Hu, B., Hao. Y., Xi, X., Wei, L., Ratanamahatana, C.A.: The UCR time series classification/clustering page (2011). http://www.cs.ucr.edu/~eamonn/time_series_data/
18. Liao, T.W.: Clustering of time series data - a survey. Pattern Recogn. **38**(11), 1857–1874 (2005)
19. Lines, J., Davis, L.M., Hills, J., Bagnall, A.: A shapelet transform for time series classification. In: KDD'12, pp. 289–297 (2012)
20. Mörchen, F.: Time series feature extraction for data mining using DWT and DFT. Technical report, Philipps Univeristy Marburg (2003)
21. Nanopoulos, A., Alcock, R., Manolopoulos, Y.: Feature-based classification of time-series data. In: Mastorakis, N., Nikolopoulos, S.D. (eds.) Information Processing and Technology, pp. 49–61. Nova Science (2001)
22. Rakthanmanon, T., Keogh, E.: Fast shapelets: a scalable algorithm for discovering time series shapelets. In: SIAM DM'13 (2013)

23. Ramsay, J., Silverman, B.: Functional Data Analysis. Springer, New York (2005)
24. Shannon, C.E.: A mathematical theory of communication. Bell System Technical Journal (1948)
25. Stefan, A., Athitsos, V., Das, G.: The move-split-merge metric for time series. Trans. Knowl. Data Eng. **25**, 1425–1438 (2013)
26. Vilalta, R., Rish, I.: A decomposition of classes via clustering to explain and improve naive bayes. In: Lavrač, N., Gamberger, D., Todorovski, L., Blockeel, H. (eds.) ECML 2003. LNCS (LNAI), vol. 2837, pp. 444–455. Springer, Heidelberg (2003)
27. Wang, X., Mueen, A., Ding, H., Trajcevski, G., Scheuermann, P., Keogh, E.: Experimental comparison of representation methods and distance measures for time series data. Data Min. Knowl. Disc. **26**(2), 275–309 (2013)
28. Xi, X., Keogh, E.J., Shelton, C.R., Wei, L., Ratanamahatana, C.A.: Fast time series classification using numerosity reduction. In: ICML'06, pp. 1033–1040 (2006)
29. Xing, Z., Pei, J., Yu, P.S., Wang, K.: Extracting interpretable features for early classification on time series. In: SDM'11, pp. 247–258 (2011)
30. Yamada, Y., Suzuki, E., Yokoi, H., Takabayashi, K.: Decision-tree induction from time-series data based on a standard-example split test. In: ICML'03, pp. 840–847 (2003)
31. Ye, L., Keogh, E.J.: Time series shapelets: a novel technique that allows accurate, interpretable and fast classification. Data Min. Knowl. Disc. **22**(1–2), 149–182 (2011)
32. Zhang, H., Ho, T.-B., Lin, M.-S.: A non-parametric wavelet feature extractor for time series classification. In: Dai, H., Srikant, R., Zhang, Ch. (eds.) PAKDD 2004. LNCS (LNAI), vol. 3056, pp. 595–603. Springer, Heidelberg (2004)

A Classification Based Scoring Function for Continuous Time Bayesian Network Classifiers

Daniele Codecasa and Fabio Stella[✉]

DISCo, Università degli Studi di Milano-Bicocca,
Viale Sarca 336, 20126 Milano, Italy
{codecasa,stella}@disco.unimib.it

Abstract. Continuous time Bayesian network classifiers are designed for analyzing multivariate streaming data when time duration of events matters. New continuous time Bayesian network classifiers are introduced while their conditional log-likelihood scoring function is developed. A learning algorithm, combining conditional log-likelihood with Bayesian parameter estimation is developed. Classification accuracies achieved on synthetic data by continuous time and dynamic Bayesian network classifiers are compared. Results show that conditional log-likelihood scoring combined with Bayesian parameter estimation outperforms marginal log-likelihood scoring in terms of classification accuracy. Continuous time Bayesian network classifiers are applied to post-stroke rehabilitation.

Keywords: Continuous time Bayesian networks · Multivariate streaming data · Conditional log-likelihood · Structural learning

1 Introduction

Streaming data are relevant to *finance* for high frequency trading [5], *computer science* for system error logs, web search query logs, network intrusion detection, social networks [24] and temporal semantic [16], and *engineering* for image, audio and video processing [30]. They are also important for analyzing GPS data, as shown in [14] and [4] where buses and animals paths are analyzed. Streaming data are becoming increasingly important in medicine for patient monitoring and continuous time diagnosis [10] including the study of firing patterns of neurons [28]. Finally, they are becoming relevant in biology where time course data [1] allow the reconstruction of gene regulatory networks, to model the evolution of infections, and to learn and analyze metabolic networks [29].

Dynamic Bayesian networks (DBNs) [6] and hidden Markov models (HMMs) [20] offer a natural way to represent and analyze streaming data. However, DBNs are concerned with discrete time and thus suffer from several limitations, due to the fact that it is not clear how timestamps should be discretized. In the case where a too slow sampling rate is used, the data will be poorly represented; while a too fast sampling rate rapidly makes learning and inference prohibitive.

A. Appice et al. (Eds.): NFMCP 2013, LNAI 8399, pp. 35–50, 2014.
DOI: 10.1007/978-3-319-08407-7_3, © Springer International Publishing Switzerland 2014

Furthermore, it has been pointed out [12] that when allowing long term dependencies it is required to condition on multiple steps into the past; thus, choosing a too fast sampling rate will increase the number of such steps that need to be conditioned on.

Continuous time Bayesian networks (CTBNs) [17], continuous time noisy-or (CT-NOR) [23], Poisson cascades [24] and Poisson networks [21] together with the piecewise-constant conditional intensity model (PCIM) [12] are interesting models to represent and analyze continuous time processes. CT-NOR and Poisson cascades are devoted to model event streams while they require the modeler to specify a parametric form for temporal dependencies. This aspect significantly impacts performance and the problem of model selection in CT-NOR and Poisson cascades has not been addressed yet. This limitation is overcome by PCIMs which perform structure learning to model how events in the past affect future events of interest. CTBNs are continuous time homogeneous Markov models which allow to represent joint trajectories of discrete finite variables.

In this paper we consider the problem of *temporal classification*, where data stream measurements are available over a period of time in history, while the class is expected to occur in the future. This kind of problem can be addressed by discrete and continuous time models. Discrete time models include dynamic latent classification models [31], a specialization of the latent classification model (LCM) [13], and DBNs [6]. Continuous time models, as continuous time Bayesian network classifiers (CTBNCs) [25], have overcome the problem of timestamps discretization. The main contributions of the paper are:

- definition of new classifiers from the class of CTBNCs,
- development of the conditional log-likelihood scoring function for CTBNCs,
- performance comparison of CTBNCs learned with the conditional log-likelihood score to CTBNCs learned with marginal log-likelihood score and to DBN classifiers.

The paper is organized as follows; Sect. 2 is devoted to notations and definitions. New classifiers are introduced and analyzed in Sect. 3. Section 4 concerns numerical experiments where synthetic data sets generated from models of increasing complexity are used. In this section a real data set on post-stroke rehabilitation is analyzed. Conclusions are proposed in Sect. 5.

2   Continuous Time Classification

2.1   Continuous Time Bayesian Networks

Dynamic Bayesian networks (DBNs) model dynamic systems without representing time explicitly. They discretize time to represent a dynamic system through several time slices. In [18] the authors pointed out that *"since DBNs slice time into fixed increments, one must always propagate the joint distribution over the variables at the same rate"*. Therefore, if the system consists of processes which evolve at different time granularities and/or the obtained observations are irregularly spaced in time, the inference process may become computationally intractable.

Continuous time Bayesian networks (CTBNs) have overcome the limitations of DBNs by explicitly representing temporal dynamics, which allows us to recover the probability distribution over time when specific events occur. A continuous time Bayesian network (CTBN) is a probabilistic graphical model whose nodes are associated with random variables and whose state evolves continuously over time.

Definition 1. *(Continuous time Bayesian network (CTBN)) [18]. Let* **X** *be a set of random variables* $X_1, X_2, ..., X_N$. *Each* X_n *has a finite domain of values* $Val(X_n) = \{x_1, x_2, ..., x_I\}$. *A continuous time Bayesian network* $\aleph$ *over* **X** *consists of two components: the first is an initial distribution* $P_{\mathbf{X}}^0$, *specified as a Bayesian network* $\mathcal{B}$ *over* **X**. *The second is a continuous transition model, specified as:*

- *a directed (possibly cyclic) graph* $\mathcal{G}$ *whose nodes are* $X_1, X_2, ..., X_N$; $Pa(X_n)$ *denotes the parents of* X_n *in* $\mathcal{G}$.
- *a conditional intensity matrix,* $\mathbf{Q}_{X_n}^{Pa(X_n)}$, *for each variable* $X_n \in \mathbf{X}$.

Given the random variable X_n, the *conditional intensity matrix* (CIM) $\mathbf{Q}_{X_n}^{Pa(X_n)}$ consists of a set of intensity matrices, one intensity matrix

$$\mathbf{Q}_{X_n}^{pa(X_n)} = \begin{bmatrix} -q_{x_1}^{pa(X_n)} & q_{x_1x_2}^{pa(X_n)} & \cdot & q_{x_1x_I}^{pa(X_n)} \\ q_{x_2x_1}^{pa(X_n)} & -q_{x_2}^{pa(X_n)} & \cdot & q_{x_2x_I}^{pa(X_n)} \\ \cdot & & \cdot & \cdot \\ q_{x_Ix_1}^{pa(X_n)} & q_{x_Ix_2}^{pa(X_n)} & \cdot & -q_{x_I}^{pa(X_n)} \end{bmatrix},$$

for each instantiation $pa(X_n)$ of the parents $Pa(X_n)$ of node X_n, where $q_{x_i}^{pa(X_n)} = \sum_{x_j \neq x_i} q_{x_ix_j}^{pa(X_n)}$ is the rate of leaving state x_i for a specific instantiation $pa(X_n)$ of $Pa(X_n)$, while $q_{x_ix_j}^{pa(X_n)}$ is the rate of arriving to state x_j from state x_i for a specific instantiation $pa(X_n)$ of $Pa(X_n)$. Matrix $\mathbf{Q}_{X_n}^{pa(X_n)}$ can equivalently be summarized by using two types of parameters, $q_{x_i}^{pa(X_n)}$ which is associated with each state x_i of the variable X_n when its parents are set to $pa(X_n)$, and $\theta_{x_ix_j}^{pa(X_n)} = \frac{q_{x_ix_j}^{pa(X_n)}}{q_{x_i}^{pa(X_n)}}$ which represents the probability of transitioning from state x_i to state x_j, when it is known that the transition occurs at a given instant in time.

Example 1. Figure 1 shows a part of the drug network introduced in [18]. It contains a cycle, indicating that whether a person is hungry (H) depends on how full his/her stomach (S) is, which depends on whether or not he/she is eating (E), which in turn depends on whether he/she is hungry.

Assume that E and H are binary variables with states *no* and *yes* while the variable S can be in one of the following states; *full*, *average* or *empty*. Then, the variable E is fully specified by the [2×2] CIM matrices $\mathbf{Q}_E^n$, and $\mathbf{Q}_E^y$, the

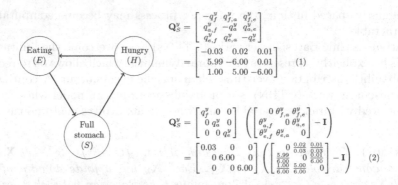

$$\mathbf{Q}_S^y = \begin{bmatrix} -q_f^y & q_{f,a}^y & q_{f,e}^y \\ q_{a,f}^y & -q_a^y & q_{a,e}^y \\ q_{e,f}^y & q_{a,a}^y & -q_e^y \end{bmatrix}$$

$$= \begin{bmatrix} -0.03 & 0.02 & 0.01 \\ 5.99 & -6.00 & 0.01 \\ 1.00 & 5.00 & -6.00 \end{bmatrix} \qquad (1)$$

$$\mathbf{Q}_S^y = \begin{bmatrix} q_f^y & 0 & 0 \\ 0 & q_a^y & 0 \\ 0 & 0 & q_e^y \end{bmatrix} \left(\begin{bmatrix} 0 & \theta_{f,a}^y & \theta_{f,e}^y \\ \theta_{a,f}^y & 0 & \theta_{a,e}^y \\ \theta_{e,f}^y & \theta_{e,a}^y & 0 \end{bmatrix} - \mathbf{I} \right)$$

$$= \begin{bmatrix} 0.03 & 0 & 0 \\ 0 & 6.00 & 0 \\ 0 & 0 & 6.00 \end{bmatrix} \left(\begin{bmatrix} 0 & \frac{0.02}{0.03} & \frac{0.01}{0.03} \\ \frac{5.99}{6.00} & 0 & \frac{0.01}{6.00} \\ \frac{1.00}{6.00} & \frac{5.00}{6.00} & 0 \end{bmatrix} - \mathbf{I} \right) \qquad (2)$$

Fig. 1. A part of the drug network and the two equivalent parametric representations of $\mathbf{Q}_S^y$ where $\mathbf{I}$ is the identity matrix.

variable S is fully specified by the $[3 \times 3]$ CIM matrices $\mathbf{Q}_S^n$ and $\mathbf{Q}_S^y$, while the variable H is fully specified by the $[2 \times 2]$ CIM matrices $\mathbf{Q}_H^f$, $\mathbf{Q}_H^a$ and, $\mathbf{Q}_H^e$.

If the hours are the units of time, then a person who has an empty stomach ($S=empty$) and is eating ($E=yes$) is expected to stop having an empty stomach in 10 min ($\frac{1.00}{6.00}$ h). The stomach will then transition from state $empty$ ($S=empty$) to state $average$ ($S=average$) with probability $\frac{5.00}{6.00}$ and to state $full$ ($S=full$) with probability $\frac{1.00}{6.00}$. Equation 1 is a compact representation of the CIM while Eq. 2 is useful because it explicitly represents the transition probability value from state x to state x', i.e. $\theta_{xx'}^{pa(X)}$.

CTBNs allow two types of evidence, namely *point evidence* and *continuous evidence*, while HMMs and DBNs allow only point evidence. *Continuous evidence* is the knowledge of the states of a set of variables $\mathbf{X}$ throughout an entire half-closed interval of time $[t_1, t_2)$: $\mathbf{Z}^{[t_1,t_2)} = \mathbf{z}^{[t_1,t_2)}$, where $\mathbf{Z}^{[t_1,t_2)} = (X_1^{[t_1,t_2)}, X_2^{[t_1,t_2)}, ..., X_k^{[t_1,t_2)})$ while $\mathbf{z}^{[t_1,t_2)} = (x_1^{[t_1,t_2)}, x_2^{[t_1,t_2)}, ..., x_k^{[t_1,t_2)})$.

Inference in CTBNs can be performed by exact and approximate algorithms. *Full amalgamation* [18] allows exact inference by generating an exponentially-large matrix representing the transition model over the entire state space. Exact inference in CTBNs is known to be intractable, and thus different approximate algorithms have been proposed. In [17] the authors introduced the *Expectation Propagation* algorithm (EP), while in [22] an optimized variant of EP is presented. Alternatives are offered by sampling based inference algorithms, such as *importance sampling* algorithm [8] and *Gibbs sampling* algorithm [7].

Given the data set $\mathcal{D}$, parameter learning is based on *marginal log-likelihood estimation*. It takes into account the *imaginary counts* of the hyperparameters (i.e. $\alpha_x^{pa(X)}$, $\alpha_{xx'}^{pa(X)}$ and, $\tau_x^{pa(X)}$):

$$q_x^{pa(X)} = \frac{\alpha_x^{pa(X)} + M[x \mid pa(X)]}{\tau_x^{pa(X)} + T[x \mid pa(X)]}; \theta_{xx'}^{pa(X)} = \frac{\alpha_{xx'}^{pa(X)} + M[x, x' \mid pa(X)]}{\alpha_x^{pa(X)} + M[x \mid pa(X)]} \qquad (3)$$

where $M[x, x' \mid pa(X)]$, $M[x \mid pa(X)]$ and $T[x \mid pa(X)]$ are the *sufficient statistics*. $M[x, x' \mid pa(X)]$ is the count of transitions from state x to state x' for node X when the state of its parents $Pa(X)$ is set to $pa(X)$. $M[x \mid pa(X)] = \sum_{x' \neq x} M[x, x' \mid pa(X)]$ is the count of transitions leaving state x of node X when the state of its parents $Pa(X)$ is set to $pa(X)$. Finally, $T[x \mid pa(X)]$ represents the time spent in state x by the variable X when the state of its parents $Pa(X)$ is set to $pa(X)$.

Learning the structure of a CTBN from a given data set $\mathcal{D}$ has been addressed as an optimization problem over possible CTBN structures [19]. It consists of finding the structure $\mathcal{G}$ which maximizes the following Bayesian score:

$$\text{score}\,(\mathcal{G} : \mathcal{D}) = \ln P(\mathcal{D}|\mathcal{G}) + \ln P(\mathcal{G}). \tag{4}$$

However, the search space of this optimization problem is significantly simpler than that of BNs or DBNs. Indeed, it is known that learning the optimal structure of a BN is NP-hard, while the same does not hold true in the context of CTBNs where all edges are across time and thus represent the effect of the current value of one variable on the next value of the other variables. Therefore, no acyclicity constraints arise, and it is possible to optimize the parent set for each variable of the CTBN independently. This allow a polynomial structural learning algorithm with respect to the number of variables and the dimension of the dataset, once fixed the maximum number of parents.

2.2   Continuous Time Bayesian Network Classifiers

Continuous time Bayesian network classifiers (CTBNCs) [25] are a specialization of CTBNs. They allow polynomial time classification of a static class, while for CTBNs general inference is intractable [18]. Classifiers from this class explicitly represent the evolution in continuous time of the set of random variables X_n, $n = 1, 2, ..., N$ which are assumed to depend on the static class node Y.

Definition 2. *(Continuous time Bayesian network classifier (CTBNC))*[1]. *A continuous time Bayesian network classifier is a pair* $C = \{\aleph, P(Y)\}$ *where* $\aleph$ *is a CTBN model with attribute nodes* $X_1, X_2, ..., X_N$, Y *is the class node with marginal probability* $P(Y)$ *on states* $Val(Y) = \{y_1, y_2, ..., y_K\}$, $\mathcal{G}$ *is the graph of the CTBNC, such that the following conditions hold:*

– $Pa(Y) = \emptyset$, the class variable Y is associated with a root node;
– Y is fully specified by $P(Y)$ and does not depend on time.

Given a data set $\mathcal{D}$ with no missing data, a CTBNC is learned by maximizing the score (4) subjected to the constraints listed in Definition 2. However, exact learning requires to set in advance the maximum number of parents k for the

---

[1] This definition differs from the one proposed in [25]. In fact, this definition does not require the CTBNC graph to be connected. Therefore, it allows structural learning algorithms to naturally perform feature selection.

nodes $X_1, X_2, ..., X_N$ [17] and thus in the case where k is not small a considerable computational effort is required to find the graph structure $\mathcal{G}^*$ which maximizes the score (4). In such a case we resort to hill-climbing or to the continuous time naive Bayes (CTNB).

Definition 3. *(Continuous time naive Bayes (CTNB)) [25]. A continuous time naive Bayes classifier is a continuous time Bayesian network classifier $\mathcal{C} = \{\aleph, P(Y)\}$ such that $Pa(X_n) = \{Y\}$, $n = 1, 2, ..., N$.*

Example 2. Figure 2a depicts the structure of a CTBNC to diagnose eating disorders from the eating process (Fig. 1). An example of the eating process is shown in Fig. 2b.

According to [25] a CTBNC $\mathcal{C} = \{\aleph, P(Y)\}$ classifies a stream of continuous time evidence $\mathbf{z} = (x_1, x_2, ..., x_N)$ for the attributes $\mathbf{Z} = (X_1, X_2, ..., X_N)$ over J contiguous time intervals, i.e. a stream of continuous time evidence $\mathbf{Z}^{[t_1, t_2)} = \mathbf{z}^{[t_1, t_2)}$, $\mathbf{Z}^{[t_2, t_3)} = \mathbf{z}^{[t_2, t_3)}$, ..., $\mathbf{Z}^{[t_{J-1}, t_J)} = \mathbf{z}^{[t_{J-1}, t_J)}$, by selecting the value y^* for the class Y which maximizes the posterior probability $P(Y | \mathbf{z}^{[t_1, t_2)}, \mathbf{z}^{[t_2, t_3)}, ..., \mathbf{z}^{[t_{J-1}, t_J)})$, which is proportional to

$$P(Y) \prod_{j=1}^{J} q_{x_{m_j}^j x_{m_j}^{j+1}}^{pa(X_{m_j})} \prod_{n=1}^{N} exp\left(-q_{x_n^j}^{pa(X_n)} \delta_j\right), \tag{5}$$

where:

- $\delta_j = t_j - t_{j-1}$ is the length of the jth time interval of the stream $\mathbf{z}^{[t_1, t_2)}, \mathbf{z}^{[t_2, t_3)}, ..., \mathbf{z}^{[t_{J-1}, t_J)}$ of continuous time evidence;
- $q_{x_n^j}^{pa(X_n)}$ is the parameter associated with state x_n^j, in which the variable X_n was during the jth time interval, given the state of its parents $pa(X_n)$ during the jth time intervals;
- $q_{x_m^j x_m^{j+1}}^{pa(X_m)}$ is the parameter associated with the transition from state x_m^j, in which the variable X_m was during the jth time interval, to state x_m^{j+1}, in which the variable X_m will be during the $(j + 1)^{th}$ time interval, given the state of its parents $pa(X_m)$ during the jth and the $(j + 1)^{th}$ time intervals.

The learning algorithm for the CTNB model, based on marginal log-likelihood maximization, and the inference algorithm for CTBNCs are described in [25].

3   Max-k Classifiers

3.1   Definitions

Structural learning for CTBNs is polynomial with respect to the number of variables and the size of the data set, once fixed the maximum number of parents (i.e. k). Nevertheless, increasing k rapidly brings to considerable computational efforts, while it implies more data is necessary to learn the node's parameter values conditioned on possible parents' instantiations. To overcome these limitations we propose the following instances from the class of CTBNCs: the Max-k Augmented CTNB (Max-k ACTNB) and the Max-k CTBNC (Max-k CTBNC) [3].

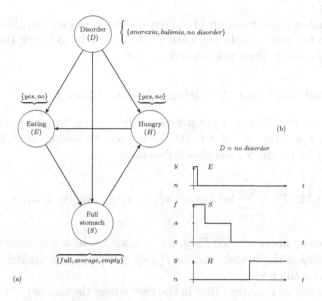

Fig. 2. CTBNC to diagnose eating disorders (a) observing the eating process (b).

Definition 4. *(Max-k Continuous Time Bayesian Network Classifier). A max-k continuous time Bayesian network classifier is a couple $\mathcal{M} = \{\mathcal{C}, k\}$, where $\mathcal{C}$ is a continuous time Bayesian network classifier $\mathcal{C} = \{\aleph, P(Y)\}$ such that the number of parents $|Pa(X_n)|$ for each attribute node X_n is bounded by a positive integer k. Formally, the following condition holds; $|Pa(X_n)| \le k$, $n = 1, 2, ..., N$, $k > 0$.*

Definition 5. *(Max-k Augmented Continuous Time Naive Bayes). A max-k augmented continuous time naive Bayes classifier is a max-k continuous time Bayesian network classifier such that the class node Y belongs to the parents set of each attribute node X_n, $n = 1, 2, ..., N$. Formally, the following condition holds; $Y \in Pa(X_n)$, $n = 1, 2, ..., N$.*

ACTNB constrains the class variable Y to be a parent of each node X_n, $n = 1, 2, ..., N$. In this way it tries to compensate for relevant dependencies between nodes which could be excluded to satisfy the constraint on the maximum number of parents k.

3.2   Learning

Learning a CTBNC from data consists of learning a CTBN where a specific node, i.e. the class node Y, does not depend on time. In such a case, the learning algorithm runs, for each attribute node X_n, $n = 1, 2, ..., N$, a local search procedure to find its optimal set of parents, i.e. the set of parents which maximizes a given score function. Furthermore, for each attribute node X_n, $n = 1, 2, ..., N$, no more than k parents are selected. The structural learning algorithm proposed in [19] uses the Bayesian score function (4) based on the marginal log-likelihood. This algorithm can be easily adapted to learn a CTBNC by introducing the constraint that the class node Y must not depend on time.

3.3 Log-likelihood and Conditional Log-likelihood

Scores based on log-likelihood are not the only scoring functions which can be used to learn the structure of a CTBN classifier. Following what is presented and discussed in [9], the log-likelihood function:

$$LL(\mathcal{M} \mid \mathcal{D}) = \sum_{i=1}^{|\mathcal{D}|} \log P_{\aleph}(y_i \mid \mathbf{x}_i^1, ..., \mathbf{x}_i^{J_i}) + \log P_{\aleph}(\mathbf{x}_i^1, ..., \mathbf{x}_i^{J_i}) \tag{6}$$

consists of two components; $\log P_{\aleph}(y_i \mid \mathbf{x}_i^1, ..., \mathbf{x}_i^{J_i})$, which measures the classification *capability* of the model, and $\log P_{\aleph}(\mathbf{x}_i^1, ..., \mathbf{x}_i^{J_i})$, which models the dependencies between the nodes.

In [9] the authors remarked that in the case where the number of the attribute nodes X_n, $n = 1, 2, ..., N$ is large, the contribution, to the log-likelihood function (6), of $\log P_{\aleph}(\mathbf{x}_i^1, ..., \mathbf{x}_i^{J_i})$ overwhelms the contribution of $\log P_{\aleph}(y_i \mid \mathbf{x}_i^1, ..., \mathbf{x}_i^{J_i})$. However, the contribution of $\log P_{\aleph}(\mathbf{x}_i^1, ..., \mathbf{x}_i^{J_i})$ is not directly related to the classification accuracy achieved by the classifier. Therefore, to improve the classification performance, in [9] it has been suggested to use the *conditional log-likelihood* as scoring function. In such a case the maximization of the conditional log-likelihood results in maximizing the classification performance of the model without paying specific attention to the discovery of the existing dependencies between the attribute nodes X_n, $n = 1, 2, ..., N$.

In the case where continuous time Bayesian network classifiers are considered, the conditional log-likelihood can be written as follows:

$$CLL(\mathcal{M} \mid \mathcal{D}) = \sum_{i=1}^{|\mathcal{D}|} \log P_{\aleph}(y_i \mid \mathbf{x}_i^1, ..., \mathbf{x}_i^{J_i}) \tag{7}$$

$$= \sum_{i=1}^{|\mathcal{D}|} \log \left(\frac{P_{\aleph}(\mathbf{x}_i^1, ..., \mathbf{x}_i^{J_i} \mid y_i) P_{\aleph}(y_i)}{P_{\aleph}(\mathbf{x}_i^1, ..., \mathbf{x}_i^{J_i})} \right)$$

$$= \sum_{i=1}^{|\mathcal{D}|} \log \left(P_{\aleph}(y_i) \right) + \sum_{i=1}^{|\mathcal{D}|} \log \left(P_{\aleph}(\mathbf{x}_i^1, ..., \mathbf{x}_i^{J_i} \mid y_i) \right) -$$

$$- \sum_{i=1}^{|\mathcal{D}|} \log \left(\sum_{y'} P_{\aleph}(y') P_{\aleph}(\mathbf{x}_i^1, ..., \mathbf{x}_i^{J_i} \mid y') \right).$$

Conditional log-likelihood scoring consists of three terms: *class probability term* (8), *posterior probability term* (9), and *denominator term* (10).

The *class probability term* is estimated from the learning data set $\mathcal{D}$ as follows:

$$\sum_{i=1}^{|\mathcal{D}|} \log\left(P_\aleph(y_i)\right) = \sum_y M[y]\log(\theta_y) \tag{8}$$

where θ_y represents the parameter associated with the probability of class y.
From (5) it is possible to write the following:

$$P_\aleph(\mathbf{x}^1,...,\mathbf{x}^J \mid y) = \prod_{j=1}^{J} q_{x_{m_j}^j x_{m_j}^{j+1}}^{pa(X_{m_j})} \prod_{n=1}^{N} exp\left(-q_{x_n}^{pa(X_n)}\delta_j\right)$$

$$= \prod_{j=1}^{J} q_{x_{m_j}^j}^{pa(X_{m_j})} \theta_{x_{m_j}^j x_{m_j}^{j+1}}^{pa(X_{m_j})} \prod_{n=1}^{N} exp\left(-q_{x_n}^{pa(X_n)}\delta_j\right)$$

Therefore, the *posterior probability term* can be estimated as follows:

$$\sum_{i=1}^{|\mathcal{D}|} \log\left(P_\aleph(\mathbf{x}_i^1,...,\mathbf{x}_i^{J_i} \mid y_i)\right) = \sum_{n=1}^{N} \sum_{x_n, pa(X_n)} M[x_n \mid pa(X_n)]\log\left(q_{x_n}^{pa(X_n)}\right)$$

$$- q_{x_n}^{pa(X_n)}T[x_n \mid pa(X_n)] + \sum_{x_n' \neq x_n} M[x_n x_n' \mid pa(X_n)]\log(\theta_{x_n x_n'}^{pa(X_n)}). \tag{9}$$

The *denominator term*, because of the sum, cannot be decomposed further.
The sufficient statistics allow us to write the following:

$$\sum_{i=1}^{|\mathcal{D}|} \log\left(\sum_{y'} P_\aleph(y')P_\aleph(\mathbf{x}_i^1,...,\mathbf{x}_i^{J_i} \mid y')\right) =$$

$$= \log\left(\sum_{y'} \theta_{y'} \prod_{n=1}^{N} \prod_{x_n, pa'(X_n)} (q_{x_n}^{pa'(X_n)})^{M[x_n|pa'(X_n)]}\right.$$

$$\left. exp(-q_{x_n}^{pa'(X_n)}T[x_n \mid pa'(X_n)]) \prod_{x_n' \neq x_n} (\theta_{x_n x_n'}^{pa'(X_n)})^{M[x_n x_n'|pa'(X_n)]}\right) \tag{10}$$

where $pa(X_n) = \{\pi_n \cup y\}$, $pa'(X_n) = \{\pi_n \cup y'\}$, while π_n is the instantiation of the non-class parents of the attribute node X_n.

The use of the conditional log-likelihood scoring function to learn continuous time Bayesian network classifiers is analyzed. Unfortunately, no closed form solution exists to compute the optimal value of the model's parameters, i.e. those parameters values which maximize the conditional log-likelihood (7). Therefore, the approach introduced and discussed in [11] is followed. The scoring function is computed by using the conditional log-likelihood, while parameter values are obtained by using the Bayesian approach as described in [19].

4   Numerical Experiments

Considering the Bayesian score (4), the prior distribution of the structure (i.e. $\ln P(\mathcal{G})$) becomes less relevant with the increase of the data set dimension. In the case where the data set dimension tends to infinity, the Bayesian score is equivalent to the marginal log-likelihood (MLL) score (i.e. $MLLscore(\mathcal{G} : \mathcal{D}) = \ln P(\mathcal{D}|\mathcal{G})$). For a fair comparison of the classification performance achieved when using the conditional log-likelihood (CLL) score (7), which does not use a graphs structure penalization term, the MLL score is used instead of the Bayesian score.

The performance of CTBNCs, namely CTNB, $k = 2$ ACTNB, $k = 2$ CTBNC, $k = 3$ CTBNC, and $k = 4$ CTBNC, is compared to that of DBNs by exploiting synthetic data sets. Classifiers are associated with a suffix related to the scoring function which has been used for learning. Suffix MLL is associated with marginal log-likelihood scoring while suffix CLL is associated with conditional log-likelihood scoring. DBNs are implemented by using the MATLAB Bayesian Nets toolbox [15]. Because of the computational effort to deal with DBNs, it was necessary to force a sampling ratio to discretize the trajectories that generates no more then 50 time slices per trajectory. In the same way the generated data sets were too big to allow the structural learning of DBNs. For this reason two naive Bayes models are used for comparison: the first one allows intra-slice naive Bayes relationships, while the second allows extra-slice relationships.

Numerical experiments for performance estimation and comparison of classifiers are implemented with 10-fold cross validation.

Numerical experiments on CTBNCs have been performed using the CTBNCToolkit [2], an open source Java toolkit developed by the authors to address and solve the problem of continuous time multivariate trajectory classification.

4.1   Synthetic Data Sets

Accuracy, learning and inference time of different CTBNCs are compared on synthetic data sets generated by sampling from models of increasing complexity. Data sets consist of $1,000$ trajectories with average length ranging from 300 (CTNBs) to 1,400 ($k = 4$ CTBNCs). Analyzed model structures are CTNB, $k = 2$ ACTNB, $k = 2$ CTBNC, $k = 3$ CTBNC, and $k = 4$ CTBNC (Fig. 3). For each structure, different assignments of parameter values (q parameters) are sampled in a given interval. Each pair, (*structure, parameter assignments*), is used to generate a learning data set.

Performance is analysed on *full data sets* (100 %) and *reduced data sets*, i.e. when the number and the length of trajectories are reduced to: 80 %, 60 %, 40 %, and 20 %. Accuracy values on *full data sets* (100 % data sets) are summarized in Table 1, while Fig. 4 depicts how accuracy behaves when reduced data sets (80 %, 60 %, 40 %, 20 %) are used for learning.

DBNs are outperformed by all continuous time models, while the CLL scoring seems to perform better, or at least to be not inferior, than the MLL scoring (see Fig. 4(a)). Figure 4(b) shows that CLL scoring outperforms the MLL scoring on the *reduced data sets*.

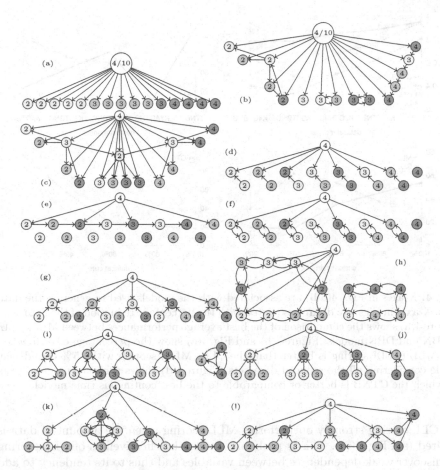

Fig. 3. CTNB (a), max-2 ACTNB (b,c), max-2 CTBNC (d–f), max-3 CTBNC (g–i) and, max-4 CTBNC (j–l) tested structures. Numbers associated with nodes represent the cardinality of the corresponding variables. White nodes are associated with classes, while the grey tonalities of the other nodes represent the width of the interval used to sample the q parameters.

Table 1. Classifier's average accuracy value with respect to different categories of the data set generating model, 10-fold cross validation over 100 % data sets. Bold characters are associated with the best model with 95 % confidence.

Test	$k=2$ CTNB	$k=2$ ACTNB (MLL)	$k=2$ ACTNB (CLL)	$k=2$ CTBNC (MLL)	$k=3$ CTBNC (CLL)	$k=3$ CTBNC (MLL)	$k=4$ CTBNC (CLL)	$k=4$ CTBNC (MLL)	CTBNC (CLL)	DBN-NB1	DBN-NB2
CTNB	**0.95**	**0.95**	0.93	0.93	0.92	0.93	0.82	0.93	0.64	0.80	0.81
K2ACTNB	0.78	0.89	**0.92**	0.76	**0.92**	0.76	0.85	0.76	0.72	0.62	0.63
K2CTBNC	0.68	0.84	**0.86**	**0.85**	0.86	**0.85**	0.76	**0.85**	0.60	0.48	0.50
K3CTBNC	0.49	0.65	0.63	0.66	0.63	**0.79**	0.75	**0.79**	0.64	0.32	0.33
K4CTBNC	0.64	0.74	0.79	0.69	0.79	0.76	**0.94**	0.79	0.90	0.40	0.40

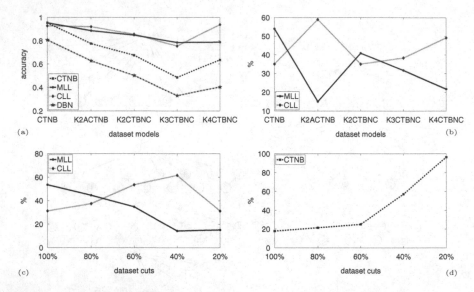

Fig. 4. X-axes in Figs. 4(a,b) are associated with the models used to generate the data sets. X-axes in Figs. 4(c,d) are associated with the percentage reduction of the data sets. Figure 4(a) shows the comparison of the best average performances between MLL, CLL, CTBN and DBN models. Figure 4(b) and Fig. 4(c) show the percentage of tests when the MLL (CLL) scoring is better than the CLL (MLL) scoring with 95 % confidence, while considering all the reduced data sets. Figure 4(d) shows the percentage of tests in which the CTNB is better or comparable to the best continuous time model.

CLL scoring strongly outperforms MLL scoring when the amount of data is limited (see Fig. 4(c)). This is probably due to the effectiveness of CLL scoring to discover weak dependences between variables and thus to its tendency to add the class variable as a parent of all nodes which are useful for the classification task. On the contrary, when the amount of data is too low, CLL scoring tends to overfit by learning classifiers which are too complex for the available amount of data. In these cases, MLL scoring achieves poor accuracy, while simple models like CTNB are the best option (see Fig. 4(d)).

To make the tests feasible DBN tests are made by using a discretization rate that reduces the number of data rows. For this reason it is impossible to exactly compare the time performances between continuous time classifiers and DBNs. Nevertheless, it is clear that dealing with discrete time models requires more computational effort than working in continuous time. Learning and inference times are summarized in Fig. 5. Numerical experiments have been performed on Intel(R) Xeon(R) CPU X5670 2.93 GHz, 15 Gb RAM. Inference time is almost the same for all the continuous time classifiers (Fig. 5b), while learning time varies across classifiers because of the different values of parent bounds (Fig. 5a). There is not a clear relation between learning time required by MLL and CLL. Theoretically, CLL should be more expensive to compute and should require little additional time than MLL, but probably due to the hill climbing algorithm

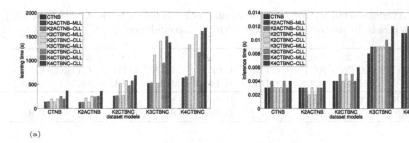

(a)   (b)

Fig. 5. Average learning (a) and inference (b) time for each model; in order: CTNB, k = 2ACTNB-MLL, k = 2ACTNB-CLL, k = 2CTBNC-MLL, k = 2CTBNC-CLL, k = 3CTBNC-MLL, k = 3CTBNC-CLL, k = 4CTBNC-MLL, k = 4CTBNC-CLL. X-axis is associated with the models used to generate the data sets.

and the structures induced by CLL it happens that often learning with CLL is faster than learning with MLL. Since CTNB does not require structural learning, but only parameter learning, it is the fastest model to learn.

4.2   Post-stroke Rehabilitation Data Set

In [27] the authors proposed a movement recognition system to face the automatic post-stroke rehabilitation problem. The idea is to provide the patient with a system capable of recognizing movements and to inform him/her about the correctness of the rehabilitation exercise performed. The authors focused on upper limb post-stroke rehabilitation and provided a data set of 7 rehabilitation exercises. For each exercise 120 multivariate trajectories are recorded by using 29 sensors working with a frequency of 30 Hz [26]. Each movement is addressed separately as a classification problem. We focused our the attention on 2, and 6 class problems where classes are associated with the same number of trajectories. The binary problem requires to identify if the movement is correct or not, while the 6 class problem introduces different degrees of correctness and errors of movements.

CTBNCs performances are better or comparable to Dynamic Time Warping (DTW) performances obtained in [27][2], even after the great simplification due to the original state space discretization. Accuracy values achieved by almost all CLL classifiers are better than accuracy values achieved by the corresponding MLL classifiers (Table 2). For the 6 class classification problem, in the case where no information about variable dependency is available (i.e. the links between the class and the other variables), CLL always outperforms MLL. Performance achieved by CTBNCs learned with CLL are robust with respect to the choice of the imaginary count values, while the same does not apply to MLL.

---

[2] DTW and Open End DTW (OE-DTW) obtained 0.99 accuracy values over the 2 class data set, while DTW obtained 0.88 and OE-DTW obtained 0.87 accuracy values over the 6 class data set [27].

Table 2. Average accuracy for the post-stroke rehabilitation data set (10 fold CV). Bold characters indicate the best models with 95 % confidence.

# classes	CTNB	$k = 2$ ACTNB (MLL)	$k = 2$ ACTNB (CLL)	$k = 2$ CTBNC (MLL)	$k = 2$ CTBNC (CLL)	$k = 3$ CTBNC (MLL)	$k = 3$ CTBNC (CLL)	$k = 4$ CTBNC (MLL)	$k = 4$ CTBNC (CLL)
2 classes	0.98	0.97	**0.99**	0.87	0.85	0.87	0.92	0.87	0.95
6 classes	**0.91**	**0.91**	**0.89**	0.81	**0.88**	0.81	**0.88**	0.81	**0.88**

$k = 2$ ACTNB, learned with CLL scoring, implements the optimal trade-off between the continuous time models in terms of time and accuracy. Indeed, for both 2 and 6 class classification problems, the $k = 2$ ACTNB model when learned with CLL, achieves the highest accuracy value and is the fastest to learn because of the small number of parents.

5   Conclusions

Continuous time Bayesian network classifiers are a model that can analyze multivariate trajectories for classification purposes of a static variable. Before the contributions in this paper only the inference algorithm was introduced. In the paper for the first time the structural learning of CTBNCs was introduced. A conditional log-likelihood scoring function has been developed to learn continuous time Bayesian network classifiers. A learning algorithm for CTBNCs is designed by combining conditional log-likelihood scoring with Bayesian parameter learning. New classifier models from the class of CTBNCs have been introduced.

Numerical experiments, on synthetic and real world streaming data sets, confirm the effectiveness of the proposed approach for CTBNCs learning. In particular, conditional log-likelihood scoring outperforms marginal log-likelihood scoring in terms of the accuracy achieved by CTBNCs. This behaviour becomes more and more evident as the amount of the available streaming data becomes scarce.

Future research directions are focused on extending CTBNCs to solve clustering problems and to overcome the memoryless assumption.

Acknowledgements. The authors would like to acknowledge the many helpful suggestions of the anonymous reviewers, who helped to improve the paper clarity and quality. The authors would like to thank Project Automation S.p.A. for funding the Ph.D. programme of Daniele Codecasa.

References

1. Barber, D., Cemgil, A.: Graphical models for time-series. IEEE Signal Process. Mag. **27**(6), 18–28 (2010)
2. Codecasa, D., Stella, F.: CTBNCToolkit: continuous time Bayesian network classifier toolkit, arXiv:1404.4893v1[cs.AI] (2014)

3. Codecasa, D., Stella, F.: Conditional log-likelihood for continuous time Bayesian network classifiers. In: International Workshop NFMCP held at ECML-PKDD2013 (2013)
4. Costa, G., Manco, G., Masciari, E.: Effectively grouping trajectory streams. In: Appice, A., Ceci, M., Loglisci, C., Manco, G., Masciari, E., Ras, Z.W. (eds.) NFMCP 2012. LNCS (LNAI), vol. 7765, pp. 94–108. Springer, Heidelberg (2013)
5. Dacorogna, M.: An introduction to high-frequency finance. AP (2001)
6. Dean, T., Kanazawa, K.: A model for reasoning about persistence and causation. Comput. Intell. **5**(2), 142–150 (1989)
7. El-Hay, T., Friedman, N., Kupferman, R.: Gibbs sampling in factorized continuous-time markov processes. In: McAllester, D.A., Myllym, P. (eds.) Proceedings of the 24th Conference on UAI, pp. 169–178. AUAI (2008)
8. Fan, Y., Shelton, C.: Sampling for approximate inference in continuous time bayesian networks. In: 10th International Symposium on Artificial Intelligence and Mathematics (2008)
9. Friedman, N., Geiger, D., Goldszmidt, M.: Bayesian network classifiers. Mach. Learn. **29**(2), 131–163 (1997)
10. Gatti, E., Luciani, D., Stella, F.: A continuous time Bayesian network model for cardiogenic heart failure. Flex. Serv. Manuf. J. **24**(4), 496–515 (2012)
11. Grossman, D., Domingos, P.: Learning bayesian network classifiers by maximizing conditional likelihood. In: Proceedings of the 21st International Conference on Machine Learning, pp. 361–368. ACM (2004)
12. Gunawardana, A., Meek, C., Xu, P.: A model for temporal dependencies in event streams. In: Shawe-Taylor, J., Zemel, R., Bartlett, P., Pereira, F., Weinberger, K. (eds.) Advances in Neural Information Processing Systems, pp. 1962–1970. Morgan Kaufmann, Burlington (2011)
13. Langseth, H., Nielsen, T.: Latent classification models. Mach. Learn. **59**(3), 237–265 (2005)
14. Masciari, E.: Trajectory Clustering via Effective Partitioning. In: Andreasen, T., Yager, R.R., Bulskov, H., Christiansen, H., Larsen, H.L. (eds.) FQAS 2009. LNCS(LNAI), vol. 5822, pp. 358–370. Springer, Heidelberg (2009)
15. Murphy, K.: The bayes net toolbox for matlab. Comput. Sci. Stat. **33**(2), 1024–1034 (2001)
16. Nanni, M., Pedreschi, D.: Time-focused clustering of trajectories of moving objects. J. Intell. Inf. Syst. **27**(3), 267–289 (2006)
17. Nodelman, U., Koller, D., Shelton, C.: Expectation propagation for continuous time bayesian networks. In: Proceedings of the 21st Conference on UAI, pp. 431–440. Edinburgh, Scotland, UK (July 2005)
18. Nodelman, U., Shelton, C., Koller, D.: Continuous time bayesian networks. In: Proceedings of the 18th Conference on UAI, pp. 378–387. Morgan Kaufmann (2002)
19. Nodelman, U., Shelton, C., Koller, D.: Learning continuous time bayesian networks. In: Proceedings of the 19th Conference on UAI, pp. 451–458 (2002)
20. Rabiner, L.: A tutorial on hidden markov models and selected applications in speech recognition. Proc. IEEE **77**(2), 257–286 (1989)
21. Rajaram, S., Graepel, T., Herbrich, R.: Poisson-networks: A model for structured point processes. In: Proceedings of the 10th International Workshop on Artificial Intelligence and Statistics (2005)
22. Saria, S., Nodelman, U., Koller, D.: Reasoning at the right time granularity. In: UAI, pp. 326–334 (2007)

23. Simma, A., Goldszmidt, M., MacCormick, J., Barham, P., Black, R., Isaacs, R., Mortier, R.: Ct-nor: Representing and reasoning about events in continuous time. In: Proceedings of the 24th Conference on UAI, pp. 484–493. AUAI (2008)
24. Simma, A., Jordan, M.: Modeling events with cascades of poisson processes. In: Proceedings of the 26th Conference on UAI, pp. 546–555. AUAI (2010)
25. Stella, F., Amer, Y.: Continuous time bayesian network classifiers. J. Biomed. Inform. **45**(6), 1108–1119 (2012)
26. Tormene, P., Giorgino, T.: Upper-limb rehabilitation exercises acquired through 29 elastomer strain sensors placed on fabric. release 1.0 (2008)
27. Tormene, P., Giorgino, T., Quaglini, S., Stefanelli, M.: Matching incomplete time series with dynamic time warping: an algorithm and an application to post-stroke rehabilitation. Artif. Intell. Med. **45**(1), 11–34 (2009)
28. Truccolo, W., Eden, U., Fellows, M., Donoghue, J., Brown, E.: A point process framework for relating neural spiking activity to spiking history, neural ensemble, and extrinsic covariate effects. J. Neurophysiol. **93**(2), 1074–1089 (2005)
29. Voit, E.: A First Course in Systems Biology. Garland Science, New York (2012)
30. Yilmaz, A., Javed, O., Shah, M.: Object tracking: a survey. ACM Comput. Surv. **38**(4), 45 (2006)
31. Zhong, S., Langseth, H., Nielsen, T.: Bayesian networks for dynamic classification. Technical report (2012)

Trajectory Data Pattern Mining

Elio Masciari[1]([envelope]), Gao Shi[2], and Carlo Zaniolo[2]

[1] ICAR-CNR, Naples, Italy
masciari@icar.cnr.it,
[2] UCLA, Los Angeles, USA
{shi,zaniolo}@cs.ucla.edu

Abstract. In this paper, we study the problem of mining for frequent trajectories, which is crucial in many application scenarios, such as vehicle traffic management, hand-off in cellular networks, supply chain management. We approach this problem as that of mining for frequent sequential patterns. Our approach consists of a partitioning strategy for incoming streams of trajectories in order to reduce the trajectory size and represent trajectories as strings. We mine frequent trajectories using a sliding windows approach combined with a counting algorithm that allows us to promptly update the frequency of patterns. In order to make counting really efficient, we represent frequent trajectories by prime numbers, whereby the Chinese reminder theorem can then be used to expedite the computation.

1 Introduction

In this paper, we address the problem of extracting frequent patterns from trajectory data streams. Due to its many applications and technical challenges, the problem of extracting frequent patterns has received a great deal of attention since the time it was originally introduced for transactional data [1,12] and later adressed for dynamic datasets in [3,13,23]. For trajectory data the problem was studied in [9,11,26]. Trajectories are data logs recording the time and the position of moving objects (or groups of objects) that are generated by a wide variety of applications. Examples include GPS systems [9], supply chain management [10], vessel classification by satellite images [20]. For instance, consider moving vehicles, such as cars or trucks where personal or vehicular mobile devices produce a digital traces that are collected via a wireless network infrastructures. Merchants and services can benefit from the availability of information about frequent routes crossed by such vehicles. Indeed, a very peculiar type of trajectory is represented by stock market. In this case, space information can be assumed as a linear sequence of points whose actual values has to be evaluated w.r.t. preceding points in he sequence in order to estimate future fluctuations. Such a wide spectrum of pervasive and ubiquitous sources and uses guarantee an increasing availability of large amounts of data on individual trajectories, that could be mined for crucial information. Therefore, due to the large amount of trajectory streams generated every day, there is a need for analyzing them efficiently in

A. Appice et al. (Eds.): NFMCP 2013, LNAI 8399, pp. 51–66, 2014.
DOI: 10.1007/978-3-319-08407-7_4, © Springer International Publishing Switzerland 2014

order to extract useful information. The challenge posed by data stream systems and data stream mining is that, in many applications, data must be processed continuously, either because of real time requirements or simply because the stream is too massive for a store-now & process-later approach. However, mining of data streams brings many challenges not encountered in database mining, because of the real-time response requirement and the presence of bursty arrivals and concept shifts (i.e., changes in the statistical properties of data). In order to cope with such challenges, the continuous stream is often divided into windows, thus reducing the size of the data that need to be stored and mined. This allows detecting concept drifts/shifts by monitoring changes between subsequent windows. Even so, frequent pattern mining over such large windows remains a computationally challenging problem requiring algorithms that are faster and lighter than those used on stored data. Thus, algorithms that make multiple scans of the data should be avoided in favor of single-scan, incremental algorithms. In particular, the technique of partitioning large windows into slides (a.k.a. panes) to support incremental computations has proved very valuable in DSMS [24] and will be exploited in our approach. We will also make use of the following key observation: in real world applications there is an obvious difference between the problem of (i) finding new association rules, and (ii) verifying the continuous validity of existing rules. In order to tame the size curse of point-based trajectory representation, we propose to partition trajectories using a suitable regioning strategy. Indeed, since trajectory data carry information with a detail not often necessary in many application scenarios, we can split the search space in regions having the suitable granularity and represent them as simple strings. The sequence of regions (strings) define the trajectory traveled by a given object. Regioning is a common assumption in trajectory data mining [9,20] and in our case it is even more suitable since our goal is to extract typical routes for moving objects as needed to answer queries such as: *which are the most used routes between Los Angeles and San Diego?* thus extracting a pattern showing every point in a single route is useless.

The partitioning step allow us to represent a trajectory as string where each substring encodes a region, thus, our proposal for incremental mining of frequent trajectories is based on an efficient algorithm for frequent string mining. As a matter of fact, the extracted patterns can be profitably used in systems devoted to traffic management, human mobility analysis and so on. Although a real-time introduction of new association rules is neither sensible nor feasible, the on-line verification of old rules is highly desirable for two reasons. The first is that we need to determine immediately when old rules no longer holds to stop them from pestering users with improper recommendations. The second is that every window can be divided in small panes on which the search for new frequent patters execute fast. Every pattern so discovered can then be verified quickly. Therefore, in this paper we propose a fast algorithm, called *verifier* henceforth, for verifying the frequency of previously frequent trajectories over newly arriving windows. To this end, we use sliding windows, whereby a large window is partitioned into smaller panes [24] and a response is returned promptly at the end of each slide (rather than at the end of each large window). This also leads to a more efficient computation since the frequency of the trajectories in the whole window can be computed incrementally by counting trajectories in the new incoming (and old expiring) panes.

Our approach in a nutshell. As trajectories flow we partition the incoming stream in windows, each window being partitioned in slides. In order to reduce the size of the input trajectories we pre-process each incoming trajectory in order to obtain a smaller representation of it as a sequence of regions. We point out that this operation is well suited in our framework since we are not interested in point-level movements but in trajectory shapes instead. The regioning strategy we exploit uses PCA to better identify directions along which we should perform a more accurate partition disregarding regions not on the principal directions. The rationale for this assumption is that we search for frequent trajectories so it is unlikely that regions far away from principal directions will contribute to frequent patterns (in the following we will use frequent patterns and frequent trajectories as synonym). The sequence of regions so far obtained can be represented as a string for which we can exploit a suitable version of well known frequent string mining algorithms that works efficiently both in terms of space and time consumption. We initially mine the first window and store the frequent trajectories mined using a tree structure. As windows flow (and thus slides for each window) we continuously update frequency of existing patterns while searching for new ones. This step requires an efficient method for counting (a.k.a verification). Since trajectory data are ordered we need to take into account this feature. We implement a novel verifier that exploits prime numbers properties in order to encode trajectories as numbers and keeping order information, this will allow a very fast verification since searching for the presence of a trajectory will result in simple (inexpensive) mathematical operations.

Remark. In this paper we exploit techniques that were initially introduced in some earlier works [27–29]. We point out that in this work we improved those approaches in order to make them suitable for sequential pattern mining. Moreover, validity of data mining approaches rely on their experimental assessment. In this respect the experiments we performed confirmed the validity of the proposed approach.

2 Related Work

Mining trajectory data is an active research area and many interesting proposals exist in the literature. In [32] an algorithm for sequential pattern mining is introduced. The algorithm *TrajPattern* mines patterns by a process that identifies the top-k most important patterns with respect to a suitable measure called NM. The algorithm exploits a min-max property since the well known Apriori property no longer holds for that measure. A general formal statement for the trajectory pattern mining is provided in [9] where trajectory pattern are characterized in terms of both space (introducing the concept of regions of interest) and time (considering the duration of movements).

Trajectory mining has been deeply investigated in recent years. In [21] trajectory clustering has been explored. Clustering is performed using a two-phase algorithm that first partitions the trajectory using the MDL principle and then clusters the trajectory segments using a line-segment clustering algorithm. In [25] a technique for defining and maintaining micro clusters of moving objects. In [14] a filter-refinement approach has been used for discovering convoys in trajectory databases. In particular various trajectory simplification techniques are studied along with different query processing strategies.

In [22] a partition end detect framework is presented for trajectory outlier detection. In particular a hybrid approach distance-based and density-based is exploited to identify anomalies.

Also, in this paper we have borrowed many techniques from traditional frequent pattern mining, both for stored and streaming data. Many algorithms have been proposed for mining of frequent itemsets [1,15,33], but due to space limitations we will only discuss those that are most relevant to this paper.

For instance, Pei et al. [30] and Zaki et al. [33], present efficient algorithms, Closet and Charm, respectively, to mine closed frequent itemsets; an itemset is closed if none of its proper supersets has the same support as it has.

Han et al. [31], introduced an efficient data structure, called *fp-tree*, for compactly storing transactions given a minimum support threshold. Then they proposed an efficient algorithm (called FP-growth) for mining an *fp-tree* [31].

We borrow the double-conditionalization technique from Mozafari et al. [29] which used the so-called *fp-tree* data structure and the original conditionalization idea from [31]. This situation has motivated a considerable amount of research interest in online frequent itemsets mining as well [2,4,7]. Lee et al. [18] propose generating k-candidate sets from (k-1)-candidate sets, i.e., without verifying their frequency. This avoids extra passes over the data and, according to the authors, does not result in too many additional candidate sets. Chi et al. [7] propose the Moment algorithm for maintaining closed frequent itemsets, whereas Cats Tree [2] and CanTree [4] support the incremental mining of frequent itemsets.

There has been also a significant amount of work on counting itemsets (rather than finding the frequent ones). Hash-based counting methods, originally proposed in Park et al. [5], are in fact used by many of the above-mentioned frequent itemsets algorithms [1,5,33], whereas Brin et al. [6], proposed a dynamic algorithm, called DIC, for efficiently counting itemsets' frequencies.

3   Trajectory Size Reduction

For transactional data a tuple is a collection of features. Instead, a trajectory is an ordered set (i.e., a sequence) of timestamped points. We assume a standard format for input trajectories, as defined next. Let P and T denote the set of all possible (spatial) positions and all timestamps, respectively. A trajectory Tr of length n is defined as a finite sequence $s_1, \cdots, s_n$, where $n \geq 1$ and each s_i is a pair (p_i, t_i) where $p_i \in P$, $t_i \in T$ and $t_i < t_{i+1}$. We assume that P and T are discrete domains; however this assumption does not affect the validity of our approach. In order to deal with these intrinsically redundant data, a viable approach is to partition the space into regions in order to map the initial locations into discrete regions labeled with a timestamped symbol. The problem of finding a suitable partitioning for both the search space and the actual trajectory is a core problem when dealing with spatial data. Every technique proposed so far somehow deals with regioning and several approaches have been proposed such as partitioning of the search space in several regions of interest (*RoI*) [9] and trajectory partitioning (e.g., [21]) by using polylines. In this section, we describe the application of Principal Component Analysis (*PCA*) [16] in order to obtain a better partitioning. Indeed, *PCA* finds *preferred* directions for data being analyzed. We denote as *preferred* directions the (possibly imaginary) axes where the majority of the trajectories lie. Once we detect the preferred directions we perform a partitioning of the search space along these directions. Many tools have been implemented for computing PCA vectors such as [16], in our framework due to the streaming nature of data we exploited an incremental PCA (IPCA) algorithm proposed in [16]. method based on the idea of a singular value decomposition (SVD) updating algorithm, namely an SVD updating-based IPCA (SVDU-IPCA) algorithm. For this SVDU-IPCA algorithm, it has been mathematically proved that the approximation error is bounded. The latter is a relevant feature since the quality of regioning heavily relies on the quality of IPCA results. Due to space limitations, instead of giving a detailed description of the mathematical steps implemented in our prototype, we will present an illustrating (real life) example, that will show the main features of the approach.

Example 1. Consider the set of trajectories depicted in Fig. 1(a) regarding bus movements in the Athens metropolitan area. There are several trajectories close to the origin of the axes so it is difficult to identify the most interesting areas for analysis.

In order to properly assign regions we need to set a suitable level of granularity by defining the initial size s of each region, i.e., its diameter. We assume for the sake of simplicity squared regions and store the center of each region. The initial size s (i.e. the size of regions along principal directions) of each region should be set according to the domain being analyzed. In order to keep an intuitive semantics for regions of interest we partition the search space into square regions along the directions set by the eigenvalues returned by IPCA. Since the region

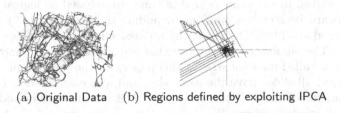

(a) Original Data   (b) Regions defined by exploiting IPCA

Fig. 1. Trajectory Pre-Elaboration steps

granularity will affect further analysis being performed the choice of region size s is guided by *DBScan* an unsupervised density based clustering algorithm. The output of the regioning step is depicted in Fig. 1(b).

Definition 1 (Dense Regions). *Let T be a set of trajectories, and X_I and Y_I the axes defined by IPCA, $C = \{C_1, C_2, \cdots, C_n\}$ a set of regions obtained with density based algorithm (DBScan) laying on X_I and Y_I, the regions defined by C_i's boundaries are* Dense.

More in detail, we denote as dense regions the one that both lay on the principal *and* belongs to a dense cluster, thus the region size is the cluster diameter.

4   Frequent Trajectories Mining

The regioning schema presented in previous section allows a compact representation of trajectories by the sequences of regions crossed by each trajectory, i.e., as a set of strings, where each substring encodes a region. It is straightforward to see that this representation transforms the problem of searching frequent information in a (huge) set of multidimensional points into the problem of searching frequent (sub)strings in a set of strings representing trajectories. We point out that our goal is to mine frequent trajectories tackling the "where is" problem, i.e., we are interested in movements made by objects disregarding time information (such as velocity). Moreover, since the number of trajectories that could be monitored in real-life scenarios is really huge we need to work on successive portions of the incoming stream of data called *windows*. Let $T = \{T_1, \cdots, T_n\}$ be the set of regioned trajectories to be mined belonging to the current window; T contains several trajectories where each trajectory is a sequence of regions. Let $S = \{S_1, \cdots, S_n\}$ denotes the set of all possible (sub)trajectories of T. The *frequency* of a (sub)trajectory S_i is the number of trajectories in T that contain S_i, and is denoted as $Count(S_i, T)$. The *support* of S_i, $sup(S_i, T)$, is defined as its frequency divided by the total number of trajectories in T. Therefore, $0 \leq sup(S_i, T) \leq 1$ for each S_i. The goal of frequent trajectories mining is to find all such S_i whose support is at least some given minimum support threshold α. The set of frequent trajectories in T is denoted as $\mathcal{F}_\alpha(T)$. We consider in this paper frequent trajectories mining over a data stream, thus T is defined as a sliding window W over the continuous stream. Each window either contains the same number of trajectories (count based or physical window), or contains all trajectories arrived in the same period of time (time-based or logical window). T moves forward by a certain amount by adding the new slide (δ^+) and dropping the expired one (δ^-). Therefore, the successive instances of T are shown as $W_1, W_2, \cdots$. The number of trajectories that are added to (and removed from) each window is called its *slide size*. In this paper, for the purpose of simplicity, we assume that all slides have the same size, and also each window consists of the same number of slides. Thus, $n = |W|/|\mathcal{S}|$ is the number of slides (a.k.a. panes) in each window, where $|W|$ denotes the window size and $|\mathcal{S}|$ denotes the size of the slides.

Mining Trajectories in W. As we obtain the string representation of trajectories, we focus on the string mining problem. In particular, given a set of input strings, we want to extract the (unknown) strings that obey certain frequency constraints. The frequent string mining problem can be formalized as follows. *Given a set T of input strings and a given frequency threshold α, find the set S_F s.t. $\forall s \in S_F, count(s, T) > \alpha$.*

Many proposals have been made to tackle this problem [8,17]. We exploit in this paper the approach presented in [17]. The algorithm works by searching for frequent strings in different databases of strings. In our paper we do not have different databases, we have different windows instead. We first briefly recall the basic notions needed for the algorithm. More details can be found in [8,17].

The suffix array SA of a string s is an array of integers in the range $[1..n]$, which describes the lexicographic order of the n suffixes of s. The suffix array can be computed in linear time [17]. In addition to the suffix array, we define the inverse suffix array SA^{-1}, which is defined $for all 1 \leq i \leq n$ by $SA^{-1}[SA[i]] = i$. The LCP table is an array of integers which is defined relative to the suffix array of a string s. It stores the length of the longest common prefix of two adjacent suffixes in the lexicographically ordered list of suffixes. The LCP table can be calculated in $O(n)$ from the suffix array and the inverse suffix array. The ω-interval is the longest common prefix of the suffixes of s. The algorithm is reported in Fig. 2 and its features can be summarized as follows.

Function *extractStrings* arranges the input strings in the window W_i in a string S^{aux} consisting of the concatenation of the strings in W_i, using # as a separation symbol and $ as termination symbol. Functions *buildSuffixes* and *buildPrefixes* compute respectively the suffixes and prefixes of S^{aux} and store them using SA and LCP variables. Function *computeRelevantStrings* first computes the number of times that a string s occurs in W_i and then subtracts so called correction terms which take care of multiple occurrences within the same string of W_i as defined in [17]. The output frequent strings are arranged in a tree structure that will be exploited for incremental mining purposes as will be explained in the next section.

Method: MineFrequentStrings
Input: A window slide S of the input trajectories;
Output: A set of frequent strings S_F.
Vars:
A string S^{aux};
A suffix array SA;
A prefix array LCP.
1: $S^{aux} = extractStrings(S)$;
2: $SA = buildSuffixes(S^{aux})$;
3: $LCP = buildPrefixes(S^{aux})$;
4: $S_F = computeRelevantStings(W_0, SA, LCP)$
5: **return** S_F;

Fig. 2. The frequent string mining algorithm

Incremental Mining of Frequent Trajectories. As the trajectory stream flows we need to incrementally update the frequent trajectories pattern so far computed (that are inserted in a *Trajectory Tree (TT)*). Our algorithm always maintains a union of the frequent trajectories of all slides in the current window W in TT, which is guaranteed to be a superset of the frequent pattern over W. Upon arrival of a new slide and expiration of an old one, we update the true count of each pattern in TT, by considering its frequency in both the expired slide and the new slide. To assure that TT contains all patterns that are frequent in at least one of the slides of the current window, we must also mine the new slide and add its frequent patterns to TT. The difficulty is that when a new pattern is added to TT for the first time, its true frequency in the whole window is not known, since this pattern was not frequent in the previous $n-1$ slides. To address this problem, we uses an auxiliary array *(aux)* for each new pattern in the new slide. The *aux* array stores the frequency of a pattern in each window starting at a particular slide in the current window. The key point is that this counting can either be done eagerly (i.e., immediately) or lazily. Under the laziest approach, we wait until a slide expires and then compute the frequency of such new patterns over this slide and update the aux arrays accordingly. This saves many additional passes through the window. The pseudo code for the algorithm is given in Fig. 3. At the end of each slide, it outputs all patterns in TT whose frequency at that time is $\geq \alpha \hat{n} |S|$. However we may miss a few patterns due to lack of knowledge at the time of output, but we will report them as delayed when

---

Method: **IncrementalMaintenance**
Input: A trajectory stream T.
Output: A trajectory pattern tree T_T.
Vars:
A window slide S of the input trajectories;
An auxiliary array aux;
A trajectory tree TT'
1: **For Each** New Slide S_{new}
2:      $updateFrequencies(TT, S)$;
3:      $TT' = MineFrequentStrings(S_{new})$;
4:      **For Each** trajectory $t \in TT \cap TT'$
5:          $annotateLast(S_{new}, t)$;
6:      **For Each** trajectory $t \in TT' \setminus TT$
7:          $update(TT, t)$;
8:          $annotateFirst(S_{new}, t, t.aux)$;
9: **For Each** Expiring Slide S_{exp}
10:      **For Each** trajectory $t \in TT$
11:          $conditionalUpdateFrequencies(S_{exp}, t)$;
12:          $conditionalUpdate(t.aux)$;
13:          **if** t has existed since arrival of S
14:              $delete(t.aux)$;
15:          **if** t no longer frequent in any of the current slides
16:              $delete(t)$;

Fig. 3. The incremental miner algorithm

other slides expire. The algorithm starts when the first slide has been mined and its frequent trajectories are stored in TT.

Herein, function *updateFrequencies* updates the frequencies of each pattern in TT if it is present in S. As the new frequent patterns are mined (and stored in TT'), we need to annotate the current slide for each pattern as follows: if a given pattern t already existed in TT we annotate S as the last slide in which t is frequent, otherwise (t is a new pattern) we annotate S as the first slide in which t is frequent and create an auxiliary array for t and start monitoring it. When a slide expires (denote it S_{exp}) we need to update the frequencies and the auxiliary arrays of patterns belonging to TT if they were present in S_{exp}. Finally, we delete auxiliary array if pattern t has existed since arrival of S and delete t, if t is no longer frequent in any of the current slides.

A very fast verifier for trajectories. In the following, we first define the verifier notion and propose our novel verifier for trajectories data.

Definition 2. *Let T be a trajectories database, P a given set of arbitrary patterns, and min_{freq} a given minimum frequency. A function f is called a verifier if it takes T, P and min_{freq} as input and for each pattern $p \in P$ returns one of the following results: (a) p's true frequency in T if it has occurred at least min_{freq} times or otherwise; (b) reports that it has occurred less than min_{freq} times (frequency not required in this case).*

It is important to notice the subtle difference between verification and simple counting. In the special case of $min_{freq} = 0$ a verifier simply counts the frequency of all $p \in P$, but in general if $min_{freq} > 0$, the verifier can skip any pattern whose frequency will be less than min_{freq}. This early pruning can be done by the Apriori property or by visiting more than $|T| - min_{freq}$ trajectories. Also, note that verification is different (and weaker) from mining. In mining the goal is to find all those patterns whose frequency is at least min_{freq}, but verification simply verifies counts for a given set of patterns, i.e., verification does not discover additional patterns. The challenge is to find a verification algorithm, which is faster than both mining and counting algorithms, since the algorithm for extracting frequent trajectories will benefit from this efficiency. In our case the verifier needs to take into account the sequential nature of trajectories so we need to count really fast while keeping the right order for the regions being verified. To this end we exploit an encoding scheme for regioned trajectories based on some peculiar features of prime numbers.

5   Encoding Paths for Efficient Counting and Querying

A great problem with trajectory sequential pattern mining is to control the exponential explosion of candidate trajectory paths to be modeled because keeping information about ordering is crucial. Indeed, our regioning step heavily reduces the dataset size that we have to deal with. Since our approach is stream oriented we also need to be fast while counting trajectories and (sub)paths. To this end, prime numbers exhibit really nice features that for our goal can be summarized in the following two theorems. They have also been exploited for similar purposes for RFID tag encodings [19], but in that work the authors did not provide a solution for paths containing cycles as we do in our framework.

Theorem 1 (The Unique Factorization Theorem). *Any natural number greater than 1 is uniquely expressed by the product of prime numbers.*

As an example consider the trajectory $T_1 = ABC$ crossing three regions A,B,C. We can assign to regions A, B and C respectively the prime numbers 3,5,7 and the position of A will be the first $(pos(A) = 1)$, the position of B will be the second $(pos(B) = 2)$, and the position of C will be the third $(pos(C) = 3)$. Thus the resulting value for T_1 (in the following we refer to it as P_1) is the product of the three prime numbers, $P_1 = 3 * 5 * 7 = 105$ that has the property that does not exist the product of any other three prime numbers that gives as results 105.

As it is easy to see this solution allows to easily manage trajectories since containment and frequency count can be done efficiently by simple mathematical operations. Anyway, this solution does not allow to distinguish among ABC, ACB, BAC, BCA, CAB, CBA, since the *trajectory number* (i.e., the product result) for these trajectories is always 105. To this end we can exploit another fundamental theorem of arithmetics.

Theorem 2 (Chinese Remainder Theorem). *Suppose that $n_1, n_2, \cdots, n_k$ are pairwise relatively prime numbers. Then, there exists W (we refer to it as witness) between 0 and $N = n_1 \cdot n_2 \cdots n_k$ solving the system of simultaneous congruences: $W\%n_1 = a_1$, $W\%n_2 = a_2$, ..., $W\%n_k = a_k$*[1].

Then, by Theorem 2, there exists W_1 between 0 and $P_1 = 3*5*7 = 105$. In our example, the witness W_1 is 52 since $52\%3 = 1 = pos(A)$ and $52\%5 = 2 = pos(B)$ and $52\%7 = 3 = pos(C)$. We can compute W_1 efficiently using the extended Euclidean algorithm. From the above properties it follows that in order to fully encode a trajectory (i.e., keeping the region sequence) it suffices to store two numbers, its prime number product (which we refer to as its trajectory number) and its witness. In order to assure that no problem will arise in the encoding phase and witness computation we assume that the first prime number we choose for encoding is greater than the trajectory size. So for example if the trajectory length is 3 we encode it using prime numbers 5,7,11. A devil's advocate may argue that multiple occurrences of the same region leading to cycles violates the injectivity of the encoding function. To this end the following example will clarify our strategy.

Dealing with Cycles. Consider the following trajectory $T_2 = ABCAD$, we have a problem while encoding region A since it appears twice, in the first and fourth position. We need to assure that the encoding value of A is such that we can say that both $pos(A) = 1$ and $pos(A) = 4$ hold (we do not want two separate encoding value since the region is the same and we are interested in the order difference). Assume that A is encoded as $(41)_5$ (i.e., 41 on base 5, we use 5 base

---

[1] The % is the classical modulo operation that computes the remainder of the division.

since the trajectory length is 5) this means that A occurs in positions 1 and 4. The decimal number associated to it is $A = 21$, and we chose as the encoding for $A = 23$ that is the first prime number greater than 21. Now we encode the trajectory using $A = 23$, $B = 7$, $C = 11$, $D = 13$ thus obtaining $P_2 = 23023$ and $W_2 = 2137$ (since the remainder we need for A is 21). As it easy to see we are still able to properly encode even trajectories containing cycles. As a final notice we point out that the above calculation is made really fast by exploiting a parallel algorithm for multiplication. We do not report here the pseudo code for the encoding step explained above due to space limitations. Finally, one may argue that the size of prime numbers could be large, however in our case it is bounded since the number of regions is small as confirmed by several empirical studies [14] (always less than a hundred of regions for real life applications we investigated).

Definition 3 (Region Encoding). *Given a set $R = \{R_1, R_2, \cdots, R_n\}$ of regions, a function enc from R to $\mathcal{P}$ (the positive prime numbers domain) is a region encoding function for R.*

Definition 4 (Trajectory Encoding). *Let $T_i = R_1, R_2 \cdots R_n$ be a regioned trajectory. A trajectory encoding $(E(T_i))$ is a function that associates T_i with a pair of integer numbers $\langle P_i, W_i \rangle$ where $P_i = \prod_{1..n} enc(R_i)$ is the trajectory number and W_i is the witness for P_i.*

Once we encode each trajectory as a pair $E(T)$ we can store trajectories in a binary search tree making the search, update and verification operations quite efficient since at each node we store the $E(T)$ pair. It could happen that there exists more than one trajectory encoded with the same value P but different witnesses. In this case, we store once the P value and the list of witnesses saving space for pointers and for the duplicate P values. Consider the following set of trajectories along with their encoding values (we used region encoding values: $A = 5$, $B = 7$, $C = 11$, $D = 13$, $E = 15$): $(ABC, \langle 385, 366 \rangle)$, $(ACB, \langle 385, 101 \rangle)$, $(BCDE, \langle 15015, 3214 \rangle)$, $(DEC, \langle 2145, 872 \rangle)$. ABC and ACB will have the same P value (385) but their witnesses are $W_1 = 366$ and $W_2 = 101$, so we are still able to distinguish them.

6 Experimental Results

In this section we will show the experimental results for our algorithms. We used the GPS dataset [34] (this dataset being part of *GeoLife*project). It records a broad range of users outdoor movements, thus, the dataset allows a severe test for our frequent sequential pattern mining. In order to compare the effectiveness of our approach we compared it with the *T-Patterns* system described in [9]. In particular since *T-Patterns* does not offer streaming functionalities we compare our system using a single large window and compare the extracted patterns w.r.t. the regioning performed. More in detail we compare our results w.r.t. the Static and Dynamic regioning offered by *T-Patterns* on window sizes of 10,000, 20,000, 50,000, 100,000.

Comparison Against Static RoI. In the following, we compare our algorithm against *T-Patterns* with static RoI by measuring the execution times, the number of extracted regions and the number of extracted patterns for a given support value. Table 2(a) and (b) summarize respectively the resultsobtained on sets of 10,000 up to 100,000 trajectories extracted for the GPS dataset with 0.5 % and 1 % as min support value. Table 2(a) shows that when the number of input trajectories increases the execution times linearly increases and our execution time is lower than *T-Patterns*. This can be easily understood since we exploit a really fast frequent miner. A more interesting result is the one on number of extracted regions. As explained in previous sections, we exploit *PCA* and we focus on regions along principal directions, this allow us to obtain less regions. As a consequence having a smaller number of regions allows more patterns to be extracted as confirmed in Table 1. The intuition behind this result is that when considering a smaller number of regions this imply a greater number of trajectories crossing those regions. The above features are confirmed by the results reported in Table 2 for 1 % minimum support value (obviously it will change the execution times and number of patterns while the number of extracted regions is the same as in the to previous table). Interestingly enough, the execution times for our algorithm slightly decrease as the min support value increases and this is due to the advantage we get from the verification strategy.

Comparison Against Dynamic RoI. In the following, we compare our algorithm against *T-Patterns* with dynamic RoI by measuring the execution times, the number of extracted regions and the number of extracted patterns for a given support value. Tables below summarize respectively the results obtained on sets of 10,000 up to 100,000 trajectories extracted for the GPS dataset with 0.5 % and 1 % as min support value. Also for this comparison, Table 3 shows, for 0.5 %

Table 1. Performances comparison with min support value 0.5 % against static ROI

Our Algorithm			T-Patterns		
Times	# regions	# patterns	Times	# regions	# patterns
1.412	94	62	4.175	102	54
2.115	98	71	6.778	107	61
3.876	96	77	14.206	108	67
7.221	104	82	30.004	111	73

Table 2. Performances comparison with min support value 1 % against static ROI

Our Algorithm			T-Patterns		
Times	# regions	# patterns	Times	# regions	# patterns
1.205	94	41	4.175	102	37
2.003	98	50	6.778	107	43
3.442	96	59	14.206	108	49
6.159	104	65	30.004	111	58

Table 3. Performances comparison with min support value 0.5 % against dynamic RoI

Our Algorithm			T-Patterns		
Times	# regions	# patterns	Times	# regions	# patterns
1.412	94	62	4.881	106	56
2.115	98	71	7.104	111	66
3.876	96	77	15.306	112	69
7.221	104	82	302.441	115	75

Table 4. Performances comparison with min support value 1 % against dynamic RoI

Our Algorithm			T-Patterns		
Times	# regions	# patterns	Times	# regions	# patterns
1.205	94	41	5.002	105	40
2.003	98	50	7.423	108	46
3.442	96	59	15.974	113	53
6.159	104	65	32.558	116	60

minimum support value, that when the number of input trajectories increases the execution times linearly increases and our execution time is better than *T-Patterns*. The other improvements obtained with our algorithm have the same rationale explained above. These features are confirmed by the results reported in Table 4 for 1 % minimum support value.

Mining Algorithm Performances. In this section we report the results we ran to test the performances of the proposed incremental mining algorithm for large sliding windows. At the best of our knowledge our algorithm is the first proposal for dealing with frequent pattern mining on trajectory streams so we do not have a "gold" standard to compare with, however the results obtained are really satisfactory since the running times are almost insensitive to the window size. Indeed, some of the approaches discussed in previous section pursue the same goal such as [3, 11, 13, 23], so we decided to compare our results with the approach presented in [3] (referred in the following as *IMFP*). We recall that the algorithm goal is maintaining frequent trajectories over large sliding windows. Indeed, the results shown in Table 5(a) show that the delta-maintenance based approach presented here is scalable with respect to the window size. Finally, we report the total number of frequent pattern as windows flow (the results shown in Table 5(b)). They are computed using a window size of 10,000 trajectories) for a minimum support value of 0,1 %. Indeed we report the total number of patterns that have been frequent wether they are still frequent or not, this information is provided to take into account the concept shift for data being monitored. The results in Table 5(b) shows that after 200 windows being monitored the number of patterns that resulted frequent in some window is more than doubled this means that the users habits heavily changed during the two years period. The results reported in Table 5(a) and (b) confirm that our performances are

Table 5. Mining algorithm results

Our Times	IMFP Times	Windows size		# Window	# Our Patterns	# IMFP Patterns
773	1,154	10,000		1	85	67
891	1,322	25,000		10	106	92
1,032	1,651	50,000		20	125	104
1,211	1,913	100,000		50	156	121
1,304	2,466	500,000		100	189	143
2,165	2,871	1,000,000		200	204	158
	(a)				(b)	

better than the ones obtained by running *IMFP* both in terms of running times and patterns quality expressed as number of frequent patterns found.

7   Conclusion

In this paper we tackled the problem of frequent pattern extraction from trajectory data by introducing a very fast algorithm to verify the frequency of a given set of sequential patterns. The fast verifier has been exploited in order to solve the sequential pattern mining problem under the realistic assumption that we are mostly interested in the new/expiring patterns. This delta-maintenance approach effectively mines very large windows with slides, which was not possible before. In summary we have proposed an approach highly efficient, flexible, and scalable to solve the frequent pattern mining problem on data streams with very large windows. Our work is subject to further improvements in particular we will investigate: (1) further improvements to the regioning strategy; (2) refining the incremental maintenance to deal with maximum tolerance for delays between slides.

References

1. Agrawal, R., Srikant, R.: Fast algorithms for mining association rules in large databases. In: VLDB (1994)
2. Cheung, W., Zaiane, O.R.: Incremental mining of frequent patterns without candidate generation or support. In: DEAS (2003)
3. Cheung, W., Zaïane, O.R.: Incremental mining of frequent patterns without candidate generation or support constraint. In: IDEAS, pp. 111–116 (2003)
4. Leung, C., et al.: Cantree: A tree structure for efficient incremental mining of frequent patterns. In: ICDM (2005)
5. Park, J.S., et al.: An effective hash-based algorithm for mining association rules. In: SIGMOD (1995)
6. Brin, S., et al.: Dynamic itemset counting and implication rules for market basket data. In: SIGMOD (1997)
7. Chi, Y., et al.: Moment: Maintaining closed frequent itemsets over a stream sliding window (2004)
8. Fischer, J., Heun, V., Kramer, S.: Optimal string mining under frequency constraints. In: Fürnkranz, J., Scheffer, T., Spiliopoulou, M. (eds.) PKDD 2006. LNCS (LNAI), vol. 4213, pp. 139–150. Springer, Heidelberg (2006)

9. Giannotti, F., Nanni, M., Pinelli, F., Pedreschi, D.: Trajectory pattern mining. In: KDD, pp. 330–339 (2007)
10. Gonzalez, H., Han, J., Li, X., Klabjan, D.: Warehousing and analyzing massive RFID data sets. In: ICDE, p. 83 (2006)
11. Hai, P.N., Poncelet, P., Teisseire, M.: GET_MOVE: an efficient and unifying spatio-temporal pattern mining algorithm for moving objects. In: Hollmén, J., Klawonn, F., Tucker, A. (eds.) IDA 2012. LNCS, vol. 7619, pp. 276–288. Springer, Heidelberg (2012)
12. Han, J., Pei, J., Yin, Y.: Mining frequent patterns without candidate generation. In: SIGMOD (2000)
13. Hernández-León, R., Hernández-Palancar, J., Carrasco-Ochoa, J.A., Martínez-Trinidad, J.F.: A novel incremental algorithm for frequent itemsets mining in dynamic datasets. In: Ruiz-Shulcloper, J., Kropatsch, W.G. (eds.) CIARP 2008. LNCS, vol. 5197, pp. 145–152. Springer, Heidelberg (2008)
14. Jeung, H., Yiu, M.L., Zhou, X., Jensen, C.S., Shen, H.T.: Discovery of convoys in trajectory databases. PVLDB 1(1), 1068–1080 (2008)
15. Jiang, N., Gruenwald, L.: Research issues in data stream association rule mining. SIGMOD Rec. 35(1), 14–19 (2006)
16. Jolliffe, I.T.: Principal Component Analysis. Springer Series in Statistics (2002)
17. Kügel, A., Ohlebusch, E.: A space efficient solution to the frequent string mining problem for many databases. Data Min. Knowl. Discov. 17(1), 24–38 (2008)
18. Lee, C., Lin, C., Chen, M.: Sliding window filtering: an efficient method for incremental mining on a time-variant database (2005)
19. Lee, C.-H., Chung, C.-W.: Efficient storage scheme and query processing for supply chain management using RFID. In: SIGMOD08, pp. 291–302 (2008)
20. Lee, J.-G., Han, J., Li, X., Gonzalez, H.: TraClass: trajectory classification using hierarchical region-based and trajectory-based clustering. PVLDB 1(1), 1081–1094 (2008)
21. Lee, J.-G., Han, J., Whang, K.-Y.: Trajectory clustering: a partition-and-group framework. In: SIGMOD07, pp. 593–604 (2007)
22. Lee, J.-G., Han, J., Li, X.: Trajectory outlier detection: A partition-and-detect framework. In: ICDE, pp. 140–149 (2008)
23. Leung, C.K., Khan, Q.I., Li, Z., Hoque, T.: Cantree: a canonical-order tree for incremental frequent-pattern mining. Knowl. Inf. Syst. 11(3), 287–311 (2007)
24. Li, J., Maier, D., Tufte, K., Papadimos, V., Tucker, P.A.: No pane, no gain: efficient evaluation of sliding-window aggregates over data streams. SIGMOD Rec. 34(1), 39–44 (2005)
25. Li, Y., Han, J., Yang, J.: Clustering moving objects. In: KDD, pp. 617–622 (2004)
26. Liu, Y., Chen, L., Pei, J., Chen, Q., Zhao, Y.: Mining frequent trajectory patterns for activity monitoring using radio frequency tag arrays. In: PerCom, pp. 37–46 (2007)
27. Masciari, E.: Trajectory clustering via effective partitioning. In: Andreasen, T., Yager, R.R., Bulskov, H., Christiansen, H., Larsen, H.L. (eds.) FQAS 2009. LNCS, vol. 5822, pp. 358–370. Springer, Heidelberg (2009)
28. Masciari, E.: Warehousing and querying trajectory data streams with error estimation. In: DOLAP, pp. 113–120 (2012)
29. Mozafari, B., Thakkar, H., Zaniolo, C.: Verifying and mining frequent patterns from large windows over data streams. In: ICDE, pp. 179–188 (2008)
30. Pei, J., Han, J., Mao, R.: CLOSET: an efficient algorithm for mining frequent closed itemsets. In: ACM SIGMOD Workshop on Research Issues in Data Mining and Knowledge Discovery (2000)

31. Pei, J., Han, J., Mortazavi-Asl, B., Pinto, H., Chen, Q., Dayal, U., Hsu, M.: Prefixspan: mining sequential patterns by prefix-projected growth. In: ICDE, pp. 215–224 (2001)
32. Yang, J., Hu, M.: TrajPattern: mining sequential patterns from imprecise trajectories of mobile objects. In: Ioannidis, Y., et al. (eds.) EDBT 2006. LNCS, vol. 3896, pp. 664–681. Springer, Heidelberg (2006)
33. Zaki, M.J., Hsiao, C.: CHARM: an efficient algorithm for closed itemset mining. In: SDM (2002)
34. Zheng, Y., Li, Q., Chen, Y., Xie, X.: Understanding mobility based on gps data. In: UbiComp 2008, pp. 312–321 (2008)

Process Mining to Forecast the Future of Running Cases

Sonja Pravilovic[1,2], Annalisa Appice[1]([✉]), and Donato Malerba[1]

[1] Dipartimento di Informatica, Università degli Studi di Bari Aldo Moro,
via Orabona, 4, 70126 Bari, Italy
[2] Montenegro Business School, Mediterranean University,
Vaka Djurovica b.b., Podgorica, Montenegro
{sonja.pravilovic,annalisa.appice,donato.malerba}@uniba.it

Abstract. Processes are everywhere in our daily lives. More and more information about executions of processes are recorded in event logs by several information systems. Process mining techniques are used to analyze historic information hidden in event logs and to provide surprising insights for managers, system developers, auditors, and end users. While existing process mining techniques mainly analyze full process instances (cases), this paper extends the analysis to running cases, which have not yet completed. For running cases, process mining can be used to notify future events. This forecasting ability can provide insights for check conformance and support decision making. This paper details a process mining approach, which uses predictive clustering to equip an execution scenario with a prediction model. This model accounts for recent events of running cases to predict the characteristics of future events. Several tests with benchmark logs investigate the viability of the proposed approach.

1 Introduction

Contemporary systems, ranging from high-tech and medical devices to enterprise information systems and e-business infrastructures, record massive amounts of events by making processes visible. Each event has a name and additional mandatory characteristics that include the timestamp (i.e. exact date and time of occurrence), the lifecycle transition state (i.e. whether the event refers to a task having been started, completed) and the resource (i.e. name of the originator having triggered the event). In addition, each event may be characterized by further optional characteristics, such as cost, location, outcome, which are specific for the process. Process mining techniques [10] can be used to analyze event logs, in order to extract, modify and extend process models, as well as to check conformance with respect to defined process models.

Thus far, several process mining techniques have been used in the discovery, conformance and enhancement of a variety of business processes [11]. They are mainly used in an off-line fashion and rarely for operational decision support. Historical full cases (i.e. instances of the process which have already completed) are

A. Appice et al. (Eds.): NFMCP 2013, LNAI 8399, pp. 67–81, 2014.
DOI: 10.1007/978-3-319-08407-7_5, © Springer International Publishing Switzerland 2014

analyzed, while running cases (i.e. instances of the process which have not completed yet) are rarely processed on-line. However, a new perspective has emerged recently. van der Aalst et al. [12] demonstrate that process mining techniques are not necessarily limited to the past, but can also be used for the present and the future. To make this philosophy concrete, they have already presented a set of approaches, which can be used very well for operational decision support. In particular, they propose to mine predictive process models from historic data, use them to estimate both the remaining processing time and the probability of a particular outcome for running cases [13].

Embracing this philosophy, we consider another predictive task and detail a process mining approach to predict future events of running cases. This forecasting service can be used to check conformance and recommend appropriate actions. In particular, we can check whether the observed event fits the predicted behavior. The moment the case deviates, an appropriate actor can be alerted in real time. Similarly, we can use predictions to notify recommendations proposing/describing activity elements (e.g. resource, activity name, cost), which will comply the process model.

In this paper, we transform the task of event forecasting for running cases into a predictive clustering task [2], where the target variables are the characteristics of future events expected in running cases, while the predictors are characteristics of recent events up to a certain time window. This transformation technique is usually known as "time delay embedding" [9] and frequently used in stream data mining, where it is also called sliding window model [5]. Historical cases, transformed with the Sliding Window model, can be processed off-line so that a predictive clustering tree (PCT) [2] can be mined for the predictive aim. A PCT is a tree structured model, which generalizes decision tree by predicting many labels of an examples at once. In this study, it allows us to reveal in advance characteristics of future events based on characteristics of recent time-delayed event elements. The PCT can be used to predict on-line event elements of any new running case.

This paper is organized as follows. Related work is discussed in Sect. 2. Section 3 introduces some notations and revises the predictive clustering technique as well as the Sliding Window model used. Section 4 describes the event-based forecasting approach proposed. Section 5 describes the empirical evaluation of our approach using benchmark case studies. Section 6 concludes the paper.

2   Related Work

A great deal of work has been done in the area of process mining in the recent years, particularly in the field of discovering process models based on observed events. An overview of the recent work in this domain is reported in [11]. The most part of existing approaches focus on discovering *descriptive* process models based on historic information and do not support users at run-time. However, the *predictive* task has recently started to receive attention in this area.

The topic of the time prediction is the most related to this work. To address this topic, Dongen et al. [4] have used the non-parametric regression, in order to predict the remaining time until the completion of a case. van der Aalst et al. [13] have proposed to annotate a transition system with information about the occurrence of specific event or the time until a particular event. Dumas et al. [8] have described a lazy-learning method for calculating log-based recommendations which account for a specific optimization goal (e.g. minimizing cycle time, or maximizing profit). These recommendations can be used to guide the user to select the next work item.

All methods, reported above, are tested for the prediction of the completion time. However, methods in [8,13] can be also used to make hypotheses on the probable name of the activity of the next event(s). In the matter of this event prediction task, Goedertier et al. [6] have proposed to learn first-order classification rules, in order to predict whether, given the state of a process instance, a particular state transition can occur. Buffet et al. [3] have adapted the Bayesian analysis, in order to label the task of a running case by accounting for the trace of events, already observed in the case. In both methods, the prediction is made for a single characteristic, in general, the activity name.

The approach presented in this paper differs from existing approaches since the complex nature of events is accounted for, so that a prediction can be made for the several characteristics of the event. A learning phase is defined, so that an interpretable predictive process model is computed off-line. The prediction is always made on-line according to the history-dependent behavior of the running case and the presence of a predictive model.

3   Basics

In this section, we introduce some basic concepts to address the task of event forecasting. We describe event logs, as well as introduce the ideas behind predictive clustering and Sliding Window model.

3.1   Event Log

The basic assumption is that the log contains information about activities executed for specific cases of a certain process type, as well as their durations.

Definition 1 (Event, Characteristic). *Let $\mathcal{E}$ be the event domain for a process $\mathcal{P}$. An event ϵ ($\epsilon \in \mathcal{E}$) is characterized by a set of mandatory characteristics, that is, the event corresponds to an activity, has a timestamp which represents date and time of occurrence, is triggered by a resource, refers to a specific lifecycle transition state which can be started or completed. Additionally, it can be characterized by variable process-specific optional characteristics such as cost, location, outcome and so on. Optional characteristics may also be away in an event.*

Table 1. A fragment of the event log *repair* [11]. The symbol (*) identifies the optional characteristics which are specific for the process.

id	name	timestamp	resource	lifecycle	phone type*	defect type*	defect fixed*	number repairs*
1	Register	1970-01-02T12:23	System	complete	-	-	-	-
1	Analyze Defect	1970-01-02T12:23	Tester3	start	-	-	-	-
1	Analyze Defect	1970-01-02T12:30	Tester3	complete	T2	6	-	-
1	Repair (Complex)	1970-01-02T12:31	SolverC1	start	-	-	-	-
1	Repair (Complex)	1970-01-02T12:49	SolverC1	complete	-	-	-	-
1	Test Repair	1970-01-02T12:49	Tester3	start	-	-	-	-
1	Test Repair	1970-01-02T12:55	Tester3	complete	-	-	-	-
1	Inform User	1970-01-02T13:10	System	complete	-	-	-	-
1	Archive Repair	1970-01-02T13:10	System	complete	-	-	true	0
2	Register	1970-01-01T11:09	System	complete	-	-	-	-
2	Analyze Defect	1970-01-01T11:09	Tester2	start	-	-	-	-
2	...	...	...	...	...	...	...	...
... ...		...		...		...	...	...

An event log is a set of events. Each event in the log is linked to a particular trace and globally unique. A trace in a log represents a process instance (e.g. a customer order or an insurance claim) also referred to as case. Time is non-decreasing in each case in the log.

Definition 2. *A case C is a finite sequence of events $e \in \mathcal{E}$ such that each event occurs only once (i.e. for $1 \leq i < j \leq |C|$: $\epsilon(i) \neq \epsilon(j)$) and time is non-decreasing (i.e. $time(\epsilon(i)) \leq time(\epsilon(j))$) . A log $\mathcal{L}$ is a bag of cases.*

A fragment of event log is reported in Table 1. A case containing nine events is shown. Each event has the mandatory characteristics and several optional characteristics. We note that the value may also lack in an event for an optional characteristic.

3.2 Predictive Clustering Trees

The task of mining predictive clustering trees is now formalized. *Given*

- a descriptive space $\mathbf{X} = \{X_1, X_2, \ldots X_m\}$ spanned by m independent (or predictor) variables X_j,
- a target space $\mathbf{Y} = \{Y_1, Y_2, \ldots, Y_q\}$ spanned by q dependent (or target) variables Y_j,
- a set $\mathcal{T}$ of training examples, $(\mathbf{x_i}, \mathbf{y_i})$ with $\mathbf{x_i} \in \mathbf{X}$ and $\mathbf{y}_i \in \mathbf{Y}$.

Find a tree structure τ which represents:

- A set of hierarchically organized clusters on T such that for each $u \in T$, a sequence of clusters $C_1, C_2, \ldots, C_k$ exist for which $u \in C_{i_r}$ and the containment relation $C_1 \supseteq C_2 \supseteq \ldots \supseteq C_k$ is satisfied. Clusters $C_1, C_2, \ldots, C_k$ are associated to the nodes $t_1, t_2, \ldots, t_k$, respectively, where each $t_i \in \tau$ is a direct

child of $t_{i-1} \in \tau$ $(j = 1, \ldots, k)$, t_1 is the root of the structure τ and t_k is a leaf of the structure.
- A predictive piecewise function $f : \mathbf{X} \to \mathbf{Y}$, defined according to the hierarchically organized clusters. In particular,

$$\forall u \in \mathbf{X}, \ f(u) = \sum_{t_i \in leaves(\tau)} D(u, t_i) f_{t_i}(u), \tag{1}$$

where $D(u, t_i) = \begin{cases} 1 \text{ if } u \in C_i \\ 0 \text{ otherwise} \end{cases}$ and $f_{t_i}(u)$ is a (multi-target) prediction function associated to the leaf t_i. It includes a categorical value for a discrete attribute, a numeric value for a continuous attribute.

Clusters are identified according to both the descriptive space and the target space $\mathbf{X} \times \mathbf{Y}$. This is different from what is commonly done in predictive modeling and classical clustering, where only one of the spaces is typically considered. This general formulation of the problem allows us to have the prediction model mining phase which can consider multiple target variables at once. This is the case of predicting the "several" characteristics of the next event in a case.

The construction of a PCT is not very different from the construction of standard decision tree (see, for example, the C4.5 algorithm [7]): at each internal node t, a test has to be selected according to a given evaluation function. The main difference is that for a PCT, the best test is selected by maximizing the (inter-cluster) variance reduction over the target space $\mathbf{Y}$, defined as:

$$\Delta_Y(C, \mathcal{P}) = Var_{\mathbf{Y}}(C) - \sum_{C_i \in \mathcal{P}} \frac{|C_i|}{|C|} Var_{\mathbf{Y}}(C_i), \tag{2}$$

where C is the cluster associated with t and $\mathcal{P}$ defines the partition $\{C_1, C_2\}$ of C. The partition is defined according to a Boolean test on a predictor variable of the descriptive space $\mathbf{X}$. By maximizing the variance reduction, the cluster homogeneity is maximized, improving at the same time the predictive performance. $Var_{\mathbf{Y}}(C)$ is the variance of $\mathbf{Y}$ in the cluster C. It is computed as the average of variances of target variables $Y_j \in \mathbf{Y}$, that is, $Var_{\mathbf{Y}}(C) = \sum_{Y_j \in \mathbf{Y}} var_{Y_j}(C)$.

3.3 Sliding Window Model

A Sliding Window model is the simplest model to consider the recent data of a running case and run queries over the data of the recent past only. Originally defined for data stream mining, this type of window is similar to the first-in, first-out data structure. When the event ϵ_i is acquired and inserted in the window, the latest event ϵ_{i-w} is discarded (see Fig. 1). w is the size of the window.

Definition 3 (Sliding Window model). *Let w be the window size of the model. A Sliding Window model views a case $\mathcal{C}$ as a sequence of overlapping windows of events,*

$$\mathcal{C}(1 \to w), \ \mathcal{C}(2 \to w + 1), \ \ldots, \ \mathcal{C}(i - w + 1 \to i), \ \ldots, \tag{3}$$

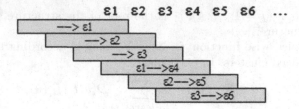

Fig. 1. Sliding Window model of a running case with window size $w = 4$.

where $C(i - w + 1 \to i)$ is the series of the w events ϵ_{i-w+1}, ϵ_{i-w+2}, ..., ϵ_{i-1}, ϵ_i of the case C with $time(\epsilon_{i-j-1}) \leq time(\epsilon_{i-j})$ (for all $j = 0, \ldots, w - 2$).

By considering the Sliding Window model, the window $C(i - w + 1 \to i)$ defines the recent history of the event ϵ_i.

4 Framework for PCT-based Event Forecasting

We use the Sliding Window model to transform our event-based forecasting problem into a predictive clustering problem where the target variables are the characteristics of the next event in the case. A PCT is learned off-line from an event log which records full cases and used on-line to predict the next event expected in a running case (see Fig. 2).

Let $\mathcal{L}$ be the event log which records full cases of a process $\mathcal{P}$. In the off-line phase, $\mathcal{L}$ is transformed in a training set $\mathcal{T}$. This transformation is performed by using the Sliding Window model. Let w be the window size. Each training case $C \in \mathcal{L}$ is transformed in a bag $training(C, w)$ of training examples. This bag collects a training example for each event ϵ_i of C so that the training example

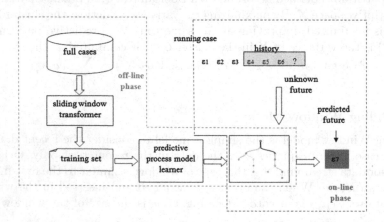

Fig. 2. Event forecasting framework. In the off-line phase, the Sliding Window model is used to transform the event forecasting task in a predictive clustering task: a PCT is learned from historical full traces of a process. In the on-line phase, the future event of a running case is predicted from its known past.

$\mathbf{xy}(\mathcal{C}(i - w + 1 \rightarrow i))$ is generated based upon the Sliding Window history of the event. Formally,

$$training(\mathcal{C}, w) = \{\mathbf{xy}(\mathcal{C}(i - w + 1 \rightarrow i)) | i = 1, \ldots, |\mathcal{C}|\}, \tag{4}$$

where $|\mathcal{C}|$ denotes the length of the case $\mathcal{C}$, and

$$\mathcal{T} = \bigcup_{\mathcal{C} \in \mathcal{T}} training(\mathcal{C}, w). \tag{5}$$

Each characteristic of an event is transformed into a variable. The target space $\mathbf{Y}$ is populated with variables originated from the newest event in the Sliding Window, while the descriptive space $\mathbf{X}$ is populated with variables generated from the oldest $w - 1$ time-delayed events in the Sliding Window. The window size influences the size of the descriptive space $\mathbf{X}$. The longer the window history, the higher the number of the descriptive variables considered for the predictive clustering task. The timestamp is used when generating the descriptive space only. It is transformed into the time (in seconds) gone by the beginning of the case. When an optional characteristic lacks in the related event, the associated variable assumes the value "none" in the training example. An example of this Sliding Window transformation of a case is reported in Example 1.

Example 1 (Sliding Window transformation). Let us consider the case 1 of the event log reported in Table 1. The Sliding Window transformation of this case generates nine examples, one for each event in the case. Each timestamp is transformed into the time (in seconds) gone by the beginning of the case. By considering $w = 3$, the following training examples are generated:

      none, 0,none, none, none, none, none, none,
(1) none, 0,none, none, none, none, none, none,
      Register, System, complete,none, none, none, none

      none, 0,none, none, none, none, none, none,
(2) Register, 0, System, complete,none, none, none, none,
      Analyze Defect, Tester3, start, none, none, none, none

      Register,0,System, complete,none, none, none, none,

(3) Analyze Defect,0,Tester3,start,none, none, none, none,
      Analyze Defect, Tester3, complete, T2, 6, none, none

      . . .

      TestRepair, 1320, Test3, complete, none, none, none, none,
(9) Inform User, 2820, System, complete, none, none, none, none,
      Active Repair, System, complete, none, none, true,0

```
conceptname2 in {none,Inform_User,Test_Repair}
+--yes: lifecycletransition2 = start
      +--yes: orgresource2 in {Tester1,Tester2,Tester5}
            +--yes: orgresource1 in {SolverS1,SolverS2,SolverS3}
                  +--yes: orgresource2 = Tester2
                        +--yes: orgresource1 = SolverS1
                              +--yes: [Tester2,Test_Repair,complete,none,none,false,1]
                              +--no:  [Tester2,Test_Repair,complete,none,none,true,1]
```

Fig. 3. A fragment of PCT learned from the event log *repair*.

In this case, the descriptive variables are generated from the characteristics of the oldest two $(w-1)$ time-delayed events, while the target variables (in italics) are generated from the characteristics of the last event of the window.

Let τ be the PCT learned from $\mathcal{T}$. It generalizes an event-based predictive process model for the process $\mathcal{P}$ (see Fig. 3). In the on-line phase, this process model can be used to predict characteristics of the next event in each new running case of $\mathcal{P}$. The prediction is that produced by traversing τ based on the characteristics of the past $w-1$ time delayed events in the case. The selected leaf contains predictions of characteristics for next event.

5   Empirical Study

The event forecasting framework has been implemented in Java. It supports both the off-line (Fig. 4a) and the on-line phase (Fig. 4b) described in this paper. It processes event logs in the XES[1] format [14], that is, the XML-based standard for event logs. The predictive clustering tree learner is CLUS[2] [2].

5.1   Event Log Description

We consider the event logs recorded for four different processes[3] (see details in Table 2). These logs contain the mandatory information about activity, resource, lifecycle and time as well as process-specific optional characteristics.

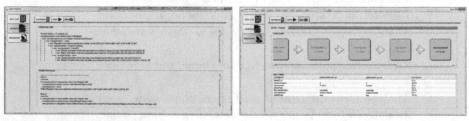

(a) Off-line learning   (b) On-line prediction

Fig. 4. EventPredictor: off-line (a) and on-line (b) activity.

---

[1] http://www.xes-standard.org/

[2] http://dtai.cs.kuleuven.be/clus/

[3] http://www.processmining.org/event_logs_and_models_used_in_book

Table 2. Event log description.

event log	nr. cases	nr. events	case statistics			nr. target characteristics
			min length	max length	avg length	
reviewing	100	3730	11	92	37.3	9
repair	1104	11855	4	25	10.7	7
teleclaims	3512	46138	5	18	10.7	5
lfull	1391	7539	5	17	5.4	1

The event log *reviewing* handles reviews for a journal. Each paper is sent to three different reviewers. The reviewers are invited to write a report. However, reviewers often do not respond. As a result, it is not always possible to make a decision after a first round of reviewing. If there are not enough reports, then additional reviewers are invited. This process is repeated until a final decision can be made (accept or reject). The optional characteristics are Result by Reviewer A (accept, reject), Result by Reviewer B (accept, reject), Result by Reviewer C (accept, reject), Result by Reviewer X (accept, reject), accepts (0, 1, 2, 3, 4, 5) and rejects (0, 1, 2, 3, 4, 5).

The event log *repair* is about a process to repair telephones in a company. The process starts by registering a telephone device sent by a customer. After registration, the telephone is sent to the Problem Detection Department where the defect is analyzed and classified. Once the problem is identified, the telephone is sent to the Repair Department which has two teams: one of the team fixes simple defects and the other fixes the complex defects. Once the repair employer finishes, the device is sent to the quality Assurance department. If the telephone is not repaired, it is sent again to the Repair department. Otherwise the case is archived and the telephone is sent to the customer. The company tries to fix a defect a limited number of times. The optional characteristics are defect type (ten categories), phone type (three categories), defect fixed (true or false) and number of repairs (0, 1, 2, 3).

The event log *teleclaim* contains an event log describing the handling of claims in an insurance company. It consists of 3512 cases (claims) and 46138 events. The process deals with the handling of inbound phone calls, whereby different types of insurance claims (household, car, etc.) are lodged over the phone. The process is supported by two separate call centers operating for two different organizational entities. Both centers are similar in terms of incoming call volume and average total call handling time, but different in the way call center agents are deployed. After the initial steps in the call center, the remainder of the process is handled by the back-office of the insurance company. It is noteworthy that this log is difficult to mine; the alpha algorithm fails to extract the right process model [11]. The optional characteristics are outcome (B insufficient information, S insufficient information, not liable, processed, rejected) and location (Brisbane and Sydney).

The event log *lfull* contains an event log describing the handling of the tickets. Events concern with the following activities: register request, examine thoroughly, examine casually, check ticket, decide, reinitiate request, pay compensation, and reject request.

5.2   Goal and Experimental Set-up

The goal of this experimental study is to investigate the performance of the PCT-based process models. The Sliding Window transformer is compared to the Landmark one [1], that is, a different kind of stream data model. For each event, the Landmark goes from the starting time point of the case to the present time. Therefore, the descriptive characteristics are aggregated on the Landmark time. For this aggregation scope, we transform each categorical characteristic into n numeric descriptive variable (one variable for each distinct value of the characteristic domain). Each aggregated variable measures the frequency of the value over the Landmark. Similarly, we transform each numeric characteristic (such as the time) into a numeric descriptive variable that sums values in the Landmark.

The performance of a learner is evaluated in terms of several metrics, which include predictive accuracy, model complexity, and learning time. These performance measures are estimated by using the 10-fold cross validation of cases in a log and by varying the window size between two and the maximum length of a case in the log. In particular, the accuracy is measured in terms of the classification accuracy for events of the test running cases. The model complexity is measured in terms of the number of leaves in the learned trees. The computation time is measured in seconds by running experiments on an Intel (R) Core(TM) i7-2670QM CPU @ 2.20 GHz server running the Windows 7 Professional.

5.3   Results and Discussion

For each event log in this study, the predictive accuracy is plotted in Fig. 5(a–d), the model complexity is plotted in Fig. 6(a–d), while the learning time is plotted in Fig. 7(a–d). The predictive accuracy is averaged on the target space. The Sliding Window transformer is used by varying the window size w. The analysis of the performance of the Sliding Window transformer, as well as the comparison between the Sliding Window transformer and the Landmark transformer deserve several considerations. First of all, results show that, in general, the predictive accuracy, as well as the model complexity and the learning time increase when enlarging the size w of the Sliding Window transformer. In any case, both the predictive accuracy and the model complexity metrics reach a (near) stable pick when this size is greater than 6. Second, although the time spent to learn a PCT-based process model is always below 7 s for reviewing and teleclaim cases, 4 s for repair cases, and 0.6 s for lfull cases, it grows-up continuously and linearly with the window size. To reduce the computation effort, it is appropriate to diminish the window size as more as possible. This study indicates that accurate predictions can be already produced by limiting the predictive analysis on the

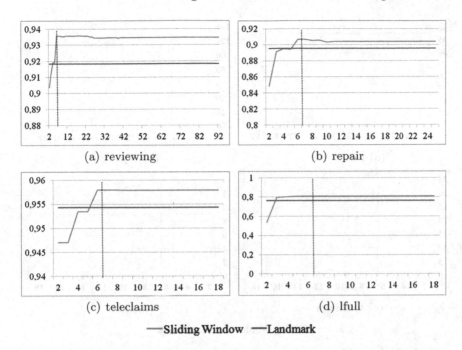

(a) reviewing   (b) repair

(c) teleclaims   (d) lfull

——Sliding Window ——Landmark

Fig. 5. The predictive accuracy averaged on the target space. The dotted vertical line indicates the accuracy when the baseline Sliding Window size $w = 6$ is used.

five time-delayed past events. This recommends the choice $w = 6$ that, in all logs, guarantees the trade-off between the highest predictive accuracy and the lowest learning time cost when the Sliding Window transformer is used. In addition, we can also observe that when Sliding Window transformer is used with $w = 6$ the accuracy, averaged on the target space, is always high, above 90 % for all logs. This valuable predictive ability of the proposed approach is confirmed by analysing the accuracy metric as it is computed for each target characteristic individually (see Tables 4–6). Finally, the comparison between transformers supports our preference towards the Sliding Window transformer, in order to deal with temporal-defined events. In fact, the accuracy achieved with the Sliding Window transformer with $w \geq 6$ is always better than the accuracy with the Landmark.

Detailed accuracy results with $w = 6$ are reported in Tables 4–6. This finer analysis shows that our predictive model is able to predict the majority of characteristics of an event with an accuracy that is greater than 92 %. Low accuracy is observed only when predicting resource of events of repair cases. However, a deeper analysis of this process reveals that repair resource domain includes the values SolverC1, SolverC2, SolverC3, SolverS1, SolverS2, SolverS3, System, Tester1, Tester2, Tester3, Tester4, Tester5, Tester6. By grouping SolverC1, SolverC2 and SolverC3 in a SolverC category, SolverS1, SolverS2, SolverS3 in a SolverS category, as well as Tester1, Tester2, Tester3, Tester4, Tester5, Tester6

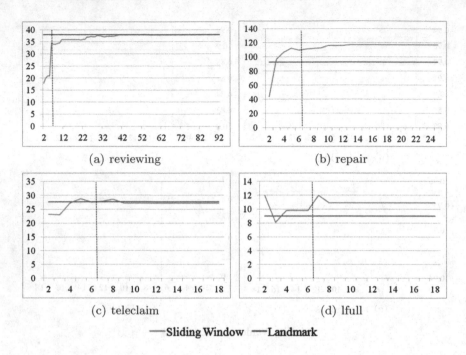

(a) reviewing   (b) repair

(c) teleclaim   (d) lfull

——Sliding Window ——Landmark

Fig. 6. The complexity of the predictive model (number of rules per model). The dotted vertical line indicates the complexity of the model when the baseline Sliding Window size $w = 6$ is used.

Table 3. reviewing.xes $w = 6$ (Sliding Window vs Landmark): rows 1-2 collect the accuracy metric for each target variable; row 3 collects the p-value of the pair-wise Wilcoxon tests comparing the accuracy of both the Sliding Window transformer and the Landmark transformer when they are used to compute the predictive process model.

	resource	name	lifecycle	ResultA	ResultB	ResultC	ResultX	accepts	rejects	avg
Sliding	**.79**	.82	1.0	.99	.99	.99	.93	**.96**	**.96**	**.94**
Landmark	.71	. 82	.99	.98	.98	.98	.92	.92	.92	.92
p-value	.0019	.4922	1	.625	.625	.9219	.3223	.0019	.00339	.0039

in a Tester category, the accuracy for predicting the resource passes from to 0.66 to 0.92. This means that the learned model is able, at least, to identify the resource category accurately although it is not very accurate when identifying the individual resource within the specific resource category. In addition, the pair-wise Wilcoxon tests, performed to compare the accuracy of both the Sliding Window transformer and the Landmark transformer, show that Sliding Window based PCTs are frequently statistically better than Landmark-based PCTs (when the hypothesis of equal performance is rejected with p-value 0.05).

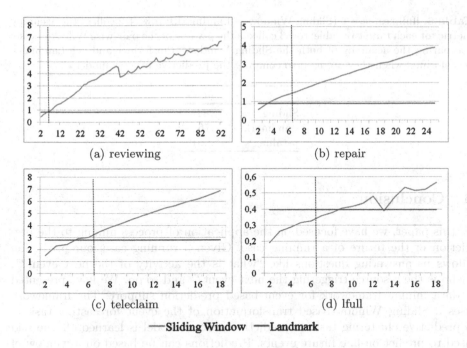

(a) reviewing

(b) repair

(c) teleclaim

(d) lfull

——Sliding Window ——Landmark

Fig. 7. Learning time (seconds). The dotted vertical line indicates the learning time when the baseline Sliding Window size $w = 6$ is used

Table 4. repair.xes, $w = 6$ (Sliding Window vs Landmark): rows 1-2 collect the accuracy metric for each target variable; row 3 collects the p-value of the pair-wise Wilcoxon tests comparing the accuracy of both the Sliding Window transformer and the Landmark transformed when they are used to compute the predictive process model.

	resource	name	lifecycle	defectType	phoneType	defectFixed	nrRepairs	avg
Sliding	**.66**	**.93**	**.99**	**.92**	**.93**	.96	.97	**.91**
Landmark	.57	.92	.98	.91	.92	**.97**	**.98**	.89
p-value	.0019	.0019	.4316	.0019	.0058	.0039	.0019	.0019

Table 5. teleclaims.xes, $w = 6$ (Sliding Window vs Landmark): rows 1-2 collect the accuracy metric for each target variable; row 3 collects the p-value of the pair-wise Wilcoxon tests comparing the accuracy of both the Sliding Window transformer and the Landmark transformed when they are used to compute the predictive process model.

	resource	name	lifecycle	outcome	location	avg
Sliding	**.96**	**.90**	1.0	.97	**.96**	**.96**
Landmark	.60	.89	1.0	.96	.95	.95
p-value	.0273	.0019	1.0	.1602	.0273	.0019

Table 6. lfull.xes, $w = 6$ (Sliding Window vs Landmark): rows 1-2 collect the accuracy metric for each target variable; row 3 collects the p-value of the pair-wise Wilcoxon tests comparing the accuracy of both the Sliding Window transformer and the Landmark transformed when they are used to compute the predictive process model.

	name
Sliding	**.81**
Landmark	.76
p-value	.0019

6   Conclusion

In this paper, we have focused on the application of process mining to the prediction of the future of a running case. Given a running case, our prediction allows us answering questions like "what is the activity of the next event?", "who is the resource triggering the next event?" and so on. We have presented a data mining framework for event-based prediction support. The framework uses a Sliding Window-based transformation of the event forecasting task in a predictive clustering task. A predictive process model is learned off-line and used to predict on-line future events. Predictions can be based on a window of time-delayed events of a running case. In the future, we would like to extend this study by using our predictions to check conformance of running cases and recommend appropriate actions. We also plan to extend this study by using alternative data stream learning schema, like the relevant event selection [15] for training, in order to be able of selecting the most relevant older data to be taken into account when learning the forecasting model.

Acknowledgments. This work fulfills the research objectives of the PON 02_00563_34 70993 project "VINCENTE - A Virtual collective INtelligenCe ENvironment to develop sustainable Technology Entrepreneurship ecosystems" funded by the Italian Ministry of University and Research (MIUR). The authors wish to thank Gianluca Giorgio and Luca Nardulli for their support in developing the framework, anonymous reviewers of the workshop paper for their useful suggestions to improve the manuscript.

References

1. Aggarwal, C.C. (ed.): Data Streams: Models and Algorithms. Advances in Database Systems, vol. 31. Springer, Heidelberg (2007)
2. Blockeel, H., De Raedt, L., Ramon, J.: Top-down induction of clustering trees. In: ICML 1998, pp. 55–63. Morgan Kaufmann (1998)
3. Buffett, S., Geng, L.: Using classification methods to label tasks in process mining. J. Softw. Maint. Evol. **22**(67), 497–517 (2010)
4. van Dongen, B.F., Crooy, R.A., van der Aalst, W.M.P.: Cycle time prediction: when will this case finally be finished? In: Meersman, R., Tari, Z. (eds.) OTM 2008, Part I. LNCS, vol. 5331, pp. 319–336. Springer, Heidelberg (2008)

5. Gaber, M.M., Zaslavsky, A., Krishnaswamy, S.: Mining data streams: a review. ACM SIGMOD Rec. **34**(2), 18–26 (2005)
6. Goedertier, S., Martens, D., Baesens, B., Haesen, R., Vanthienen, J.: Process mining as first-order classification learning on logs with negative events. In: ter Hofstede, A.H.M., Benatallah, B., Paik, H.-Y. (eds.) BPM Workshops 2007. LNCS, vol. 4928, pp. 42–53. Springer, Heidelberg (2008)
7. Quinlan, R.J.: C4.5: Programs for Machine Learning. Morgan Kauffmann, San Mateo (1993)
8. Schonenberg, H., Weber, B., van Dongen, B.F., van der Aalst, W.M.P.: Supporting flexible processes through recommendations based on history. In: Dumas, M., Reichert, M., Shan, M.-C. (eds.) BPM 2008. LNCS, vol. 5240, pp. 51–66. Springer, Heidelberg (2008)
9. Takens, F.: Detecting strange attractors in turbulence. In: Rand, D., Young, L.-S. (eds.) Dynamical Systems and Turbulence, Warwick 1980, vol. 898, pp. 366–381. Springer, Heidelberg (1981)
10. van der Aalst, W., et al.: Process mining manifesto. In: Daniel, F., Barkaoui, K., Dustdar, S. (eds.) BPM Workshops 2011, Part I. LNBIP, vol. 99, pp. 169–194. Springer, Heidelberg (2012)
11. van der Aalst, W.M.P.: Process Mining: Discovery, Conformance and Enhancement of Business Processes. Springer, Heidelberg (2011)
12. van der Aalst, W.M.P., Pesic, M., Song, M.: Beyond process mining: from the past to present and future. In: Pernici, B. (ed.) CAiSE 2010. LNCS, vol. 6051, pp. 38–52. Springer, Heidelberg (2010)
13. van der Aalst, W.M.P., Schonenberg, M.H., Song, M.: Time prediction based on process mining. Inf. Syst. **36**(2), 450–475 (2011)
14. Verbeek, H.M.W., Buijs, J.C.A.M., van Dongen, B.F., van der Aalst, W.M.P.: XES, XESame, and ProM 6. In: Soffer, P., Proper, E. (eds.) CAiSE Forum 2010. LNBIP, vol. 72, pp. 60–75. Springer, Heidelberg (2011)
15. Žliobaitė, I.: Combining time and space similarity for small size learning under concept drift. In: Rauch, J., Raś, Z.W., Berka, P., Elomaa, T. (eds.) ISMIS 2009. LNCS, vol. 5722, pp. 412–421. Springer, Heidelberg (2009)

Classification, Clustering
and Pattern Discovery

A Hybrid Distance-Based Method and Support Vector Machines for Emotional Speech Detection

Vladimer Kobayashi[✉]

Department of Mathematics, Physics, and Computer Science,
University of the Philippines Mindanao Mintal, Davao City, Philippines
vladimer.kobayashi@upmin.edu.ph

Abstract. We describe a novel methodology that is applicable in the detection of emotions from speech signals. The methodology is useful if we can safely ignore sequence information since it constructs static feature vectors to represent a sequence of values; this is the case of the current application. In the initial feature extraction part, the speech signals are cut into 3 speech segments according to relative time interval process. The speech segments are processed and described using 988 acoustic features. Our proposed methodology consists of two steps. The first step constructs emotion models using principal component analysis and it computes distances of the observations to each emotion models. The distance values from the previous step are used to train a support vector machine classifier that can identify the affective content of a speech signal. We note that our method is not only applicable for speech signal, it can also be used to analyse other data of similar nature. The proposed method is tested using four emotional databases. Results showed competitive performance yielding an average accuracy of at least 80 % on three databases for the detection of basic types of emotion.

Keywords: Emotion recognition from speech · Support vector machines · Speech segment-level analysis

1 Introduction

The advent of Ubiquitous Computing research has paved the way to the development of systems that are more accustomed and able to respond in a timely manner according to human needs and behaviour [1,2]. Computers and other machines have been successfully integrated to every aspect of life. To be truly practical machines must not only "think" but also "feel" since meaningful experiences are communicated through changes in affective states or emotions. At the center of all of these is the type of data that we will handle to proceed with the computational task. In a typical scenario we deal with data types commonly encountered in signal processing since emotions are either overtly expressed through voice signal or covertly through other physiological signals.

A. Appice et al. (Eds.): NFMCP 2013, LNAI 8399, pp. 85–99, 2014.
DOI: 10.1007/978-3-319-08407-7_6, © Springer International Publishing Switzerland 2014

These signals can be captured through the use of sophisticated sensors. The challenge not only lies on the collection and pre-processing part but also on the development of methods adapted to the type of data we wish to analyse.

During the past decade we have seen the explosion of studies that try to extract human emotions from various physiological signals. By far the most carefully studied signal for this purpose is the speech signal [3,4]. It is commonly accepted that it is relatively easy for humans to identify (to a certain extent) the emotion of another person based on the voice, although, we are still on the process to understand how we manage to do it. There are salient features in speech signal which we implicitly distinguish and process that enable us to detect certain types of emotions. The central idea of many researches is to automate the process of detecting emotions which is crucial to the creation of emotion-aware technologies and systems [4]. For this purpose researchers have been using techniques from machine learning to construct classifier models that can achieve this task.

In this paper, as a first step, we also deal with speech signal. We argue that this kind of signal is the most convenient and reasonable to deal with since in real setting it can be easily captured. Unlike other signals such as ECG and EEG, speech signal is not particularly troublesome to collect and can be recorded anywhere and at any time. Also, many studies were published about speech processing thus we can try the features proposed in those studies in this work. The true nature of a speech signal is dynamic, a sequence of values indexed by time, however, the approach that we follow is to represent it as a single static feature vector. This approach is influenced by the fact that the sequence information is not essential for emotion recognition [5].

The objective of this work is the proposal of a novel approach that extracts emotions from speech signal. Our proposed method consisted of two steps. The first step is the creation of *emotion models* and the computation of deviations or distances of the speech signal "parts" from each of the emotion models. The first step is reminiscent of the method called Soft Modelling of Class Analogies (SIMCA) [6] although we do not attempt to classify the speech signals in this step. The second step is to use the distance values computed in the previous step to construct a classifier that can detect the over-all emotional content of a given speech signal. In contrast with other techniques, we obtained additional knowledge such as the importance of different variables and the degree of separation of the speech signal components from each emotion category. Another advantage of our technique is the tremendous flexibility it offers. Among other things, the modeller has the freedom to use different sets of features in both the first and second steps and adjust the underlying methods to make it robust to noise.

As primary application of our proposed approach we deal with the problem of detecting basic common types of emotion. The emotions are of interest in the analysis of telephone conversation and diagnosis of certain medical disorders. We tested our approach using four emotional speech databases. The results showed the effectiveness of our approach based from the analysis on the four databases.

The rest of the paper is organized as follows: Sect. 2 discusses prior works related to this study. Section 3 provides description of the four speech databases to which we tested our approach. Section 4 discusses the pre-processing and the extraction of speech acoustic features. Section 5 elaborates on our proposed approach. The results of our experiments are presented in Sect. 6. Finally, we close the paper in Sect. 7.

2   Related Studies

The work of Bi *et al.* (2007) [7] made use of the technique called decision templates ensemble algorithm to combine classifiers built on segment-level feature sets. They segmented an utterance according to *relative time intervals* (RTI) scheme where an utterance is cut at fixed relative positions [8]. Features were extracted at both segment and utterance levels. They described four strategies to train the base classifiers depending on the feature vectors (either segment or utterance level feature vectors) used in the training and testing phase. Using the *Berlin Database of Emotional Speech* (see next part for details on the speech datasets) the best accuracy was at 80.5 % obtained by using segment level feature vectors in the training phase and the utterance level feature vectors in the testing phase with 3-segment length. It should be noted that in their experiments they conducted separate analyses for male and female speakers which could have somewhat influenced the accuracy since previous studies demonstrated that categorizing speakers according to gender could improve classification performance.

Shami and Kamel (2005) [9] also used segment-based approach to the recognition of emotions in speech. In their work they combined features extracted from both segment and utterance levels. They constructed segment classifiers based from support vector machines (SVM) and K nearest neighbors (K-NN). The output of the classifiers were in the form of vector of posteriori class probabilities which can be thought of as degrees of membership of the segments to each emotion classes. To obtain decision in the utterance level they proposed aggregation techniques to combine the posterior class probabilities. For a certain utterance, the class probabilities of its segments were combined together with the information about the utterance itself such as duration. Simple aggregation techniques namely mean, product, and maximum were employed to combine the posterior probabilities from segments. The method was tested using the *KISMIT* speech database which contains 1002 American English utterances obtained from 3 female speakers with 5 affective communicative intents. Affective intents were acted and strongly expressed. The accuracy was reported at 87 % using K-NN.

In a paper of Pan *et al.* (2012) [10], they investigated different combinations of features in conjunction with support vector machines that could yield maximum detection of emotion in the *Berlin Database of Emotional Speech*. The study focused on only three emotions, namely, 'happy', 'sad' and 'neutral'. The best combination of features were mel-frequency cepstrum coefficients (MFCCs), mel-energy spectrum dynamic coefficients (MEDCs) and energy. The performance using three emotion classes reached as high as 95 %.

3   Speech Databases

We tested our proposed approach using four emotional speech databases. They are described next:

The **Berlin Database of Emotional Speech**[1] [11], also known as *Berlin EmoDB*, has been used in several studies thus we can make comparisons between our results and the previous ones. The database was constructed using 10 speakers, 5 male and 5 female, who read 10 sentences in German that have little emotional content textually but they are read in such a way to simulate emotions. Seven discrete emotions were considered, namely, 'anger', 'boredom', 'disgust', 'fear', 'joy', 'neutral', and 'sadness'. The sentence utterances are of variable lengths ranging from 1 to 4 s. There are a total of 535 sentence utterances that were evaluated in a perception test.

Surrey Audio-Visual Expressed Emotion (SAVEE) database[2] has been recorded as a pre-requisite for the development of an automatic emotion recognition system. The database consists of recordings from 4 male actors in 7 different emotions, 480 British English utterances in total. The data were recorded in a visual media lab with high quality audio-visual equipment, processed and labeled.

The **RML emotion database**[3] contains 720 audiovisual emotional expression samples that were collected at Ryerson Multimedia Lab. Six basic human emotions are expressed: 'anger', 'disgust', 'fear', 'happiness', 'sadness', 'surprise'. The RML emotion database is language and culturally background independent. The video samples were collected from eight human subjects, speaking six different languages (English, Mandarin, Urdu, Punjabi, Persian, Italian). Each video clip has a length of about 3–6 s with one emotion being expressed.

The **eNTERFACE'05 EMOTION Database**[4] is an audio-visual emotion database that can be used as a reference database for testing and evaluating video, audio or joint audio-visual emotion recognition algorithms. The final version of the database thus contains 42 subjects, coming from 14 different nationalities. Among the 42 subjects, a percentage of 81 % were men, while the remaining 19 % were women. All the experiments were driven in English.

For the four databases we considered sentence utterances as our speech signals and we only dealt with the audio aspect.

4   Pre-processing and Feature Extraction

Instead of modelling directly the sequence of values and the dynamic nature of speech signal we took a slightly different approach. We firstly cut each speech signal into 3 segments in the manner of relative time interval (RTI) process [8] and we represented each segment by a single static feature vector. We found

---

[1] http://database.syntheticspeech.de/

[2] http://personal.ee.surrey.ac.uk/Personal/P.Jackson/SAVEE/

[3] http://www.rml.ryerson.ca/rml-emotion-database.html

[4] http://www.enterface.net/enterface05/main.php?frame=emotion

out that dividing a speech utterance into 3 segments gives the optimal compromise to balance emotional content and variability. In this part, features do not directly describe the whole speech signal but rather they describe the individual segments, thus a speech segment becomes our unit of analysis.

4.1 Pre-processing

The speech signals were initially preprocessed by removing the silent parts at the beginning and end of the signals. Then they were cut into 3 non-overlapping segments. Here, we did a blind segmentation approach where no prior delimitation of word or syllable boundary has been performed. We assumed that segments may contain emotional primitives that contribute to the over-all emotional content of a speech signal. The segmentation as well as the subsequent extraction of features are illustrated in Fig. 1.

4.2 Speech Acoustic Features

Here the unit of analysis is not the utterances but the segment. Thus, we extracted segment-level features. We used the baseline feature set in the openS-MILE/openEAR software [12]. The features set is named the "emobase" set. The features are low level descriptors (LLD) which include Intensity, Loudness, 12 MFCC, Pitch (F0), Probability of voicing, F0 envelope, 8 LSF (Line Spectral Frequencies), and Zero-Crossing Rate. Delta regression coefficients are calculated from the mentioned LLD and functionals such as Maximum and Minimum values and respective relative position within input, range, arithmetic mean,

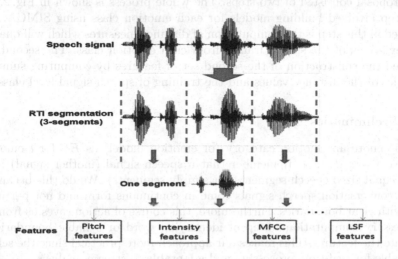

Fig. 1. Segmentation and Feature Extraction Part. An utterance is segmented into 3 non-overlapping speech segments. The features are extracted from the individual speech segments.

2 linear regression coefficients and linear and quadratic error, standard deviation, skewness, kurtosis, quartile 1–3, and 3 interquartile ranges are applied both to LLD and delta coefficients. A total of 988 features were extracted. We will no longer delve into the details of feature extraction instead we refer the reader to the manual of the openSMILE software for a complete description of the features. The features discussed here are used in the first step of our proposed approach.

4.3   Feature Selection

The sheer number of features necessitate the application of feature selection techniques to retain only the approximate essential features. The presence of irrelevant features could add unnecessary complexity to the model and may even adversely affect its performance. For this purpose we used the random forest algorithm to assign importance to each feature. First we run the random forest algorithm, next we retrieved the mean decrease in node impurity for each feature, then we ranked the features according to the values of the mean decrease in node impurity, and finally we apply a simplified best-first search. The best-first search that we used mimics a grid search methodology where we started with the top fifty features until we reached the top 500 features incremented by 10 features in each step. For each set of features we trained our classifier and measured the performance on a test set. The feature set with the lowest average test loss through cross validation was our final set.

5   Our Proposal

Our proposal consisted of two steps. The whole process is shown in Fig. 2. The first step involved building models for each emotion class using SIMCA. Also included in this step is the computation of distance measures which will quantify the deviations of the speech frames from each emotion class. The second step involved the construction of the second set of features by computing summary statistics of the distance values and the training of speech signal level classifier.

5.1   Preliminaries

Let us denote an emotion category (or emotion model) as E^j, for l emotions we have $j = 1, 2, \ldots, l$. Remember that a speech signal (mother signal) is cut into 3 equal sized speech segments (or simply segments). We do this because in actual conversation speech signals come in continuous form and not per utterance with clear boundaries. Furthermore, this course of action saves us from the unnecessary computational step of identifying word or syllable boundaries to compute the features. This makes our approach more practical since the scheme is suitable for real-time processing and adaptable to stream analysis.

Depending on the length of the speech signal the length of each of the segments may vary. We assigned emotion labels for each of the segments during training. Here we assume that *the emotion label of a segment is the same as the*

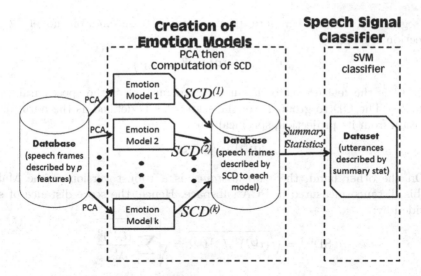

Fig. 2. Diagrammatic presentation of our proposed approach

emotion label of the mother signal where the segment came from. We represent a segment as $\mathbf{s}_i^j$, which can be interpreted as segment i that has emotion label E^j. Finally each segment is described by p (=988) features. Thus for a given segment its static single feature vector representation is $\mathbf{s}_i^j = (s_{i1}^j, s_{i2}^j, \ldots, s_{ip}^j)^T$.

5.2 First Step: Construction of Emotion Models

The motivation for this step is to reveal underlying structure for each emotion category. To achieve this task we perform Principal Component Analysis (PCA) classes on each emotion category. This way we understand better the properties of speech frames with respect to each emotion. Also, this stage will involve a second feature reduction part since we want to derive a condensed set of features that are useful to describe each group.

To obtain the emotion models, we run separate PCA on each emotion class. After we run PCA on each class E^j we will obtain matrix of scores T^j and loadings P^j. Since we run separate PCA we can now summarize each emotion class in a different subspace model (according to the PCA models). The number of retained principal components for each class is denoted by $k_j \ll p$. The relevance of the features on each model can be assessed by examining the loadings of the features on the extracted principal components.

Once we have the emotion models, we can now compute the deviations of each segments to the different models. Two deviations can be defined: the *orthogonal distance* (OD) and the *score distance* (SD).

The *orthogonal distance* is simply the Euclidean distance of a segment to an emotion model in the PCA subspace. To compute the orthogonal distance of any segment $\mathbf{s}$, we first compute its projection $\mathbf{s}^{(j)}$ on emotion model E^j. Its projection is given by

$$\mathbf{s}^{(j)} = \bar{\mathbf{s}}^j + P^j (P^j)^T (\mathbf{s} - \bar{\mathbf{s}}^j) \tag{1}$$

where $\bar{\mathbf{s}}^j$ is the feature vector mean (column means) of the speech frames in group E^j. The OD to group E^j of the segment $\mathbf{s}$ is defined as the norm of its deviation from its projection, specifically,

$$\mathrm{OD}^{(j)} = \|\mathbf{s} - \mathbf{s}^{(j)}\| \tag{2}$$

On the other hand, the *score distance* is a robust version of the Mahalanobis distance measured in PCA subspace. Hence, the score distance of $\mathbf{s}$ is provided by:

$$\mathrm{SD}^{(j)} = \sqrt{(\mathbf{t}^{(j)})^T J^{-1} \mathbf{t}^{(j)}} = \sqrt{\sum_{a=1}^{k_j} \frac{(t_a^{(j)})^2}{\lambda_a^{(j)}}} \tag{3}$$

where $\mathbf{t}^{(j)} = (P^j)^T (\mathbf{s} - \bar{\mathbf{s}}^j) = (t_1^{(j)}, t_2^{(j)}, \ldots, t_{k_j}^{(j)})^T$ is the score of $\mathbf{s}$ with respect to the E^j group, $\lambda_a^{(j)}$ for $a = 1, 2, \ldots, k_j$ stands for the largest eigenvalues in the E^j group, and J is the diagonal matrix of the eigenvalues. The advantage of SD over OD is that SD uses information about the eigenvalues.

In the usual case we need to decide which of the two distances is more appropriate for the problem, but we opted to combine the two distances by using a parameter γ. We first standardized the distance so that neither one of them dominates the other in terms of magnitude. To perform the standardization we need to apply cut-off values for each of them.

The cut-off value for the score distance, denoted by c_{SD}^j, uses the chi-square distribution since from statistics the squared score distances follow asymptotically a χ^2-distribution with k_j degrees of freedom if the projected observations are i.i.d and normally distributed. Thus, the cut-off value is set to $c_{SD}^j = \sqrt{\chi_{k_j;0.975}^2}$. To standardize we divide the score distances by the cut-off value c_{SD}^j.

The same analysis can be done in the case of orthogonal distance. We start with the following information: The scaled chi-squared distribution $g_1 \chi_{g2}^2$ approximates the unknown distribution of the squared orthogonal distances. From the information, we need to estimate the parameters g_1 and g_2, the estimation is accomplished by using the method of moments. Using the Wilson-Hilferty approximation for a chi-squared distribution the cut-off value can be derived which also implies that the orthogonal distances to the power $\frac{2}{3}$ are approximately normally distributed with mean $\mu = (g_1 g_2)^{\frac{1}{3}} (1 - \frac{2}{9 g_2})$ and variance $\sigma^2 = (2 g_1^{\frac{2}{3}}) / (9 g_2^{\frac{1}{3}})$. The estimates for $\hat{\mu}$ and $\hat{\sigma}$ are calculated by means of univariate MCD applied to the orthogonal distances of the training samples from group j. The cut-off value for the orthogonal distances then equals $c_{OD}^j = (\hat{\mu} + \hat{\sigma} z_{0.975})^{\frac{3}{2}}$ with $z_{0.975} = \Phi^{-1}(0.975)$ the 97.5 % quantile of the Gaussian distribution. The same with the score distance, to standardize we divide the orthogonal distances with the cut-off value c_{OD}^j.

Using the standardized score and orthogonal distance we can now define a combined distance and named it *scortho distance* (SCD) [6] for a given s by

$$\text{SCD}^{(j)} = \gamma \left(\frac{\text{OD}^{(j)}}{c_{OD}^j} \right) + (1 - \gamma) \left(\frac{\text{SD}^{(j)}}{c_{SD}^j} \right) \tag{4}$$

where $\gamma \in [0, 1]$. We can also consider the square-scortho distance in the following way

$$\text{SCD}^{2(j)} = \gamma \left(\frac{\text{OD}^{(j)}}{c_{OD}^j} \right)^2 + (1 - \gamma) \left(\frac{\text{SD}^{(j)}}{c_{SD}^j} \right)^2 \tag{5}$$

We can optimize the results by choosing appropriate values for the parameter γ. One suggestion is to perform cross-validation by optimizing certain criterion like test prediction accuracy.

Another advantage of building emotion models is we can identify segments which are markedly far from any of the models. We do this by utilizing the cut-off values of the computed distances. Segments which computed distances are outside the cut-off values for all the models with respect to a particular distance are termed unrepresentative segments because purportedly they do not contain any emotion. Another possibility is they may form an unknown group and after detection can lead to a new knowledge about the nature of the problem we are studying. In this paper, we are not yet going to pursue the idea of identifying unrepresentative segments.

To summarize the first step of our proposed approach, we first build emotion models using PCA and then proceed to the computation of the SCD (or the square SCD). At this point, each speech frame is represented by a vector consisting of its distances to each emotion models. Notationally, using the square SCD distances, given s and we wish to identify l types of emotion, we have

$$\mathbf{s} = (\text{SCD}^{2(1)}, \text{SCD}^{2(2)}, \dots, \text{SCD}^{2(l)}) \tag{6}$$

This new representation of the speech frames will be used in the second step.

5.3 Second Step: Speech Signal Level Classifier

The new representation of the segments obtained from the previous step is aggregated with respect to each speech signal. Remember that we cut the speech signals into segments so that we can proceed with the initial feature extraction. The aggregation is made possible by computing certain functionals or summary statistics. Suppose we have a speech signal u and the speech frames cut from it are $\{\mathbf{s}_1, \mathbf{s}_2, \mathbf{s}_3\}$. One summary statistic that we can compute is the mean. Hence, using the mean we can represent the speech signal u as

$$\mathbf{u} = (\mu_1, \mu_2, \dots, \mu_l) \tag{7}$$

where l is the number of emotion categories and

$$\mu_j = \frac{1}{3} \sum_{i=1}^{3} \text{SCD}_i^{2(j)} \tag{8}$$

where, $\text{SCD}_i^{2(j)}$ is the distance of ith segment extracted from u to emotion model E^j. The interpretation of the components is straightforward in this case: the μ_j are the mean SCD^2 of the speech frames components in u to emotion model E^j. We can also view this as the "distance" of a speech signal from each emotion category.

Aside from the mean, we can also use additional summary statistics like the standard deviation or the maximum and minimum values to capture additional information. It is important to note that if we use many statistical measures we will be increasing the size of the vector representation of the speech signals. For instance in the case of l emotions, if we consider two summary statistics then the number of vector components could be equal to $2l$. Thus it is imperative to choose the right summary statistics to capture the important characteristics of the speech signals as expressed by the deviations of their member speech segments.

Once we have represented each speech signal u as feature vector $\mathbf{u}$ in the manner we described above we can now construct a speech signal matrix $\mathbf{U}$ denoted by

$$\mathbf{U} = \begin{pmatrix} \mathbf{u}_1 \\ \mathbf{u}_2 \\ \vdots \\ \mathbf{u}_n \end{pmatrix} \tag{9}$$

where n is the total number of utterances or speech signals.

With emotion labels associated to each utterance we are now ready to train the classifier at this step. The user has the liberty to select the classification algorithm to use here. For our case, we decided to employ Support Vector Machine (SVM) technique since it has shown competitive performance in other pattern recognition problems [13,14]. For an in-depth discussion about the principles of SVM we refer the reader to [15].

6   Results

All throughout the experiments we made use of the segment level features in the first step. Initially, we have a large feature set consisting of 988 features in total. We attempted to reduce the dimension by applying a wrapper approach to feature selection using the importance measure generated from random forest algorithm. We used the square SCD as our distance to measure the deviation of each segment from the emotion models. In the second step we have aggregated the distances obtained from the first step by computing the arithmetic mean, standard deviation, minimum, maximum, range, the categories where the minimum and maximum values could be found. A total of 41 (or 36 depending on the number of emotion classes) derived features were included in the second stage. Thus, each speech signal is represented by a vector of 41 (or 36) elements. Moreover, we trained SVM classifier in the second step using the ordinary dot product kernel function. As a standard practice to get reliable estimate of the performance of each classifier we use 10 times 10-fold cross validation and computed the mean accuracies for each 10-fold and also we computed the macro F-measure. The accuracy is computed by:

$$Accu = \frac{\# \text{ of correctly classified}}{\text{total}\# \text{ of observations}} \tag{10}$$

Whereas to compute the macro F-measure we needed to compute the F-measure for each category. F-measure of class j is computed by:

$$F_j = 2 * \frac{Recall_j * Precision_j}{Recall_j + Precision_j} \tag{11}$$

recall and precision measures are calculated using the formulas:

$$Recall_j = \frac{tp_j}{tp_j + fn_j} \tag{12}$$

and

$$Precision_j = \frac{tp_j}{tp_j + fp_j} \tag{13}$$

The symbols, tp, fp, and fn stand for true positive, false positive, and false negative, respectively. Now to compute the macro F-measure for a multi-class classification problem we take the average among all the F-measures of the all the classes. Thus for a l-class classification problem the macro F-measure is given by:

$$MacroF = \frac{\sum_{j=1}^{j=l} F_j}{l} \tag{14}$$

The macro F-measure gives an indication of the global *effectiveness* of the classifier on a given data.

It is important to emphasize here that we trained the models without taking into account the gender or age or language of speakers and the results reported are the mean accuracies and mean macro F-measures on the speech signal level. We trained separate models for each databases. The whole process of train and testing is depicted in Fig. 3.

From the results shown in Table 1 we find that our approach was able to effectively identify the emotions. Particularly our approach is superior in detecting emotions "Sadness", "Anger", and "Surprise". This can be explained by the fact that the three emotions have distinctive arousal and duration characteristics as expressed in the speech signal. "Sadness" is a lowly active emotion and both "Anger" and "Surprise" are highly-active ones. "Anger" and "Surprise" are further contrasted due to the fact that "Surprise" can be detected for shorter duration. The case of emotions "Disgust" and "Happiness" is interesting because they have somewhat almost the same arousal as "Anger", nevertheless the classifier has successfully distinguished it from "Anger" many times especially in the RML and SAVEE databases. Our method was also able to distinguish speech

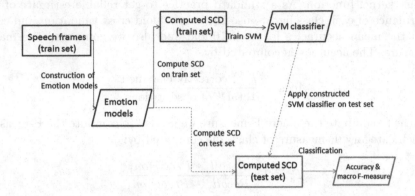

Fig. 3. Training and testing followed in the experiments

Table 1. Average precision and recall for each emotion categories in each database using our proposed approach.

Database	Emotion	Average precision	Average recall
	Fear	0.778	0.778
	Disgust	0.571	0.500
	Happiness	0.769	0.588
Berlin EmoDB	Boredom	0.929	0.867
	Neutral	0.833	0.833
	Sadness	0.889	0.941
	Anger	0.806	0.967
	Fear	0.857	0.632
	Disgust	0.778	0.737
RML	Happiness	0.789	0.833
	Sadness	0.947	0.900
	Anger	0.643	0.947
	Surprise	0.944	0.810
	Fear	0.600	0.500
	Disgust	0.657	0.767
	Happiness	0.581	0.621
eNTERFACE	Sadness	0.833	0.333
	Anger	0.636	0.700
	Surprise	0.512	0.733
	Fear	1.000	0.750
	Disgust	0.778	0.875
	Happiness	0.857	0.750
SAVEE	Neutral	1.000	0.875
	Sadness	1.000	0.750
	Anger	0.667	1.000
	Surprise	0.778	0.875

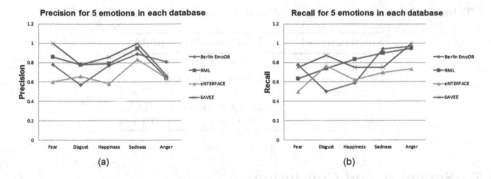

Fig. 4. Precision (a) and Recall (b) graphs for 5 emotions.

Table 2. Performance of our proposed approach on the four databases assessed using average accuracy and average macro F-measure. We also present the baseline accuracy obtained by classifying all speech signals to the majority class.

Database	Baseline accuracy	Average accuracy	Average macro F-measure
Berlin EmoDB	23.7 %	82.10 %	0.799
RML	16.70 %	81.00 %	0.810
eNTERFACE	16.70 %	60.90 %	0.600
SAVEE	16.70 %	83.90 %	0.841

signals that do not contain emotion ("Neutral"). A graphical presentation of the comparison of precision and recall among 5 emotions on the four databases is displayed in Fig. 4. We notice that although "Anger" has high recall its precision is relatively low since most misclassification for other emotions occurred here.

We present in Table 2 the mean accuracy and mean macro F-measures obtained by our approach. The table confirms our claim that our method is particularly effective. Our method reached as high as at least 80 % accuracy except for eNTERFACE database which is particularly difficult to get even an accuracy of 60 % [16].

Aside from the good detection rates we also obtained additional knowledge regarding the features that are useful in the discrimination of emotions. The information is derived from the first step of our approach. We find that although we have used an initial set of 988 features we discovered that we can construct at most 10 (latent) features from the PCA components and still capture the overall variation in each emotion model. An analysis on the loadings of the original features to the PCA components revealed that Pitch related features and MFCCs are the most influential in the detection of emotions, the other features just seemed to complement them. Lastly, in Table 3 we provide information regarding the most important features to discriminate among emotion models obtained from the Berlin EmoDB.

Table 3. Summary characteristics of each emotion models for the Berlin EmoDB.

Emotion model	Most important variables
Fear	MFCCs, LSP
Disgust	Intensity, MFCCs
Happiness	Pitch, intensity, MFCCs
Angry	ntensity, loudness, MFCCs
Sadness	Pitch, intensity, MFCCs

7   Summary and Conclusion

The methodology we have proposed in this paper has shown competitive performance in the detection of basic emotion types on the four databases. Although the results from the eNTERFACE database did not fare well with the other databases we can still put confidence on our method because it is better than the baseline accuracy and for this database it is notoriously difficult to achieve an accuracy of greater than 60 %. Aside from the good classification accuracies the methods also reveal additional knowledge regarding the features useful to extract emotions. This knowledge is desirable if we want to understand better the type of features useful for the discrimination of emotion classes.

Our proposed approach did not only achieve maximum detection rate for the emotion but also it provides added information toward the understanding of the synergy among emotions within a speech signal. By doing this, we are able to make a detailed analysis of a speech signal by examining its speech segment components.

Another confirmation we got in this study is the usefulness of using speech segments as unit of our analysis. The speech segments acts as atoms of emotions from which we can identify emotional primitives that could be used to deduce the over-all emotion of a speech signal.

In practice, our proposed approach can be implemented relatively fast. In the first step PCA can be run rapidly and the computation of distances is speedy. In the second step the complexity only depends on the complexity of the classifier. Lastly, our proposed approach offer new insights to the kind of analysis that can be done and additional tool to analyse emotional speech database.

In our future work we will investigate further the use of other features both in the first and second steps and other classifiers in the second step. We will also test our method on other speech databases especially on continuous ones.

References

1. Cowie, R., Douglas-Cowie, E., Tsapatsoulis, N., Votsis, G.N., Kollias, S.D., Fellenz, W.A., Taylor, J.G.: Emotion recognition in human-computer interaction. IEEE Sig. Process. Mag. **18**, 32–80 (2001)

2. Vogt, T., André, E., Wagner, J.: Automatic recognition of emotions from speech: a review of the literature and recommendations for practical realisation. In: Peter, C., Beale, R. (eds.) Affect and Emotion in HCI. LNCS, vol. 4868, pp. 75–91. Springer, Heidelberg (2008)
3. El Ayadi, M.M.H., Kamel, M.S., Karray, F.: Survey on speech emotion recognition: features, classification schemes, and databases. Pattern Recogn. **44**, 572–587 (2011)
4. Koolagudi, S.G., Rao, K.S.: Emotion recognition from speech: a review. Int. J. Speech Technol. **15**, 99–117 (2012)
5. Dileep, A.D., Veena, T., Sekhar, C.C.: A review of kernel methods based approaches to classification and clustering of sequential patterns, part i: sequences of continuous feature vectors. In: Kumar, P., Krishna, P.R., Raju, S.B. (eds.) Pattern Discovery Using Sequence Data Mining: Applications and Studies. IGI Global (2012)
6. Branden, K.V., Hubert, M.: Robust classification in high dimensions based on the SIMCA method. Chemometr. Intell. Lab. Syst. **79**, 10–21 (2005)
7. Bi, F., Yang, J., Yu, Y., Xu, D.: Decision templates ensemble and diversity analysis for segment-based speech emotion recognition. In: 2007 International Conference on Intelligent Systems and Knowledge Engineering (ISKE 2007). Advances in Intelligent Systems Research (2007)
8. Schuller, B., Rigoll, G.: Timing levels in segment-based speech emotion recognition. In: Ninth International Conference on Spoken Language Processing, INTERSPEECH 2006 - ICSLP, Pittsburgh, PA, USA, pp. 1818–1821. ISCA (2006)
9. Shami, M.T., Kamel, M.S.: Segment-based approach to the recognition of emotions in speech. In: Proceedings of the 2005 IEEE International Conference on Multimedia and Expo, ICME 2005, Amsterdam, The Netherlands, pp. 366–369. IEEE (2005)
10. Pan, Y., Shen, P., Shen, L.: Speech emotion recognition using support vector machine. Int. J. Smart Home **6**(2), 101–108 (2012)
11. Burkhardt, F., Paeschke, A., Rolfes, M., Sendlmeier, W.F., Weiss, B.: A database of german emotional speech. In: INTERSPEECH 2005, pp. 1517–1520. ISCA (2005)
12. Eyben, F., Wöllmer, M., Schuller, B.: Opensmile - the munich versatile and fast open-source audio feature extractor. In: Proceedings of ACM Multimedia (MM), Florence, Italy, pp. 1459–1462. ACM (2010)
13. Osuna, E., Freund, R., Girosi, F.: Training support vector machines: an application to face detection. In: CVPR '97, IEEE Computer Society, pp. 130–136 (1997)
14. Joachims, T.: Text categorization with suport vector machines: learning with many relevant features. In: Nédellec, C., Rouveirol, C. (eds.) ECML 1998. LNCS, vol. 1398, pp. 137–142. Springer, Heidelberg (1998)
15. Herbrich, R.: Learning Kernel Classifiers: Theory and Algorithms. MIT Press, Cambridge (2001)
16. Schuller, B., Zhang, Z., Weninger, F., Rigoll, G.: Using multiple databases for training in emotion recognition: to unite or to vote? In: 12th Annual Conference of the International Speech Communication Association, INTERSPEECH 2011, Florence, Italy, pp. 1553–1556. ISCA (2011)

Methods for the Efficient Discovery
of Large Item-Indexable Sequential Patterns

Rui Henriques[1,2]([✉]), Cláudia Antunes[2], and Sara C. Madeira[1,2]

[1] KDBio, Inesc-ID, Instituto Superior Técnico,
Universidade de Lisboa, Lisboa, Portugal
[2] Department of Computer Science and Engineering, IST,
Universidade de Lisboa, Lisboa, Portugal
{rmch,claudia.antunes,sara.madeira}@tecnico.ulisboa.pt

Abstract. An increasingly relevant set of tasks, such as the discovery of biclusters with order-preserving properties, can be mapped as a sequential pattern mining problem on data with item-indexable properties. An item-indexable database, typically observed in biomedical domains, does not allow item repetitions per sequence and is commonly dense. Although multiple methods have been proposed for the efficient discovery of sequential patterns, their performance rapidly degrades over item-indexable databases. The target tasks for these databases benefit from lengthy patterns and tolerate local mismatches. However, existing methods that consider noise relaxations to increase the average short length of sequential patterns scale poorly, aggravating the yet critical efficiency. In this work, we first propose a new sequential pattern mining method, IndexSpan, which is able to mine sequential patterns over item-indexable databases with heightened efficiency. Second, we propose a pattern-merging procedure, MergeIndexBic, to efficiently discover lengthy noise-tolerant sequential patterns. The superior performance of IndexSpan and MergeIndexBic against competitive alternatives is demonstrated on both synthetic and real datasets.

1 Introduction

Sequential pattern mining (SPM) has been proposed to deal efficiently with the discovery of frequent precedences and co-occurrences in itemset sequences. SPM methods can be applied to solve tasks centered on extracting order-preserving regularities, such as the discovery of flexible (bi)clusters [14]. These tasks commonly rely on a more restricted form of sequences, item-indexable sequences, which do not allow item repetitions per sequence. Illustrative examples of item-indexable databases include sequences derived from microarrays, molecular interactions, consumer ratings, ordered shoppings, tasks scheduling, among many others. However, these tasks are characterized by two major challenges. First, their hard nature, which is related with two factors: average high number of items per transaction and high data density. Second, order-preserving solutions are optimally described by lengthy noise-tolerant sequential patterns [5].

A. Appice et al. (Eds.): NFMCP 2013, LNAI 8399, pp. 100–116, 2014.
DOI: 10.1007/978-3-319-08407-7_7, © Springer International Publishing Switzerland 2014

Although existing SPM approaches can be applied over item-indexable data-bases, they suffer from two problems. First, they show inefficiencies due to the commonly observed density levels and high average transaction length of these datasets, which leads to a combinatorial explosion of sequential patterns under low support thresholds [14]. Additionally, the few dedicated methods able to discover sequential patterns in item-indexable databases [13,14] show significant memory overhead.

Second, the average length of sequential patterns is typically short. A common desirable property for the tasks formulated over these databases is the discovery of sequential patterns with a medium-to-large number of items. For instance, order-preserving patterns from ratings and biological data are only relevant above a minimum number of items. Such lengthy patterns can be discovered under very low support thresholds, aggravating the yet hard computational complexity, or by assuming local noise (violation of ordering constraints for a few transactions). However, some of existing SPM extensions to deal with noise relaxations have been tuned for different settings [6], while others can deteriorate the yet critical SPM efficiency [25]. Additionally, methods to discover colossal patterns from smaller itemsets [30] have not been extended for the SPM task.

This work proposes a new method for the efficient retrieval of lengthy order-preserving regularities based on sequential patterns discovered over item-indexable sequences. This is performed in two steps. First, we propose a new method, IndexSpan, that uses efficient data structures to keep track of the position of items per sequence and relies on fast database projections based on the relative order of items. Pruning techniques are available when the user has only interest in sequential patterns above a minimum length. Second, we propose an efficient method, MergeIndexBic, to guarantee the relevance of order-preserving regularities by deriving medium-to-large sequential patterns from multiple short sequential patterns. MergeIndexBic uses an error threshold based on the percentage of shared sequences and items among sets of sequential patterns. This is accomplished by mapping this problem in one of two tasks: discovery of maximal circuits in graphs or multi-support frequent itemset mining.

The paper is structured as follows. Section 2 introduces and motivates the task of mining sequential patterns over item-indexable databases, and covers existing contributions in the field. Section 3 describes the proposed solution based on the IndexSpan and MergeIndexBic methods. Section 4 assesses the performance of IndexSpan on both real and synthetic datasets against SPM and dedicated algorithms. The performance of MergeIndexBic against default alternatives is also validated. Finally, the implications of this work are synthesized.

2 Background

Let an item be an element from an ordered set $\mathcal{L}$. An *itemset* I is a set of non-repeated items, $I \subseteq \mathcal{L}$. A *sequence* s is an ordered set of itemsets. A sequence $a = a_1 \cdots a_n$ is a *subsequence* of $b = b_1 \cdots b_m$ ($a \subseteq b$), if $\exists_{1 \leq i_1 < \cdots < i_n \leq m} : a_1 \subseteq b_{i_1}, \ldots, a_n \subseteq b_{i_n}$. The illustrative sequence $s_1 = \{a\}, \{be\} = a(be)$ is contained in $s_2 = (ad)c(bce)$. A *sequence database* is a set of sequences $D = \{s_1, \ldots, s_n\}$.

The **coverage** Φ_s of a sequence s w.r.t. to a set of sequences D, is the set of all sequences in D with s as subsequence: $\Phi_s = \{s' \in D \mid s \subseteq s'\}$. The **support** of a sequence s in D, denoted sup_s, is its coverage size $|\Phi_s|$. Illustrating, consider the sequence database $D = \{s_1 = (bc)a(abc)d, s_2 = cad(acd), s_3 = a(ac)c\}$. For this database, we have $|\mathcal{L}| = 4$, $\Phi_{\{a(ac)\}} = \{s_1, s_2, s_3\}$, and $sup_{\{a(ac)\}} = 3$.

Given a set of sequences D and some user-specified minimum support threshold θ, a sequence $s \in D$ is *frequent* when is subsequence of at least θ sequences. The **sequential pattern mining** (SPM) problem consists of computing the set of frequent sequences, $\{s \mid sup_s \geq \theta\}$.

The set of maximal frequent sequences for the illustrative sequence database, $D = \{(bc)a(abc)d, cad(acd), a(ac)c\}$, under $\theta = 3$ is $\{a(ac), cc\}$.

Let an item-indexable sequence be a sequence without repeated items. An item-indexable sequence database is a set of item-indexable sequences.

Let $|I|$ be the length of an itemset, and $|s|$ be the length of a sequence, $\Sigma_i |s^i|$. Given a set of item-indexable sequences D, a minimum support threshold θ, and a minimum sequence length δ. The task of **SPM over item-indexable sequences**, or simply item-indexable SPM, consists of computing:

$$\{s \mid sup_s \geq \theta \wedge |s| \geq \delta\}$$

This formalization allows the definition of new methods prone to seize the properties of item-indexable sequences. Understandably, the resulting sequential patterns, referred as item-indexable sequential patterns, preserve the consistency of item ordering constraints since they do not allow for item duplicates.

2.1 Applications

From the large set of applications of item-indexable SPM[1], one prominent task is order-preserving biclustering, a form of local clustering based on frequent ordering constraints [5,13]. Order-preserving biclustering is commonly applied over biological domains for the analysis of gene expression data, networks and genomic structural variations. Illustrating, finding subsets of genes respecting orderings on the levels of expression across conditions is critical to study frequent variations, otherwise not discovered under the original levels of expression.

To compose an item-indexable database D from a real-value or discrete matrix, (X, Y), where $X = \{s_1, \ldots, s_n\}$ is a set of rows and $Y = \{y_1, \ldots, y_m\}$ is a set columns, the column indexes are linearly ordered for each transaction according to their values. Each transaction is seen as a sequence of items that correspond to column indexes. A bicluster, (I, J), a correlated subset of rows $I \subset X$ and columns $J \subset Y$, is order-preserving if the permutation of its columns J is strictly increasing across the I rows. A order-preserving bicluster can be derived from a frequent sequence s by mapping $(I, J) = (\Phi_s, \{s_i \mid i = 1 \cdots |s|\})$. Figure 1 illustrates how order-preserving biclustering can be solved using SPM.

---

[1] Detailed description of tasks available in
http://web.ist.utl.pt/rmch/software/indexspan.

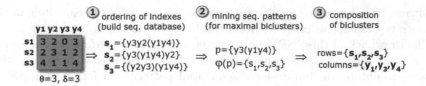

Fig. 1. Mining order-preserving biclusters from item-indexable databases.

An increasingly important application of item-indexable SPM for recommendations based on user preferences, quality assessments and questionnaires [11]. Item-indexable sequences are derived from an ordering of ratings, such as ratings of videos, hotels, shopped items, restaurants, among other products and experiences recorded in large-scale platforms (e.g. IMDb, booking.com, Amazon). Frequent precedences and co-occurrences disclose relevant priorities for different groups of users. Other applications include the discovery of order-preserving regularities in scheduling, planning, shopping, and traveling behavior [9,28].

2.2 Related Work

Two major lines of research are considered. First, we review the general SPM methods and item-indexable dedicated methods and cover their major drawbacks. Second, we gather the potentialities and limitations of the available alternatives to compose lengthy (sequential) patterns.

Efficient SPM in item-indexable databases. Although general SPM methods are not optimized to deal with item-indexable specificities, they have been the largely adopted to solve these applications [14]. Since the SPM problem proposal [1], multiple extensions and applications have been proposed, ranging from scalable implementations to alternative pattern representations. Current SPM methods can be classified into three main categories: apriori-based, pattern-growth, and early-pruning [15]. Apriori-based algorithms [22], and vertical-based variations [27], rely on join procedures to generate candidate sequences in a breadth-first manner using multiple database scans. To overcome the computational complexity of maintaining the support count for each sequence generated, the use of bitmaps or direct comparison have been proposed [3,7].

Pattern growth methods [3,17,18] avoid the costs from candidate generation by building a representation of the database and recursively traversing it to grow the frequent sequences. PrefixSpan [17], an efficient option, recursively constructs patterns by growing their prefix and by maintaining their corresponding postfix subsequences into projected databases. This guarantees a narrowed search space and avoids the generation of candidates since it only counts the frequency of local sequences. The major cost of PrefixSpan resides on database projections.

Early-pruning methods emerged more recently in the literature [7,20,26]. They adopt a sort of position induction to prune candidate sequences very early in the mining process and to avoid support counting as much as possible. These algorithms usually employ a table to track the last positions of each item in the sequence to evaluate whether the item can be appended to a given prefix.

The drawback of these SPM alternatives is that their performance does not scale for very low support thresholds, which are often required in item-indexable contexts to obtain medium-to-large sequential patterns. In fact, new methods can seize the item-indexable property, that guarantees that each item appears at most one time per item-indexable sequence, to minimize this problem.

Seizing this property, Liu and Wang [13,14] proposed an alternative SPM method that constructs a compact tree structure, OPC-Tree, where sequences sharing the same prefix are gathered and recorded in the same branch. The discovery of frequent subsequences and the association of rows with frequent subsequences are performed simultaneously. However, the memory complexity of OPC-Tree is $\Theta(n \times m^2)$, where n is the number of records and m the average number of items per transaction. Although some pruning techniques can be applied in the OPC-Tree structure, their impact is not sufficient to turn this approach scalable for medium-to-large databases.

Discovery of Lengthy (Sequential) Patterns. The existing attempts for the discovery of lengthy patterns can be synthesized according to four major directions. First direction is to rely on efficient methods able to discover sequential patterns under very low support thresholds. There are two main classes of such methods. First class incorporates look-ahead heuristics, such as the ones used by MaxMiner [4], to avoid the traversal of every frequent sequence [12]. Second class generates patterns by reducing large candidates using the monotocity property of frequency (if an itemset is frequent, then its subsets are also frequent). Constraint programming methods can make good use of monotonic constraints to effectively reduce the search space [19]. However, under very low support thresholds, there are two structural problems. First, there is a high probability of discovering precedences and co-occurrences by chance due to the small set of supporting transactions. Second, the increase of length in the number of items comes at the cost of an heightened decrease in the number of supporting transactions. This does not support the final goal since the size of order-preserving regularities is defined by the length of both item and transaction sets (as previously illustrated in Fig. 1). To overcome this drawback, the remaining directions allow for noisy patterns to increase their length without a significant decrease of support levels. These directions assume a tolerance of item mismatches observed for small subsets of the supporting transactions.

Second direction is to extend SPM to discover sequential patterns with local mismatches and gap-based relaxations [2,6,25,29]. However, such extensions either: assume the presence of very long sequences for the creation of partitions, or increase the computational complexity of the original methods, limiting even more the discovery sequential patterns in useful time.

Third direction is to rely on approximative pattern mining under specific principle for composing lengthy patterns [8]. In particular, colossal pattern mining relies on the approximative fusion of smaller patterns [30]. However, these principles have been synthesized in the context of frequent itemset mining and, to our knowledge, have not yet been extended for sequential pattern mining.

Final direction is to view patterns as biclusters and rely on dedicated biclustering merging strategies [21]. The need for merging biclusters is based on the observation that when two biclusters share a significant area it is probable that they are part of a larger coherent bicluster. The simplest criterion for merging is to rely on the overlapping area (as a percentage of the larger bicluster). Nevertheless, the existing approaches require the computation of similarities between all pairs of biclusters. Understandably, this solution is impracticable for solutions characterized by a large number of biclusters (sequential patterns).

3   Solution

The proposed solution is defined by two major methods. First, IndexSpan method for the efficient discovery of sequential patterns over dense item-indexable databases. Second, MergeIndexBic method to consider relaxations that allow local ordering violations to compose larger and noise-tolerant sequential patterns.

3.1   IndexSpan: Boosting Item-Indexable SPM

To avoid the drawbacks of existing approaches, we propose the IndexSpan method, an extension of PrefixSpan to discover sequential patterns with heightened efficiency from item-indexable databases. Comparison of existing SPM algorithms [15] shows key heuristics to turn the SPM efficient: mechanisms to reduce the support counting; narrowing of the search space; optimally sized data structure representations of the sequence database; and early pruning of candidate sequences. Seizing these properties, IndexSpan guarantees a search space as small as possible and relies on a narrow search procedure, depth-first search.

IndexSpan extends PrefixSpan [17] in order to incorporate additional efficiency gains from three principles. First, IndexSpan relies on an easily indexable and compacted version of the original sequence database. Second, it uses faster and memory-efficient database projections. A projected database only maintains a list with the IDs of the active sequences. Finally, IndexSpan relies on early-pruning techniques. IndexSpan is described in Algorithm 1.

IndexSpan considers the three following structural adaptations over the PrefixSpan algorithm. First, it maintains a simple matrix in memory that maintains the index of each item per row. This matrix is constructed at the very beginning (*lines 2–5*) and the original database is removed. Additionally, for sparse databases, this matrix is replaced by a vector of hash tables to optimize memory usage. These data structures support position induction. The idea behind is simple: if an item's last/start position precedes the current prefix/postfix position, the item can no longer appear before/after the current prefix.

Second, a projected database can be constructed with heightened efficiency by avoiding the need to update and maintain postfixes. A projected database simply maintains the identifiers of the supporting sequences for a specific prefix.

To know if a sequence is still frequent when an item is added over a specific prefix, there is only the need to compare its index against the index of the previous item as well as their lexical order for the case where the index is the same (i.e. the new item co-occurs with the last items of the pattern). In this way, database projections, the most expensive step of PrefixSpan both in terms of time and memory, are handled with heightened efficiency. The proposed projection method is described in Algorithm 1, *lines 12–19 and 28–37*.

Finally, the input minimum number of items per sequential pattern, δ, can be used to prune the search as early as possible. If the number of items of the current prefix ($|\alpha|$) plus the items of a postfix s_α (computed based on the current and last index positions) is less than δ, then the sequence identifier related with

---

Algorithm 1. IndexSpan

Input: sequence database D, minimum support θ, minimum sequence length δ
Output: set of sequential patterns S
Note: α is a sequence, D_α is the α-projected database
     (D_α simply maintains a reference to the current sequences)

```
1   mainMethod() begin
2       foreach sequence s in D /*add array of item indexes per sequence*/ do
3           foreach item c do
4               s.indexes[c] ← position(s,c);
5       α.items ← φ; α.trans ← φ;
6       indexSpan(α,D);

7   indexSpan(α,Dα) begin
8       foreach frequent item c in Dα do
9           β.items ← α.items ∪ c;//co-occurrence (c is added to the last α itemset)
10          γ.items ← α.items · c;//α precedes c (c is inserted as a new itemset)
11          //pruning and fast gathering of sup. transactions (for efficient data projection)
12          foreach sequence s in Dα do
13              currentIndex ← s.indexes[c];
14              upperIndex ← s.indexes[αn]/*αn is the last item*/;
15              if leftPositions(currentIndex)≥δ-|α| /*pruning*/ then
16                  if currentIndex > upperIndex then
17                      γ.trans ← γ.trans ∪ s.ID;
18                  else
19                      if currentIndex=upperIndex ∧ c>αn then β.trans ← β.trans∪s.ID;
20          if supβ(Dα) ≥ θ then
21              S ← S ∪ {β};
22              Dβ ← fastProjection(β,Dα);
23              indexSpan(β,Dβ);
24          if supγ(Dα) ≥ θ then
25              S ← S ∪ {γ};
26              Dγ ← fastProjection(γ,Dα);
27              indexSpan(γ,Dγ);

28  fastProjection(β,Dα) begin
29      foreach sequence s in Dα do
30          currentIndex ← s.indexes[βn];
31          upperIndex ← s.indexes[βn-1];
32          if leftPositions(currentIndex)≥δ-|α| /*pruning*/ then
33              if currentIndex > upperIndex then
34                  Dβ ← Dβ ∪ s;
35              else
36                  if currentIndex=upperIndex ∧ c > αn then Dβ ← Dβ ∪ s;
37      return Dβ;
```

the s_α postfix can be removed from the projected database since all the patterns supported by s will have a number of items below the inputted threshold.

For an optimal pruning, this assessment is performed before item indexes comparisons, which occurs in two distinct moments during the prefixSpan recursion (Algorithm 1 *lines 15* and *32*).

The efficiency gains from fast database projections and early pruning techniques, combine with the absence of memory overhead, turn IndexSpan highly attractive in comparison with the OPC-Tree peer method.

3.2 MergeIndexBic: Composing Large Item-Indexable Patterns

In real-world contexts, an ordering permutation observed among a set items can be violated for specific transactions due to the presence of noise. This can either result in a sequential pattern with a reduced set of transactions or items (if this violation turns the original sequence infrequent). In these scenarios, it is desirable to allow some of these violations. Four directions to accomplish this goal were covered in Sect 2.2, with two directions being limited with regards to their outcome and the other two directions with regards to their levels of efficiency. In this section, we propose MergeIndexBic, which makes available two efficient methods to compose lengthy sequential patterns based on merging procedures.

Merging procedures have been applied over sets of biclusters by computing the similarities (overlapping degree) between all pairs of biclusters. Remind that a sequential pattern s can be viewed as a bicluster (I, J), by mapping the supporting transactions Φ_s as the I rows and the pattern items as the J columns. In this context, merging occurs when a set of biclusters (I_k, J_k) has an overlapping area above a minimum threshold, meaning that a new bicluster (I', J') is composed with $I' = \cup_k I_k$ and $J' = \cup_k J_k$. This new bicluster can be mapped back as a sequential pattern, for order-preserving tasks where biclustering is not the ultimate goal. This strategy is reliable as it considers both the set of shared items and the set of shared transactions to perform the merging step, leading to an effective identification and allowance of local ordering violations. A global view of this step is provided in Fig. 2, where three larger biclusters are derived from subsets of biclusters satisfying the minimum overlapping constraint.

In this illustration, three major steps are considering: (1) mapping sequential patterns as sets of rows and columns; (2) discovering candidates for merging; and (3) recovering the new sequential patterns. In particular, we consider the overlapping criteria for the discovery of merging candidates to be the shared percentage of the larger bicluster. When multiple biclusters have overlapping areas, two criteria can be consider: to merge all rows and columns if all pairs of biclusters satisfy the considered overlapping criterion (relaxed setting) and to merge all rows and columns by comparing the shared area among all with the area of the larger bicluster (restrictive setting). We propose two procedures to efficiently deal with the merging of biclusters, one for each setting.

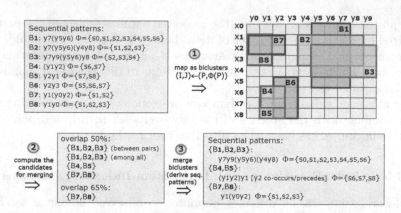

Fig. 2. Composing lengthy sequential patterns from smaller patterns by merging biclusters: considering the influence of both the set of items and the set of transactions.

Maximal Circuits. The first proposed procedure, MergeCycle, is the combination of a graph search method with several heuristics to guide the search space exploration. Consider μ to be the overlapping degree between two biclusters. Since the overlapping degree is typically defined as the number of shared elements by the larger bicluster, heuristics can be defined assuming that biclusters are order by size. Consider two biclusters: a larger bicluster, (I, J), and a smaller bicluster (I', J'). If they do not satisfy $|I'| \times |J'| \leq \mu |I| \times |J|$, we do not need to compute their similarity, neither to compute the similarity for smaller biclusters than (I', J'). This is the first heuristic for pruning the search space. Second heuristic is to further prune the space by computing similarities along one dimension only. Pairs of biclusters not satisfying either $|I \cap I'| \geq \mu |I|$ or $|J \cap J'| \geq \mu |J|$ can be removed without computing the similarities for the remaining dimension. The chosen dimension is the one with average lower size among biclusters.

After computing the pairs of biclusters satisfying the overlapping threshold, a new procedure needs to be applied to verify the availability of larger candidates. Illustrating, if (B_1, B_2), (B_1, B_3) and (B_2, B_3) are candidates for merging, then (B_1, B_2, B_3) is also a candidate. For this step, we map the candidate pairs of biclusters as an unweighted undirected graph, where the nodes are the biclusters and their links are given by the pairs. Under this formulation, the larger candidates for merging are edge-disjoint cyclic subgraphs (or circuits). This procedure is illustrated in Fig. 3, where the cycles in the graph composed of candidate pairs are used to derive the three larger biclusters identified in Fig. 2.

Multi-support FIM. The second proposed merging procedure, MergeFIM, maps the task of merging biclusters as an adapted frequent itemset mining task. Let the elements of the original matrix be the available transactions, and the biclusters be the available items. Recovering the illustrative case presented in Fig. 2, the (x_2, y_2) transaction would now have three items assigned, $\{B_1, B_2, B_3\}$. In this context, the support represents the number of shared elements for a specific set of biclusters (itemset). Understandably, for this scenario,

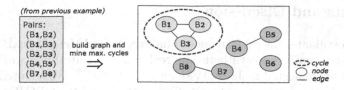

Fig. 3. MergeCycle: computing merging candidates from pair candidates as a search for edge-disjoint cycles in graphs of biclusters.

we cannot rely on a general minimum support threshold, as the minimum number of shared elements to find a candidate for merging depends on the size of the larger bicluster. For this reason, the items (biclusters) are ordered by descending order of their size. When verifying if an itemset is frequent, instead of comparing its support with a minimum support threshold, the support is compared with the minimum support of the larger bicluster ($\mu |I| \times |J|$) that corresponds to the first item (1-length prefix) of the itemset. In this way, no computational complexity is added, and we guarantee that the output itemsets correspond to sets of biclusters that are candidates for merging.

This procedure follows three major steps. The first step is to create a minimal itemset database. Empty transactions are removed and transactions with one item can be pruned for further efficiency gains. Similarly to MergeCycle procedure, MergeFIM can also rely on the proposed heuristics to reduce the search space in order to produce the pairwise similarities. In this case, transactions with two items can be removed, and an Apriori-based method can be applied with the already 2-length itemsets derived from valid pairs of biclusters.

The second step is to run the adapted frequent itemset mining task using closed itemset representations. As previously described, such adaptation allows to replace the general notion of minimum by an indexable support based on the size of larger bicluster in the context of an itemset. Note that by using closed representation we avoid subsets of items with the same number of transactions. This means that the output of the mining task is precisely the set of merging procedures that are required.

The final step is to compose the new biclusters and, optionally, to derive the respective sequential patterns when required. This procedure is illustrated in Fig. 4, and it follows the illustrative case introduced in Fig. 2. The efficiency of this procedure is based on the observation that the mapped itemset databases tend to be highly sparse.

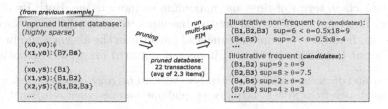

Fig. 4. MergeFIM: computing merging candidates as a frequent itemset mining task.

4   Results and Discussion

This section evaluates the performance of IndexSpan and MergeIndexBic against competitive alternatives on synthetic and real datasets. IndexSpan and MergeIndexBic were implemented in Java (JVM version 1.6.0-24)[2]. We adopted PrefixSpan[3], still considered a state-of-the-art SPM method, and the OPC-Tree method [13] as the bases of comparison. The experiments were computed using an Intel Core i5 2.30 GHz with 6 GB of RAM.

4.1   Synthetic Datasets

The generated experimental settings are described in Table 1. First, we created dense datasets (each item occurs in every sequence) by generating matrices up to 2.000 rows and 100 columns. Each sequence is derived from the ordering of column indexes for a specific row according to the generated values, as illustrated in Fig. 1 from Sect. 2. Understandably, each of the resulting item-indexable sequences contains all the items (or column indexes), which leads to a highly dense dataset. Sequential patterns were planted in these matrices by maintaining the order of values across a subset of columns for a subset of rows. The number and shape of the planted sequential patterns were also varied. For each setting we instantiated 6 matrices: 3 matrices with a background of random values, and 3 matrices with values generated according to a Gaussian distribution. The observed results are an average across these matrices. The number of supporting sequences and items for each sequential pattern followed a Uniform distribution.

Table 1. Properties of the generated dataset settings.

Matrix size (♯rows × ♯columns)	100×30	500×50	1000×75	2000×100
Nr. of hidden seq. patterns	5	10	20	30
Nr. rows for the hidden seq. patterns	$[10, 14]$	$[12, 20]$	$[20, 40]$	$[40, 70]$
Nr. columns for the hidden seq. patterns	$[5, 7]$	$[6, 8]$	$[7, 9]$	$[8, 10]$
Assumptions on the inputted thresholds	$\theta = 5\%$	$\theta = 5\%$	$\theta = 5\%$	$\theta = 5\%$
	$\delta = 3$	$\delta = 4$	$\delta = 5$	$\delta = 6$

Figure 5 compares the performance of the alternative approaches for the generated datasets in terms of time and maximum memory usage. Both PrefixSpan and OPC-Tree can be seen as competitive baselines to assess efficiency. Note that we evaluate the impact of mining sequential patterns in the absence and presence of the δ input (minimum number of items per pattern) for a fair comparison.

---

[2] Software and datasets available in: http://web.ist.utl.pt/rmch/software/indexspan/.
[3] Implementation from SPMF: http://www.philippe-fournier-viger.com/spmf/.

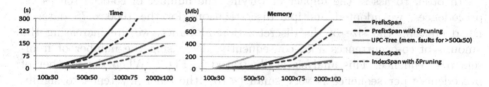

Fig. 5. Performance of alternative SPM methods for datasets with varying properties.

Two main observations can be derived from this analysis. First, the gains in efficiency from adopting fast database projections are significant. In particular, the adoption of fast projections for hard settings dictates the scalability of the SPM task. Pruning methods should also be considered in the presence of the pattern length threshold δ. Contrasting with OPC-Tree and PrefixSpan, IndexSpan guarantees acceptable levels of efficiency for matrices up to 2000 rows and 100 columns for a medium-to-large occupation of sequential patterns ($\sim$3 %–10 % of matrix total area). Second, IndexSpan performs searches with minimal memory waste. The memory is only impacted by the lists of sequence identifiers maintained by prefixes during the depth-first search. Memory of PrefixSpan is slightly hampered due to the need to maintain the projected postfixes. OPC-Tree requires the full construction of the pattern-tree before the traversal, which turns this approach only applicable for small-to-medium databases. For an allocated memory space of 2 GB, we were not able to construct OPC-Trees for input matrices with more than 40 columns.

To further assess the performance of IndexSpan, we fixed the 1000 × 75 experimental setting and varied the level of sparsity by removing specific positions from the input matrix, while preserving the planted sequential patterns. We randomly selected these positions to cause a heighten variance of length among the generated sequences. The amount of removals varied between 0 and 40 %. This analysis is illustrated in Fig. 6.

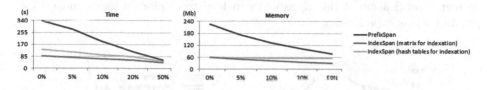

Fig. 6. Performance for varying levels of sparsity (1000 × 75 dataset).

Two main observations can be retrieved. First, to guarantee an optimal memory usage, there is the need to adopt vectors of hash tables in IndexSpan. Second, although the use of these new data structures hampers the efficiency of IndexSpan, the observable computational time is still significantly preferable over the PrefixSpan alternative.

In order to assess the impact of varying the number of co-occurrences vs. precedences, we adopted multiple discretizations for the 1000 × 75 dataset. By decreasing the size of the discretization alphabet, we are increasing the amount of co-occurrences and, consequently, decreasing the number of itemsets per sequence. This analysis is illustrated in Fig. 7. When the number of precedences per sequence is very small (<10), the efficiency tends to significantly decrease due to the exponential increase of sequential patterns. However, for the remaining discretizations, the efficiency does not strongly differ since the number of frequent patterns is identical and pattern-growth methods are able to deal with co-occurrences and precedences in similar ways (Algorithm 1 *lines 20–27*).

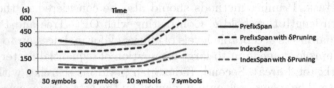

Fig. 7. Performance for varying weights of precedences vs. co-occurrences.

To evaluate the relevance of considering noise-relaxations in order to compose larger sequential patterns, we selected the 1000 × 75 setting and exchanged the order of 5 % of the items. Figure 8 traces item-indexable SPM performance for varying noise-relaxations by using MergeIndexBic with different overlapping degrees. Performance is measured using match scores based on the Jaccard index to assess: (1) to what the extent do the found patterns match with planted patterns (correctness), and (2) how well are the planted biclusters recovered (completeness). When relaxing the overlapping criteria, match scores increase, as the merging step allows for the recovery of order violations. However, this improvement in behavior is only observable until a certain overlapping threshold. The correct identification of this threshold can lead to significant gains (near 15 % points for this experiment).

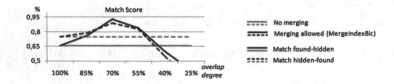

Fig. 8. Impacting of merging sequential patterns in noisy contexts.

Finally, in order to show why MergeIndexBic is needed for an efficient merging step, we maintained the previous experimental settings and compared the performance of traditional combinatorial procedures (where similarities are computed for all pairs of biclusters, and the composition of larger candidates is recursively accomplished) against the proposed MergeCycle and MergeFIM procedures. MergeFIM procedure relies on the efficient Charm[4] method for the delivery of closed frequent itemsets. Figure 9 illustrates this analysis for varying overlapping degrees. Clearly, traditional procedures do not scale. MergeFIM outperforms MergeCycle for hard scenarios where there are large candidates for merging, a case that is commonly observed for relaxed levels of overlapping.

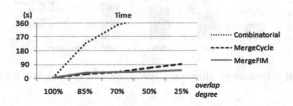

Fig. 9. Comparing the efficiency of MergeCycle and MergeFIM against peer methods.

4.2   Real Datasets

To assess the performance of the proposed approaches in real datasets, we used multiple gene expression matrices[5]: dlblc (180 items per instance, 660 instances), coloncancer (62 items per instance, 2000 instances), leukemia (38 items per instance, 7129 instances). The goal is to discover order-preserved biclusters. For this purpose, we followed the procedure described in Fig. 1 to generate the sequence databases using a discretization alphabet with 20 symbols. The positions corresponding to missing values were removed. Figure 10 compares the performance of the alternative approaches for the $\theta = 8\%$ and $\delta = 5$ thresholds. This analysis reinforces the previous observations. OPC-Tree is bounded by the size of the database. The adoption of IndexSpan strategies to deal with item-indexable sequences strongly impacts the SPM performance, and, consequently, the ability to discover order-preserving biclusters in real data.

The relevance of tolerating noise to compose larger sequences is illustrated in Fig. 11 for the leukemia dataset. Here, we computed the functional enrichment of the genes supporting each frequent sequnce, Φ_s, recurring to the GoToolBox [16]. As a measure of significance, we counted the number of overrepresented terms with Bonferroni corrected p-values below 0.01. We observe that MergeIndexBic procedure increases not only the number of significant terms, but also their relative percentage as it removes short patterns from the output.

---

[4] Implementation from SPMF: http://www.philippe-fournier-viger.com/spmf/.

[5] http://www.upo.es/eps/bigs/datasets.html
http://www.bioinf.jku.at/software/fabia/gene_expression.html

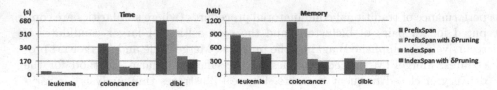

Fig. 10. Performance of SPM-based order-preserving biclustering for biological data.

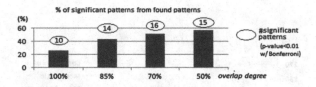

Fig. 11. Biological relevance of allowing noise using MergeIndexBic for leukemia data.

5   Conclusions

This work formalizes the task of performing noise-tolerant sequential pattern mining over item-indexable databases and motivates its relevance for a critical set of applications. The performance of existing approaches based on general SPM methods and on dedicated algorithms, such as the OPC-Tree, is discussed. To tackle the inefficiencies of existing solutions, we propose the IndexSpan algorithm. IndexSpan relies on position induction to deliver fast and memory-free database projections. Additionally, early-pruning techniques with impact on the performance of IndexSpan can be adopted to guarantee that only large sequential patterns are discovered.

Furthermore, we explore alternatives to efficiently extend IndexSpan in order to compose large sequential patterns under parameterizable noise allowance guarantees. MergeIndexBic is proposed to surpass the limited robustness and efficiency of existing options. This is done by merging sequential patterns with significant overlap on sequence items and on the supporting transactions. Pruning heuristics to avoid the computation of similarities among all pair are proposed. Efficient computation of candidates is achieved by mapping the merging task as maximal cycle discovery in undirected graphs or as multi-support frequent itemset mining.

Results on both synthetic and real datasets show the superior performance and relevance of the IndexSpan and MergeIndexBic methods.

Since the proposed item-indexable SPM relies on a prefix-growth search, it can easily accommodate principles from existing research to deliver condensed pattern representations, such as CloSpan [24], in order to reduce the complexity of the merging step; and to discover sequential patterns in distributed settings, such as MapReduce [23], in order to relax the efficiency boundaries of IndexSpan.

Acknowledgments. This is an extension of previous work [10] supported by *Fundação para a Ciência e Tecnologia* under the project D2PM (PTDC/EIA-EIA/110074/2009), project Neuroclinomics (PTDC/EIA-EIA/ 111239/2009), and PhD grant SFRH/BD/ 75924/2011.

References

1. Agrawal, R., Srikant, R.: Mining sequential patterns. In: ICDE. pp. 3–14. IEEE CS, Washington (1995)
2. Antunes, C., Oliveira, A.L.: Mining patterns using relaxations of user defined constraints. In: Knowledge Discovery in Inductive Databases (2004)
3. Ayres, J., Flannick, J., Gehrke, J., Yiu, T.: Sequential pattern mining using a bitmap representation. In: KDD. pp. 429–435. ACM, New York (2002)
4. Bayardo, R.J.: Efficiently mining long patterns from databases. SIGMOD Rec. **27**(2), 85–93 (1998)
5. Ben-Dor, A., Chor, B., Karp, R., Yakhini, Z.: Discovering local structure in gene expression data: the order-preserving submatrix problem. In: RECOMB. pp. 49–57. ACM, New York (2002)
6. Cheng, H., Yu, P.S., Han, J.: Approximate frequent itemset mining in the presence of random noise. In: Maimon, O., Rokach, L. (eds.) Soft Computing for Knowledge Discovery and Data Mining, pp. 363–389. Springer, New York (2008)
7. Chiu, D.Y., Wu, Y.H., Chen, A.L.P.: An efficient algorithm for mining frequent sequences by a new strategy without support counting. In: ICDE. p. 375. IEEE CS, Washington (2004)
8. Han, J., Cheng, H., Xin, D., Yan, X.: Frequent pattern mining: current status and future directions. Data Min. Knowl. Discov. **15**(1), 55–86 (2007)
9. Han, J., Yang, Q., Kim, E.: Plan mining by divide-and-conquer. In: ACM SIGMOD IW on Research Issues in DMKD (1999)
10. Henriques, R., Madeira, S., Antunes, C.: Indexspan: efficient discovery of item-indexable sequential patterns. In: ECML/PKDD IW on New Frontiers in Mining Complex Patterns (2013)
11. Kumar, P., Krishna, P., Raju, S.: Pattern Discovery Using Sequence Data Mining: Applications and Studies. IGI Global, Hershey (2011)
12. Lin, D.-I., Kedem, Z.M.: Pincer search: a new algorithm for discovering the maximum frequent set. In: Schek, H.-J., Saltor, F., Ramos, I., Alonso, G. (eds.) EDBT 1998. LNCS, vol. 1377, pp. 105–119. Springer, Heidelberg (1998)
13. Liu, J., Wang, W.: Op-cluster: clustering by tendency in high dimensional space. In: ICDM. p. 187. IEEE CS, Washington (2003)
14. Liu, J., Yang, J., Wang, W.: Biclustering in gene expression data by tendency. In: IEEE Computational Systems Bioinformatics Conference, pp. 182–193. IEEE (2004)
15. Mabroukeh, N.R., Ezeife, C.I.: A taxonomy of sequential pattern mining algorithms. ACM Comput. Surv. **43**(1), 3:1–3:41 (2010)
16. Martin, D., Brun, C., Remy, E., Mouren, P., Thieffry, D., Jacq, B.: Gotoolbox: functional analysis of gene datasets based on gene ontology. Genome Biol. **5**(12), 101 (2004)
17. Pei, J., Han, J., Mortazavi-Asl, B., Wang, J., Pinto, H., Chen, Q., Dayal, U., Hsu, M.C.: Mining sequential patterns by pattern-growth: the prefixspan approach. IEEE Trans. Knowl. Data Eng. **16**(11), 1424–1440 (2004)

18. Pei, J., Han, J., Mortazavi-Asl, B., Zhu, H.: Mining access patterns efficiently from web logs. In: Terano, T., Liu, H., Chen, A.L.P. (eds.) PAKDD 2000. LNCS, vol. 1805, pp. 396–407. Springer, Heidelberg (2000)
19. Raedt, L.D., Guns, T., Nijssen, S.: Constraint programming for data mining and machine learning. In: AAAI. AAAI Press (2010)
20. Salvemini, E., Fumarola, F., Malerba, D., Han, J.: FAST sequence mining based on sparse Id-lists. In: Kryszkiewicz, M., Rybinski, H., Skowron, A., Raś, Z.W. (eds.) ISMIS 2011. LNCS, vol. 6804, pp. 316–325. Springer, Heidelberg (2011)
21. Serin, A., Vingron, M.: Debi: discovering differentially expressed biclusters using a frequent itemset approach. Algorithms Mol. Biol. **6**, 1–12 (2011)
22. Srikant, R., Agrawal, R.: Mining sequential patterns: generalizations and performance improvements. In: Apers, P.M.G., Bouzeghoub, M., Gardarin, G. (eds.) EDBT 1996. LNCS, vol. 1057, pp. 3–17. Springer, Heidelberg (1996)
23. qing Wei, Y., Liu, D., shan Duan, L.: Distributed prefixspan algorithm based on mapreduce. In: Information Technology in Medicine and Education, vol. 2, pp. 901–904 (2012)
24. Yan, X., Han, J., Afshar, R.: CloSpan: mining closed sequential patterns in large datasets. In: SDM. pp. 166–177 (2003)
25. Yang, J., Wang, W., Yu, P.S., Han, J.: Mining long sequential patterns in a noisy environment. In: SIGMOD. pp. 406–417. ACM, New York (2002)
26. Yang, Z., Wang, Y., Kitsuregawa, M.: LAPIN: effective sequential pattern mining algorithms by last position induction for dense databases. In: Kotagiri, R., Radha Krishna, P., Mohania, M., Nantajeewarawat, E. (eds.) DASFAA 2007. LNCS, vol. 4443, pp. 1020–1023. Springer, Heidelberg (2007)
27. Zaki, M.J.: Spade: an efficient algorithm for mining frequent sequences. Mach. Learn. **42**(1–2), 31–60 (2001)
28. Zheng, Y., Zhang, L., Xie, X., Ma, W.Y.: Mining interesting locations and travel sequences from gps trajectories. In: WWW. pp. 791–800. ACM (2009)
29. Zhu, F., Yan, X., Han, J., Yu, P.S.: Mining Frequent Approximate Sequential Patterns. Chapman & Hall, London (2009)
30. Zhu, F., Yan, X., Han, J., Yu, P., Cheng, H.: Mining colossal frequent patterns by core pattern fusion. In: ICDE. pp. 706–715 (2007)

Mining Frequent Partite Episodes
with Partwise Constraints

Takashi Katoh[(✉)], Shin-ichiro Tago, Tatsuya Asai, Hiroaki Morikawa,
Junichi Shigezumi, and Hiroya Inakoshi

Fujitsu Laboratories Ltd., Kawasaki 211-8588, Japan
{kato.takashi_01,s-tago,asai.tatsuya,h.morikawa,j.shigezumi,
inakoshi.hiroya}@jp.fujitsu.com

Abstract. In this paper, we study the problem of efficiently mining
frequent *partite episodes* that satisfy *partwise constraints* from an input
event sequence. Through our constraints, we can extract episodes related
to events and their precedent-subsequent relations, on which we focus,
in a short time. This improves the efficiency of data mining using trial
and error processes. A partite episode of length k is of the form $P =
\langle P_1, \ldots, P_k \rangle$ for sets P_i $(1 \leq i \leq k)$ of events. We call P_i a *part* of P
for every $1 \leq i \leq k$. We introduce the *partwise constraints* for partite
episodes P, which consists of *shape and pattern constraints*. A shape
constraint specifies the size of each part of P and the length of P. A pattern constraint specifies subsets of each part of P. We then present a
backtracking algorithm that finds all of the frequent partite episodes
satisfying a partwise constraint from an input event sequence. By theoretical analysis, we show that the algorithm runs in output polynomial
time and polynomial space for the total input size. In the experiment, we
show that our proposed algorithm is much faster than existing algorithms
for mining partite episodes on an artificial and a real-world datasets.

1 Introduction

Episode Mining. One of the most important tasks in data mining is to discover
frequent patterns from time-related data. Mannila *et al.* [6] introduced *episode
mining* to discover frequent *episodes* in an event sequence. An episode is formulated as a labeled acyclic digraph in which labels corresponding to events and
arcs represent a temporal precedent-subsequent relation in an event sequence.

For subclasses of episodes [3,6], a number of efficient algorithms have been
developed. In a previous work [3], we introduced the class of *partite episodes*,
which are time-series patterns of the form $\langle P_1, \ldots, P_k \rangle$ for sets P_i $(1 \leq i \leq k)$
of events, which means that in an input event sequence, every event in P_{i+1}
follows every event in P_i for every $1 \leq i < k$. For the class of partite episode,
we presented an algorithm that finds all of the frequent episodes from an input
event sequence.

A partite episode is a richer representation of temporal relationship than
a subsequence, which represents just a linearly ordered relation in sequential

A. Appice et al. (Eds.): NFMCP 2013, LNAI 8399, pp. 117–131, 2014.
DOI: 10.1007/978-3-319-08407-7_8, © Springer International Publishing Switzerland 2014

pattern mining [1,9]. In particular, the partite episode can represent the precedent-subsequent relation between the sets of events that occur simultaneously (in any order) in a time span.

Main Problem. In the data analysis using trial and error processes, we often want to extract only episodes which include some events we focus on. For example, if we get an episode that means *buying a telescope before buying a PC*, then we will be interested in frequent patterns in the period between the two events.

In this paper, we introduce *partwise constraints* for partite episodes. A partwise constraint consists of a *shape constraint* and a *pattern constraint*. For a partite episode P, a shape constraint specifies the size of each part of P and the length of P. A pattern constraint specifies the subsets of each part of P. Through episode mining with a shape constraint that means the length of any extracted episode is at most three and both the size of the first and the third set are at most one, and a pattern constraint that means the first set includes the event *buying a PC* and the third set includes the event *buying a telescope*, we may obtain a knowledge: *some customers bought a digital camera and a photo printer after buying a telescope before buying a PC*.

We can select the episodes satisfying these constraints in post-processing steps after or during non-constraint episode mining. This approach, however, requires a very long execution time since mining algorithms often outputs a large number of frequent episodes. It is a better idea to use the constraint-based algorithm PPS introduced by Ma *et al.* [5]. PPS efficiently handles a *prefix anti-monotone constraint* [11], which means that if an episode satisfies the constraint, so does every prefix of the episode. Any shape constraint is prefix anti-monotonic, and PPS can extract partite episodes satisfying a given shape constraint in polynomial amortized time per output.

On the other hand, pattern constraints are not always prefix anti-monotonic. For example, a pattern constraint having *buying a telescope* in the first set and *buying a PC* in the third set, which is described above, is not prefix anti-monotonic one. This pattern constraint is not prefix anti-monotonic because there is an episode whose prefix does not satisfy it; It is obvious by considering the episode; *buying a telescope before buying a camera and then buying a PC* and it's prefix episodes.

If we allow an algorithm to add event to any part of the partite episode instead of adding only the tail of the episode, there is a possibility to enumerate episode satisfying any pattern constraint without generating candidate episodes. However, the efficient computation methods of episodes and their occurrence in that enumeration is non-trivial.

Main Results. Our goal is to present an algorithm that extracts frequent partite episodes satisfying a partwise constraint without post-processing steps. Then we show that our algorithm runs in $O(Nsc)$ time per partite episode and $O(Nsm)$ space, where N is the total size of an input sequence, s is the number of event types in the input sequence (alphabet size), c is a constant that depends only on a given partwise constraint, and m is the maximum size of episodes.

Finally, our experimental result shows that our proposed algorithm is 230 times faster than the straightforward algorithm which consists of an algorithm for non-constraint episodes and post-processing steps for partwise constraints.

Related Works. In addition to Ma's research, there have been many studies on episode mining. Méger and Rigotti [7] introduced a different type of constraint called *gapmax* that represents the maximum time gap allowed between two events. Tatti *et al.* [13] and Zhou *et al.* [15] introduced *closed episode mining*. Closed episode mining is another approach to reduce the number of outputs and improve the mining efficiency. Closed episode mining algorithms extract only representative patterns called *closed episodes*, whereas constraint-based algorithms such as our algorithm and Méger's algorithm extract only episodes that satisfy some given conditions on which we focus. Seipel *et al.* [12] applied episode mining to system event logs for network management. Since their class of episodes was a subclass of partite episodes, their analysis would be improved by applying the proposed algorithm in this paper.

2   Partite Episodes

In this section, we introduce partite episodes and the related notions and lemmas necessary for a later discussion. We denote the sets of all natural numbers by $\mathbf{N}$. Then we define ∞ as the special largest number such that $a < \infty$ for all $a \in \mathbf{N}$. For a set $S = \{s_1, \ldots, s_n\}$ $(n \in \mathbf{N})$, we denote the cardinality n of S by $|S|$. For sets S and T, we denote the difference set $\{s \in S \mid s \notin T\}$ by $S \setminus T$. For a sequence $X = \langle x_1, \ldots, x_n \rangle$ $(n \in \mathbf{N})$, we denote the length n of X by $|X|$, the i-th element x_i of X by $X[i]$, and the consecutive subsequence $\langle x_i, \ldots, x_j \rangle$ of X by $X[i, j]$, for every $1 \leq i \leq j \leq n$, where we define $X[i, j] = \langle \rangle$ when $i > j$.

2.1   Input Event Sequence

Let $\Sigma = \{1, \ldots, m\}$ $(m \geq 1)$ be a finite alphabet with the total order $\leq$ over $\mathbf{N}$. Each element $e \in \Sigma$ is called an *event*[1]. An *input event sequence* (an *input sequence*, for short) $\mathcal{S}$ on Σ is a finite sequence $\langle S_1, \ldots, S_\ell \rangle \in (2^\Sigma)^*$ of sets of events of length ℓ. For an input sequence $\mathcal{S}$, we define the *total size* $\|\mathcal{S}\|$ of $\mathcal{S}$ by $\sum_{i=1}^{\ell} |S_i|$. Clearly, $\|\mathcal{S}\| = O(|\Sigma|\ell)$. Without loss of generality, we can assume that every event in Σ appears at least once in $\mathcal{S}$.

2.2   Partite Episodes

Mannila *et al.* [6] and Katoh *et al.* [3] formulated an episode as a partially ordered set and a labeled acyclic digraph, respectively. In this paper, for a sub-class of episodes called *partite episodes*, we define an episode as a sequence of sets of events for simpler representations.

---

[1] Mannila *et al.* [6] originally referred to each element $e \in \Sigma$ itself as an *event type* and an occurrence of e as an *event*. However, we simply refer to both of these as *events*.

Definition 1. A *partite episode* P over Σ is a sequence $\langle P_1, \ldots, P_k \rangle \in (2^\Sigma)^*$ of sets of events $(k \geq 0)$, where $P_i \subseteq \Sigma$ is called the i-th *part* for every $1 \leq i \leq k$. For a partite episode $P = \langle P_1, \ldots, P_k \rangle$, we define the *total size* $\|P\|$ of P by $\sum_{i=1}^{k} |P_i|$. We call P *proper* when $P_i \neq \emptyset$ for every $1 \leq i \leq k$.

Definition 2. Let $P = \langle \{a_{(1,1)}, \ldots, a_{(1,m_1)}\}, \ldots, \{a_{(k,1)}, \ldots, a_{(k,m_k)}\} \rangle$ be a partite episode of length $k \geq 0$, and $Q = \langle \{b_{(1,1)}, \ldots, b_{(1,n_1)}\}, \ldots, \{b_{(l,1)}, \ldots, a_{(l,n_l)}\} \rangle$ a partite episode of length $l \geq 0$, where $m_i \geq 0$ for every $1 \leq i \leq k$ and $n_i \geq 0$ for every $1 \leq i \leq l$. A partite episode P is a sub-episode of Q, denoted by $P \sqsubseteq Q$, if and only if there exists some injective mapping $h : A \to B$ satisfying (i) $A = \bigcup_{i=1}^{k} \bigcup_{j=1}^{m_i} \{(i,j)\}$, $B = \bigcup_{i=1}^{l} \bigcup_{j=1}^{n_i} \{(i,j)\}$, (ii) for every $x \in A$, $a_x = b_{h(x)}$ holds, and (iii) for every $(p_1, q_1), (p_2, q_2) \in A$, and values $(p_1', q_1') = h((p_1, q_1))$ and $(p_2', q_2') = h((p_2, q_2))$, if $p_1 < p_2$ holds, then $p_1' < p_2'$ holds.

Let $P = \langle P_1, \ldots, P_k \rangle$ and $Q = \langle Q_1, \ldots, Q_l \rangle$ be partite episodes of length $k \geq 0$ and $l \geq 0$, respectively. We denote the partite episode $\langle P_1, \ldots, P_k, Q_1, \ldots, Q_l \rangle$ by $P \mapsto Q$, the partite episode $\langle P_1 \cup Q_1, \ldots, P_{\max(k,l)} \cup Q_{\max(k,l)} \rangle$ by $P \circ Q$, and the partite episode $\langle P_1 \setminus Q_1, \ldots, P_k \setminus Q_k \rangle$ by $P \setminus Q$, where we assume that $P_i = \emptyset$ for any $i > k$ and $Q_j = \emptyset$ for any $j > l$. In Fig. 1, we show examples of an input event sequence S and partite episodes P^i $(1 \leq i \leq 7)$.

2.3 Occurrences

Next, we introduce occurrences of episodes in an input sequence. An *interval* in an input sequence S is a pair of integer $(s,t) \in \mathbf{N}^2$ satisfying $s \leq t$. Let P be a partite episode, and $x = (s,t)$ be an interval in an input sequence $S = \langle S_1, \ldots, S_\ell \rangle$ $(\ell \geq 0)$. A partite episode P *occurs in* an interval x, if and only if $P \sqsubseteq S[s, t-1]$, where we define $S[i]$ for any index such that $i < 1$ and $i > |S|$ as $S[i] = \emptyset$.

For a partite episode P that occurs in an interval $x = (s,t)$ in an input sequence S, the interval x is a *minimum occurrence* if and only if P does not occur in both intervals $(s+1, t)$ and $(s, t-1)$. Moreover, a *minimum occurrence set* of P on S is a set $\{(s,t) \mid P \sqsubseteq S[s, t-1], P \not\sqsubseteq S[s, t-2], P \not\sqsubseteq S[s+1, t-1]\} \subseteq \mathbf{N}^2$ of all minimum occurrences of P on S. For a minimum occurrence set X of P, a *minimum occurrence list* (*mo-list*, for short) of P, denoted by $mo(P)$, is a sequence $\langle (s_1, t_1), \ldots, (s_n, t_n) \rangle \in (\mathbf{N}^2)^*$ of minimum occurrences such that $n = |X|$, $s_i < s_{i+1}$ for every $1 \leq i < n$.

A *window width* is a fixed positive integer $w \geq 1$. For an input sequence S and any $1 - w < i < |S|$, the interval $(i, i+w)$ is a *window* of width w on S. Then the frequency $freq_{S,w}(P)$ of a partite episode P is defined by the number of windows (of width w on S) in which P occurs. A *minimum frequency threshold* is any positive integer $\sigma \geq 1$. An episode P is σ-*frequent* in S if $freq_{S,w}(P) \geq \sigma$. For a minimum frequency threshold σ and a window width w, the *minimum support threshold* $\acute{\sigma}$ is the relative value $\acute{\sigma} = \sigma / (|S| + w - 1)$. By the definition of the frequency, we can show the next lemma.

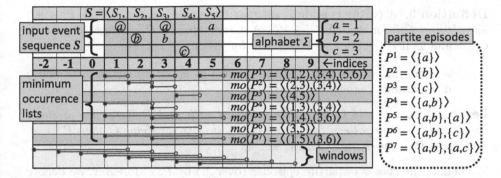

Fig. 1. An input event sequence $\mathcal{S}$ of length 5 (left top), partite episodes P^i $(1 \le i \le 7)$ (right), their minimum occurrence lists (left middle), and windows of width 4 on $\mathcal{S}$ (left bottom). In the input sequence $\mathcal{S}$, we indicate an occurrence of P^7 in the fourth window $(1,5)$ in circles. See Example 1 for details.

Lemma 1. *Let P and Q be partite episodes, $\mathcal{S}$ an input sequence, and w a window width. If $P \sqsubseteq Q$ holds, then $freq_{\mathcal{S},w}(P) \ge freq_{\mathcal{S},w}(Q)$ holds.*

Our definition of frequency is identical to the one given by Mannila [6]. With this definition, an occurrence of episode could be counted more than once, and more than two occurrences could be counted only once. Lemma 1 holds even with another definition that does not have the problems above, although we do not explain why because of the limited space.

Example 1. In Fig. 1, we show an input sequence $\mathcal{S} = \langle \{a\}, \{b\}, \{a,b\}, \{c\}, \{a\} \rangle$ of length $\ell = 5$ over an alphabet $\Sigma = \{a,b,c\}$ of events, partite episodes $P^1 = \langle \{a\} \rangle$, $P^2 = \langle \{b\} \rangle$, $P^3 = \langle \{c\} \rangle$, $P^4 = P^1 \circ P^2 = \langle \{a,b\} \rangle$, $P^5 = P^4 \mapsto P^1 = \langle \{a,b\}, \{a\} \rangle$, $P^6 = P^4 \mapsto P^3 = \langle \{a,b\}, \{c\} \rangle$, and $P^7 = P^5 \circ P^6 = \langle \{a,b\}, \{a,c\} \rangle$, and their minimum occurrence lists $mo(P^1) = \langle (1,2), (3,4), (5,6) \rangle$, $mo(P^2) = \langle (2,3), (3,4) \rangle$, $mo(P^3) = \langle (4,5) \rangle$, $mo(P^4) = \langle (1,3), (3,4) \rangle$, $mo(P^5) = \langle (1,4), (3,6) \rangle$, $mo(P^6) = \langle (3,5) \rangle$, and $mo(P^7) = \langle (1,5), (3,6) \rangle$, respectively, where $a = 1$, $b = 2$, and $c = 3$ are events. The input sequence $\mathcal{S}$ has eight windows with width 4 from $(-2,2)$ to $(5,9)$. Among these, the partite episode P^5 occurs in $(0,4)$, $(1,5)$, $(2,6)$, and $(3,7)$. Therefore, the frequency of P^5 is 4. Furthermore, the partite episode P^7 occurs in $(1,5)$, $(2,6)$, and $(3,7)$. Therefore, the frequency of P^7 is 3. We see that $P^5 \sqsubseteq P^7$ and $freq_{\mathcal{S},4}(P^5) > freq_{\mathcal{S},4}(P^7)$.

3   Partwise Constraints

In this section, we introduce *partwise constraints* for partite episodes which consist of the *shape and pattern constraints*. A *shape constraint* for a partite episode P is a sequence $C = \langle c_1, \ldots, c_k \rangle \in \mathbf{N}^*$ $(k \ge 0)$ of natural numbers. A *pattern constraint* for a partite episode P is a sequence $D = \langle D_1, \ldots, D_k \rangle \in (2^\Sigma)^*$ $(k \ge 0)$ of sets of events. We consider a pattern constraint as a partite episode.

Definition 3. A partite episode P *satisfies* the shape constraint C if $|P| \le |C|$ and $|P[i]| \le C[i]$ for every $1 \le i \le |C|$. A partite episode P *satisfies* the pattern constraint D if $|P| \ge |D|$ and $P[i] \supseteq D[i]$ for every $1 \le i \le |D|$.

By this definition, we see that a shape constraint is *anti-monotone* [10] and a pattern constraint is *monotone* [10].

Lemma 2. *Let P and Q be partite episodes such that $P \sqsubseteq Q$. For a shape constraint C and a pattern constraint D, (i) if Q is satisfying C then P is so, and (ii) if P is satisfying D then Q is so.*

We denote the classes of partite episodes (over Σ) by $\mathcal{PE}$. Moreover, we denote the classes of partite episodes satisfying a pattern constraint D by $\mathcal{PE}(D)$.

Definition 4. PARTITE EPISODE MINING WITH A PARTWISE CONSTRAINT: Given an input sequence $S \in (2^\Sigma)^*$, a window width $w \ge 1$, a minimum frequency threshold $\sigma \ge 1$, a shape constraint C, and a pattern constraint D, the task is to find all of the σ-frequent partite episodes satisfying C and D that occur in S with a window width w without duplication.

Our goal is to design an algorithm for the frequent partite episode mining problem with a partwise constraint, which we will show in the next section, in the framework of enumeration algorithms [2]. Let N be the total input size and M the number of solutions. An enumeration algorithm $\mathcal{A}$ is of *output-polynomial time*, if $\mathcal{A}$ finds all solutions in total polynomial time both in N and M. Moreover, $\mathcal{A}$ is of *polynomial delay*, if the *delay*, which is the maximum computation time between two consecutive outputs, is bounded by a polynomial in N alone.

4   Algorithm

In this section, we present an output-polynomial time and a polynomial-space algorithm PARTITECD for extracting all the frequent partite episodes satisfying partwise constraints in an input event sequence. Throughout this section, let $S = \langle S_1, \ldots, S_\ell \rangle \in (2^\Sigma)^*$ be an input event sequence over an alphabet Σ, $w \ge 1$ a window width, and $\sigma \ge 1$ the minimum frequency threshold. Furthermore, let M be the number of all solutions of our algorithm, and m the maximum size of output episode P, that is, $m = \|P\| + |P|$. Then we define the length of the input sequence $\ell = |S|$, the total size of the input sequence $N = \|S\| + \ell$, the total size of the constraints $c = |C| + \|D\| + |D|$, and the alphabet size $s = |\Sigma|$ for analysis of time and space complexity of our algorithm, where we assume $O(s) = O(N)$ and $O(m) = O(N)$.

4.1   Family Tree

The main idea of our algorithm is to enumerate all of the frequent partite episodes satisfying partwise constraints by searching the whole search space from general to specific by using depth-first search. First, we define the search space.

For a partite episode P such that $|P| \geq 1$, the *tail pattern* $tail(P)$ of P is defined by the partite episode Q such that $|Q| = \max\{1 \leq i \leq |P| \mid P[i] \neq \emptyset\}$, $Q[i] = \emptyset$ for every $1 \leq i < |Q|$, and $Q[i] = \{\max P[i]\}$ for $i = |Q|$, where $\max P[i]$ is the maximum integer in the i-th part of P. Then, for a pattern constraint D, we introduce the parent-child relationship between partite episodes satisfying D.

Definition 5. The partite episode $\perp = D$ is the *root*. The *parent* of the partite episode $P = \langle P_1, \ldots, P_k \rangle$ is defined by:

$$parent_D(P) = \begin{cases} \langle P_1, \ldots, P_{k-1} \rangle, & \text{if } |P_k| \leq 1 \text{ and } |P| > |D|, \\ P \setminus tail(P \setminus D), & \text{otherwise.} \end{cases}$$

We define the set of all children of P by $children_D(P) = \{Q \mid parent_D(Q) = P\}$. Then we define the *family tree* for $\mathcal{PE}(D)$ by a rooted digraph $\mathcal{T}(\mathcal{PE}(D)) = (V, E, \perp)$ with the root $\perp$, where the root $\perp = D$, the vertex set $V = \mathcal{PE}(D)$, and the edge set $E = \{(P, Q) \mid P = parent_D(Q), Q \neq D\}$.

Lemma 3. *The family tree* $\mathcal{T}(\mathcal{PE}(D)) = (V, E, D)$ *for* $\mathcal{PE}(D)$ *is a rooted tree with the root* D.

For any episode Q on $\mathcal{T}(\mathcal{PE}(D))$, the parent $parent_D(Q)$ is a subepisode of Q. Therefore, we can show the next lemma by Lemma 1 and Lemma 2.

Lemma 4. *Let C be a shape constraint, D a pattern constraint, and P and Q partite episodes such that $P = parent_D(Q)$. If Q is frequent then so is P. If Q is satisfying C then so is P.*

By Lemma 2 and Lemma 3, we know that a family tree $\mathcal{T}(\mathcal{PE}(D))$ contains only episodes satisfying a pattern constraint D. Thus, we can make a search tree containing only episodes that are frequent and satisfy the shape and pattern constraints by pruning episodes that do not satisfy the conditions.

Example 2. We describe the part of family trees $\mathcal{T}(\mathcal{PE}(\langle\rangle))$ and $\mathcal{T}(\mathcal{PE}(\langle\{\}, \{b\}\rangle))$ that forms the spanning trees for all partite episodes of $\mathcal{PE}(\langle\rangle)$ and $\mathcal{PE}(\langle\{\}, \{b\}\rangle)$ on an alphabet $\Sigma = \{a, b\}$ in Fig. 2. For a pattern constraint $D_1 = \langle\{\}, \{b\}\rangle$, the parent of $P^1 = \langle\{a, b\}, \{a, b\}\rangle$ is $P^1 \setminus tail(P^1 \setminus D_1) = P^1 \setminus tail((\langle\{a, b\}, \{a\}\rangle)) = P^1 \setminus \langle\{\}, \{a\}\rangle = \langle\{a, b\}, \{b\}\rangle = P^2$. For a pattern constraint $D_2 = \langle\rangle$, the parent of $\langle\{a\}, \{b\}\rangle$ is $\langle\{a\}\rangle$, because $|\{b\}| \leq 1$ and $|\langle\{a\}, \{b\}\rangle| > |D_2|$.

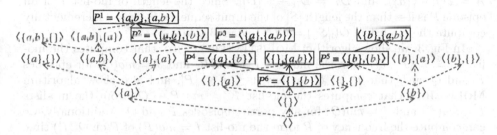

Fig. 2. The parent-child relationships on an alphabet $\Sigma = \{a, b\}$ for pattern constraints $\langle\rangle$ (dashed arrows) and $\langle\{\}, \{b\}\rangle$ (solid arrows), where $a = 1$ and $b = 2$ are events.

4.2 Pattern Expansion

Secondly, we discuss how to enumerate all children of a parent. For a pattern constraint D, a partite episode Q, and its parent P, we define the *index of the expanded part* of Q by $iex(Q) = |Q|$ if $|Q| > |P|$, and $iex(Q) = |tail(Q \setminus P)|$ otherwise. Additionally, we define $iex(Q) = 0$ for the root $Q = D$.

Let $h = iex(P)$ be the index of the expanded part of a parent episode P. We define *type-i children* of P by $children_D(P, i) = \{Q \in children_D(P) \mid iex(Q) = i, \|Q\| > \|P\|\}$ for an index $i \geq h$. By Definition 5, we can make any type-i child Q of P by adding an event $a \in \Sigma \setminus (P[i] \cup D[i])$ at $P[i]$. Moreover, by Definition 5, we see $children_D(P) = (\bigcup_{i=m}^n children_D(P, i)) \cup \{P \mapsto \langle \emptyset \rangle\}$, where $m = \max(h, 1)$, and $n = \max(h, |D|) + 1$.

Furthermore, for partite episodes P, Q, and S such that $P = parent_D(Q)$, $S = parent_D(P)$, $iex(Q) \geq 1$, and $|(Q \setminus D)[iex(Q)]| \geq 2$, we define the *uncle R* of Q by $uncle_D(Q) = S \circ tail(Q \setminus D)$.

Lemma 5. *For partite episodes P and Q such that $Q \in children_D(P, iex(P))$, there exist partite episodes R, S, and an index $i > 0$ such that $R = uncle_D(Q)$, $S = parent_D(P) = parent_D(R)$, $P, R \in children_D(S, i)$, and $P \simeq_{\bar{i}} Q \simeq_{\bar{i}} R$.*

Example 3. In Fig. 2, for a pattern constraint $D = \langle \{\}, \{b\} \rangle$, an episode $P^4 = \langle \{a\}, \{b\} \rangle$ is the parent of an episode $P^2 = \langle \{a, b\}, \{b\} \rangle$. The index $iex(P^2)$ of the expanded part of P^2 is $|tail(P^2 \setminus P^4)| = |tail(\langle \{b\}, \{\} \rangle)| = |\langle \{b\} \rangle| = 1$. Therefore, P^2 is a type-1 child of P^4. On the other hand, P^3 is a type-2 child of P^4 because $iex(P^3)$ is $|tail(\langle \{\}, \{b\} \rangle)| = 2$. Since the parent of P^4 is P^6, the uncle of P^2 is $P^6 \circ tail(P^2 \setminus D) = P^6 \circ \langle \{a\} \rangle = \langle \{a\}, \{b\} \rangle = P^5$.

4.3 Incremental Computation

To compute a type-i child Q of P and its mo-list incrementally from the parent P, we make Q as follows. Let $h = iex(P)$ be the index of the expanded part of P; (i) we make Q by $P \circ R$, and the subepisode $Q_{sub} = Q[1, i]$ of Q by $P_{sub} \circ R_{sub}$ when $i = h$, where $R = uncle_D(Q)$, $P_{sub} = P[1, i]$, and $R_{sub} = R[1, i]$; (ii) we make Q by $P_{sub} \mapsto D_H \mapsto A \mapsto D_T$ and the subepisode $Q_{sub} = Q[1, i]$ of Q by $P_{sub} \mapsto D_H \mapsto A$ when $i > h$, where $P_{sub} = P[1, h]$, $D_H = D[h + 1, i - 1]$, $A = D[i] \circ \langle \{a\} \rangle$, and $D_T = D[i + 1, |D|]$. Since the length of mo-list I of an episode P is less than the length $|S|$ of the input sequence S, we can incrementally compute the mo-list in $O(\|S\|) = O(N)$ time [3, 4, 8].

In Fig. 3, we show algorithm MoLISTINTERSECTION that computes the mo-list $K = mo(P \circ Q)$ from the mo-lists $I = mo(P)$ and $J = mo(Q)$ for episodes P and Q such that $P \simeq_{\bar{i}} Q$ for any $1 \leq i \leq |P|$, and also show algorithm MoLISTJOIN that computes the mo-list $K = mo(P \mapsto Q)$ from the mo-lists $I = mo(P)$ and $J = mo(Q)$ for any partite episodes P and Q. Additionally, we can compute the frequency of P form the mo-list $I = mo(P)$ of P on $O(|I|)$ time and $O(1)$ extra space by checking mo-list I from the head $I[1]$ to the tail $I[|I|]$.

algorithm MoListJoin(I, J)	algorithm MoListIntersection(I, J)
input: mo-lists I and J;	input: mo-lists I and J;
output: mo-list $I \mapsto J$; {	output: mo-list $I \circ J$; {
01 if $(I = \Omega)$ return J;	01 if $(I = \Omega)$ return J;
02 if $(J = \Omega)$ return I;	02 if $(J = \Omega)$ return I;
03 $i := 1; j := 1$;	03 $i := 1; j := 1$;
04 $\omega := (-\infty, +\infty); K := \langle \omega \rangle$;	04 $\omega := (-\infty, +\infty); K := \langle \omega \rangle$;
05 while $(i \le |I|$ and $j \le |J|)$ do	05 while $(i \le |I|$ and $j \le |J|)$ do
06 $\quad (s_I, t_I) := I[i]; (s_J, t_J) := J[j]$;	06 $\quad (s_I, t_I) := I[i]; (s_J, t_J) := J[j]$;
07 $\quad$ if $(t_I \le s_J)$ then $x := (s_I, t_J)$;	07 $\quad x := (\min(s_I, s_J), \max(t_I, t_J))$;
08 $\quad\quad$ if $(x \subseteq K[\|K\|])$ $K[\|K\|] := x$;	08 $\quad$ if $(x \subseteq K[\|K\|])$ $K[\|K\|] := x$;
09 $\quad\quad$ else $K := K\langle x \rangle$;	09 $\quad$ else if $(x \not\supseteq K[\|K\|])$ $K := K\langle x \rangle$;
10 $\quad\quad$ $i := i + 1$;	10 $\quad$ if $(t_I \le t_J)$ $i := i + 1$;
11 $\quad$ else $j := j + 1$; end if	11 $\quad$ if $(t_J \le t_I)$ $j := j + 1$;
12 end while	12 end while
13 if $(K = \langle \omega \rangle)$ $K := \langle \rangle$;	13 if $(K = \langle \omega \rangle)$ $K := \langle \rangle$;
14 return K;	14 return K;
}	}

Fig. 3. Algorithms MoListJoin and MoListIntersection for computing mo-lists.

4.4 Depth-First Enumeration

In Fig. 4, we describe the algorithm PartiteCD, and its subprocedure RecCD for extracting frequent partite episodes satisfying a partwise constraint. For a pattern constraint D, the algorithm is a backtracking algorithm that traverses the spanning tree $\mathcal{T}(\mathcal{PE}(D))$ based on a depth-first search starting from the root $P = D$ using the parent-child relationships over $\mathcal{PE}(D)$.

First, for an alphabet Σ, PartiteCD computes mo-lists $mo(\langle\{a\}\rangle)$ of partite episodes of size 1 for every event $a \in \Sigma$. Then PartiteCD makes the root episode $P = D$ and calls the recursive subprocedure RecCD. Finally, for an episode P the recursive subprocedure RecCD enumerates all frequent episodes that are descendants of P and satisfy the shape constraint C. In RecCD, the argument i indicates the index of the expanded part of P. RecCD incrementally makes type-i children $children(P, i)$ of P which are frequent and satisfying the shape constraint C in Lines 7–11. Then it recursively calls itself in Line 12 for the children. Finally, RecCD makes the child $P\langle\rangle$ of P in Line 15. Algorithm PartiteCD finds all of the frequent partite episodes satisfying the partwise constraints occurring in an input event sequence until all recursive calls are finished.

Theorem 1. *Algorithm* PartiteCD *runs in* $O(Nsc)$ *amortized time per output and in* $O(Nsm)$ *space.*

Proof. The main algorithm PartiteCD requires $O(N)$ time at Line 1 and $O(N(\|D\| + |D|))$ time at Line 2. It also requires $O(|\mathcal{S}|)$ time to compute the frequency of an episode P from the mo-list $I = mo(P)$ of P by checking mo-list

---

algorithm PARTITECD($\mathcal{S}, \Sigma, w, \sigma, C, D$)
input: input sequence $\mathcal{S} \in (2^\Sigma)^*$, alphabet of events Σ, window width $w > 0$, minimum frequency $1 \leq \sigma$, shape and pattern constraints C and D ,respectively;
output: frequent partite episodes satisfying the shape and pattern constraints;
01 *compute the mo-list $mo(a)$ for each event a in Σ*
    by scanning imput sequence $\mathcal{S}$ at once and store to $\Sigma_\mathcal{S}$;
02 *make the root episode $P = D$ and compute its mo-list $mo(P)$ from $\Sigma_\mathcal{S}$;*
03 **if** (*P is frequent and satisfying C*) RECCD($P, 1, \emptyset, \Sigma_\mathcal{S}, w, \sigma, C, D$);

procedure RECCD($P, i, U, \Sigma_\mathcal{S}, w, \sigma, C, D$)
output: all frequent partite episodes that are descendants of P;
04 **output** P;
05 **while** ($i \leq \max(h, |D|) + 1$) **do** // where $h = iex(P)$.
06    $V := \emptyset$;
07   **foreach** ($a \in \Sigma \setminus (P[i] \cup D[i])$) **do**
08   *make a type-i child Q of P by adding the event a at $P[i]$;*
09   *compute the mo-lists of Q and the subepisode $Q_{sub} = Q[1, i]$ by using U;*
10   **if** (*Q is frequent and satisfying C*) *store* ($Q, mo(Q), mo(Q_{sub})$) *to* V;
11   **end foreach**
12   **foreach** (*a child Q of P stored in V*) RECCD($Q, i, V, \Sigma_\mathcal{S}, w, \sigma, C, D$);
13    $i := i + 1$;
14 **end while**
15 **if** ($P \mapsto \langle \emptyset \rangle$ *is frequent and satisfying C*) RECCD($P \mapsto \langle \emptyset \rangle, i, \emptyset, \Sigma_\mathcal{S}, w, \sigma, C, D$);

Fig. 4. The main algorithm PARTITECD and a recursive subprocedure RECCD for mining frequent partite episodes satisfying shape and pattern constraints.

I from the head $I[1]$ to the tail $I[|I|]$. PARTITECD requires $O(N + \|P\|) = O(N)$ extra space for $\Sigma_\mathcal{S}$, P, and its mo-list. Thus, the main algorithm PARTITECD runs in $O(N(\|D\| + |D|)) = O(Nc)$ time and in $O(N)$ extra space. RECCD requires $O(N)$ time to incrementally make the child Q and its mo-list at Line 8–10. Since the arguments $\Sigma_\mathcal{S}$, w, σ, C, and D are constant at each iteration of RECCD, it runs in $O(N|\Sigma||D|) = O(Nsc)$ time and in $O(Ns)$ extra space to store the mo-lists of the children, and outputs exactly 1 solution. Since, the depth of recursions is at most m, and $\|D\| + |D| \leq m$ holds true, the statement holds true. □

Finally, we describe the possible improvement for PARTITECD. By using the alternating output technique [14], that is, we output episodes after Line 15 (the end of RECCD) instead of Line 4 if and only if the depth of recursions is even, our algorithm runs in $O(Nsc)$ delay.

Example 4. In Fig. 5, we show an example of PARTITECD. For the input sequence $\mathcal{S}$ and the alphabet of events Σ in Fig. 1, the window width $w = 4$, minimum frequency $\sigma = 1$, shape constraint $C = \langle +\infty, +\infty \rangle$, and pattern constraint $D = \langle \{b\}, \{b\} \rangle$, (**step 0**) PARTITECD computes the mo-lists of events a, b, and c. Then, it calls RECCD for the root episode $P^1 = D$. (**step 1**) RECCD makes the type-1 children P^2 and P^3 of P^1. Then, it computes the mo-list

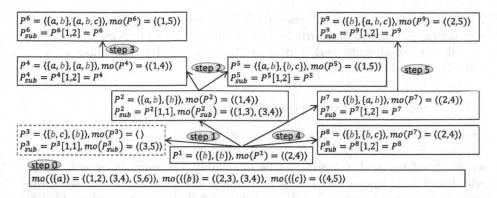

Fig. 5. Partite episodes and related arguments generated by PARTITECD from the input sequence $\mathcal{S}$ in Fig. 1. The episode P^3 in a dashed box is infrequent. See Example 4 for details.

$mo(P^2)$ of P^2 by $(mo(D[1]) \circ mo(\langle\{a\}\rangle)) \mapsto mo(D[2])$, and similarly computes the mo-lists of $P^2[1,1]$, P^3, and $P^3[1,1]$. RECCD prunes P^3 because it is infrequent. Then, it recursively calls itself for the children of P^2. **(step 2)** RECCD makes the type-2 children P^4 and P^5 of P^2. Then it incrementally computes the mo-list $mo(P^4)$ of P^4 by $mo(P^2_{sub}) \mapsto (mo(D[2]) \circ mo(\langle\{b\}\rangle))$, and similarly computes mo-lists of $P^4[1,2]$, P^5, and $P^5[1,2]$. **(step 3)** RECCD makes the type-2 child P^6 of P^4. Since the uncle of P^6 is P^5, it computes the mo-list $mo(P^6)$ of P^6 by $mo(P^4) \circ mo(P^5)$ and the mo-list $mo(P^6[1,2])$ of $P^6[1,2]$ by $mo(P^4[1,2]) \circ mo(P^5[1,2])$. **(step 4)** Similarly, RECCD makes the type-2 children P^7 and P^8 of P^1, and then **(step 5)** makes the type-2 child P^9 of P^7.

5   Experimental Results

In this section, we show that the algorithm described in Sect. 4 has a significantly better performance than a straightforward algorithm which consists of an algorithm for non-constraint episodes and post-processing steps for partwise constraints by experiments for both an artificial data set and a real-world data set.

5.1   Method

We implemented the following algorithms to compare their performance. KPER2 is a straightforward algorithm which consists of our previous algorithm [3] for non-constraint episodes and post-processing steps for partwise constraints. PPS is an algorithm which consists of Ma's algorithm [5] handling prefix anti-monotone constraints and post-processing steps for partwise constraints which are not prefix anti-monotonic. PPS efficiently handles shape constraints and a part of pattern constraints which have prefix anti-monotonic property, that is, for a partite episode P, a pattern constraint D, and an index $1 \leq i \leq |P|$, if

$P[1, i]$ does not satisfy $D[1, i]$, then PPS does not generate P. PARTITECD is our algorithm that handles partwise constraints presented in Sect. 4.

All the experiments were run on a Xeon X5690 3.47 GHz machine with 180 GB of RAM running Linux (CentOS 6.3).

5.2   Experiments for an Artificial Data Set

Data Set. An artificial data set consisted of the randomly generated event sequence $\mathcal{S} = \langle S_1, \ldots, S_\ell \rangle$ ($\ell \geq 1$) over an alphabet $\Sigma = \{1, \ldots, s\}$ ($s \geq 1$) as described below. Let $1 \leq i \leq \ell$ be any index. For every event $a \in \Sigma$, we add a into S_i independently with a probability of $p = 0.1/a$. By repeating this process for every S_i, we made a skewed data set that emulates real-world data sets having long tails such as point-of-sale data sets.

If not explicitly stated otherwise, we assume that the number of event in the input sequence $\mathcal{S}$ is $n = \|\mathcal{S}\| = 1,000,000$, the alphabet size is $s = |\Sigma| = 1,000$, the window width is $w = 20$, the minimum support threshold is $\acute{\sigma} = 0.1\,\%$, the shape constraint $C = \langle +\infty, 1, +\infty \rangle$, and the pattern constraint $D = \langle \{10\}, \emptyset, \{10\} \rangle$.

Experiments. Figure 6 shows the running time (left), and the amount of memory usage (right) of the algorithms KPER, PPS, and PARTITECD for the number of events n in input data. With respect to the total running time, we see that PARTITECD is 230 times as fast as KPER2 and 35 times as fast as PPS in this case. This difference comes from the number of episode generated by each algorithm before post-processing steps.

For the amount of memory usage, we see that PPS uses more memory than both KPAR2 and PARTITECD. The reason is that PPS is based on breadth-first search in a search space [5], whereas both KPAR2 and PARTITECD are based on depth-first search. We also see that, both the running time and the memory usage of these algorithms seem to be almost linear to the input size and must therefore scale well on large datasets.

Figure 7 shows the number of outputs before the post-processing steps (left) and the total running time (right) for pattern constraints D_i ($1 \leq i \leq 3$). In

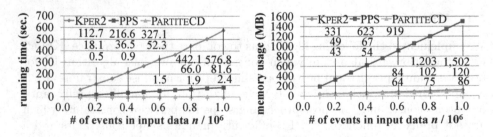

Fig. 6. Running time (left) and memory usage (right) for the number of events in input data.

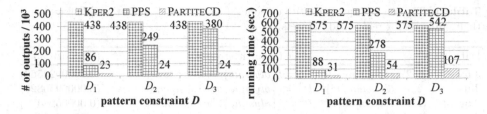

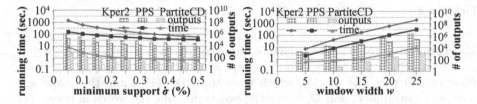

Fig. 7. The number of outputs before post-processing steps (left) and running time (rigth) for pattern constraints D_i $(1 \leq i \leq 3)$, where $D_1 = \langle \{10\}, \emptyset, \emptyset \rangle$, $D_2 = \langle \emptyset, \{10\}, \emptyset \rangle$, and $D_3 = \langle \emptyset, \emptyset, \{10\} \rangle$.

Fig. 8. Running time and number of outputs against minimum supports (left) and window width (right).

this experiment, we used a shape constraint $C = \langle +\infty, +\infty, +\infty \rangle$. We see that the number of outputs before post-processing steps of PPS is large and the running time is long if a part $\{10\}$ of the pattern constraint D appears at the tail side of D. The reason is that since PPS reduces the number of outputs by pruning based on prefix anti-monotonic property, PPS generates a large number of episodes before pruning by the constraint that appears at the tail side. On the other hand, the number of outputs of PARTITECD is almost constant for every constraint D, because PARTITECD generates only episodes that satisfy the given constraints.

Figure 8 shows the total running time and the number of outputs before the post-processing steps against minimum support $\dot{\sigma}$ (left) and window width w (right). We see that PARTITECD outperforms other algorithms both on the number of outputs and the running time for every parameter set.

5.3 Experiments for a Real-World Data Set

Data Set. The real-world data set was made from a Twitter data set. Our Twitter data set consisted of about one percent of all tweets obtained by sampling from a tweeter stream. For our Twitter data sets from July and August 2012, we extracted $1,115,726$ users whose number of tweet samples were between 15 and 25. Then, we made event sequences from the tweeted words except for the most frequent 300 words used as stop words and their tweeted times for each user. Then, we made input sequences for July and August by concatenating the event sequences for every user, where the number of events n and the alphabet

Table 1. The most frequent partite episodes P^i $(1 \leq i \leq 4)$ such that $\|P^i\| = 3$ and their supports for real-world datasets.

Data set	C	D	Most frequent episode	(Support %)
July	$\langle 1, 2 \rangle$	$\langle \{olympic\}, \emptyset \rangle$	$P^1 = \langle \{olympic\}, \{jordan, win\} \rangle$	(0.0000923563)
July	$\langle 2, 1 \rangle$	$\langle \emptyset, \{olympic\} \rangle$	$P^2 = \langle \{olympic, win\}, \{olympic\} \rangle$	(0.0000956826)
August	$\langle 1, 2 \rangle$	$\langle \{olympic\}, \emptyset \rangle$	$P^3 = \langle \{olympic\}, \{medal, gold\} \rangle$	(0.0000628452)
August	$\langle 2, 1 \rangle$	$\langle \emptyset, \{olympic\} \rangle$	$P^4 = \langle \{ceremony, closing\}, \{olympic\} \rangle$	(0.0000630654)

size s of the real-world event sequence was $n = 69,345,247$ and $s = 86,243,716$ for July, and $n = 70,218,644$ and $s = 87,621,968$ for August.

Experiments. Table 1 shows the most frequent partite episodes P^i $(1 \leq i \leq 4)$ such that $\|P^i\| = 3$ extracted from the real-world event sequences by algorithm PARTITECD with the window width $w = 86,400\,s$ (24 h), and the shape and pattern constraints C and D, where the events describe the words in tweet. For example, episode P^1 says that both the words *jordan* and *win* were tweeted after a word *olympic* was tweeted within 24 h. We can observe that the most frequent pair of words related to the word *olympic* depends on the temporal relation to the word *olympic*, and it also depends on the input data set.

On the experiment of Table 1, algorithm PARTITECD ran in 175.1 s and output 956 episodes including P^1 under the minimum support $\acute{o} = 0.00001\,\%$, the shape constraint $C = \langle 1, 2 \rangle$, and the pattern constraint $D = \langle \{olympic\}, \emptyset \rangle$. In this case, we could not extract episodes within $86,400\,s$ (24 h) without constraint because there exists a large number (more than $8,639,000$) of frequent episodes.

5.4   Summary of Experiments

Overall, we conclude that the partwise constraint reduces the number of outputs and the proposed algorithm handles the constraint much more efficiently than the straightforward algorithm using post-processing steps. In other words, by using our algorithm with our constraint, we can extract partite episodes on which we focused with lower frequency thresholds for larger input data in the same running time.

6   Conclusion

This paper studied the problem of frequent partite episode mining with partwise constraints. We presented an algorithm that finds all frequent partite episodes satisfying a partwise constraint in an input sequence. Then, we showed that our algorithm runs in the output polynomial time and polynomial space. We reported the experimental results that compare the performance of our algorithm and other straightforward algorithms. Both the theoretical and experimental results showed that our algorithm efficiently extracts episodes on which we focused. Thus, we conclude that our study improved the efficiency of data mining.

A possible future path of study will be the extension of PARTITECD for the class of proper partite episodes to reduce redundant outputs by traversing a search tree containing only proper partite episodes satisfying a pattern constraint. Our future work will also include the application of our algorithm to other real-world data sets in addition to Twitter data.

References

1. Arimura, H., Uno, T.: A polynomial space and polynomial delay algorithm for enumeration of maximal motifs in a sequence. In: Deng, X., Du, D.-Z. (eds.) ISAAC 2005. LNCS, vol. 3827, pp. 724–737. Springer, Heidelberg (2005)
2. Avis, D., Fukuda, K.: Reverse search for enumeration. Discrete Appl. Math. **65**, 21–46 (1996)
3. Katoh, T., Arimura, H., Hirata, K.: Mining frequent k-partite episodes from event sequences. In: Nakakoji, K., Murakami, Y., McCready, E. (eds.) JSAI-isAI 2009. LNCS (LNAI), vol. 6284, pp. 331–344. Springer, Heidelberg (2010)
4. Katoh, T., Hirata, K.: A simple characterization on serially constructible episodes. In: Washio, T., Suzuki, E., Ting, K.M., Inokuchi, A. (eds.) PAKDD 2008. LNCS (LNAI), vol. 5012, pp. 600–607. Springer, Heidelberg (2008)
5. Ma, X., Pang, H., Tan, K.L.: Finding constrained frequent episodes using minimal occurrences. In: ICDM, pp. 471–474 (2004)
6. Mannila, H., Toivonen, H., Verkamo, A.I.: Discovery of frequent episodes in event sequences. Data Min. Knowl. Disc. **1**(3), 259–289 (1997)
7. Méger, N., Rigotti, C.: Constraint-based mining of episode rules and optimal window sizes. In: Boulicaut, J.-F., Esposito, F., Giannotti, F., Pedreschi, D. (eds.) PKDD 2004. LNCS (LNAI), vol. 3202, pp. 313–324. Springer, Heidelberg (2004)
8. Ohtani, H., Kida, T., Uno, T., Arimura, H.: Efficient serial episode mining with minimal occurrences. In: ICUIMC, pp. 457–464 (2009)
9. Pei, J., Han, J., Mortazavi-Asl, B., Wang, J.: Mining sequential patterns by pattern-growth: the PrefixSpan approach. IEEE Trans. Knowl. Data Eng. **16**(11), 1–17 (2004)
10. Pei, J., Han, J.: Can we push more constraints into frequent pattern mining? In: KDD, pp. 350–354 (2000)
11. Pei, J., Han, J., Wang, W.: Mining sequential patterns with constraints in large databases. In: CIKM, pp. 18–25. ACM (2002)
12. Seipel, D., Neubeck, P., Köhler, S., Atzmueller, M.: Mining complex event patterns in computer networks. In: Appice, A., Ceci, M., Loglisci, C., Manco, G., Masciari, E., Ras, Z.W. (eds.) NFMCP 2012. LNCS, vol. 7765, pp. 33–48. Springer, Heidelberg (2013)
13. Tatti, N., Cule, B.: Mining closed strict episodes. Data Min. Knowl. Disc. **25**(1), 34–66 (2012)
14. Uno, T.: Two general methods to reduce delay and change of enumeration algorithms. Technical report. National Institute of Informatics (2003)
15. Zhou, W., Liu, H., Cheng, H.: Mining closed episodes from event sequences efficiently. In: Zaki, M.J., Yu, J.X., Ravindran, B., Pudi, V. (eds.) PAKDD 2010, Part I. LNCS, vol. 6118, pp. 310–318. Springer, Heidelberg (2010)

Structure Determination and Estimation of Hierarchical Archimedean Copulas Based on Kendall Correlation Matrix

Jan Górecki[1]([✉]) and Martin Holeňa[2]

[1] Department of Informatics, SBA in Karvina,
Silesian University in Opava, Karvina, Czech Republic
gorecki@opf.slu.cz
[2] Institute of Computer Science,
Academy of Sciences of the Czech Republic,
Praha, Czech Republic
martin@cs.cas.cz

Abstract. An estimation method for the copula of a continuous multivariate distribution is proposed. A popular class of copulas, namely the class of hierarchical Archimedean copulas, is considered. The proposed method is based on the close relationship of the copula structure and the values of Kendall's tau computed on all its bivariate margins. A generalized measure based on Kendall's tau adapted for purposes of the estimation is introduced. A simple algorithm implementing the method is provided and its effectiveness is shown in several experiments including its comparison to other available methods. The results show that the proposed method can be regarded as a suitable alternative to existing methods in the terms of goodness of fit and computational efficiency.

Keywords: Copula · Hierarchical Archimedean copula · Copula estimation · Structure determination · Kendall's correlation coefficient

1 Introduction

Studying relationships among random quantities is a crucial task in the field of knowledge discovery and data mining (KDDM). Having a dataset collected, the relationships among the observed variables can be studied by means of an appropriate measure of stochastic dependence. Under assumption of the multivariate continuous distribution of the variables, the famous Sklar's theorem [29] can be used to decompose the distribution in two components. While the first component describes the distributions of the univariate margins, the second component describes the copula of the distribution containing the whole information about the relationship among the variables. Thus, studying dependencies among the random variables can be restricted without any loss of generality to studying the copula.

A. Appice et al. (Eds.): NFMCP 2013, LNAI 8399, pp. 132–147, 2014.
DOI: 10.1007/978-3-319-08407-7_9, © Springer International Publishing Switzerland 2014

Despite the fact that copulas have most success in finance, they are increasingly adopted also in KDDM, where they are used due to their effective mathematical ability to capture even very complex dependence structures among variables. We can see applications of copulas in water-resources and hydro-climatic analysis [4,13,14,17,19], gene analysis [18,31], cluster analysis [3,15,26] or in evolution algorithms, particularly in the estimation of distribution algorithms [7,30]. For an illustrative example, we refer to [13], where the task for anomaly detection in climate that incorporates complex spatio-temporal dependencies is solved using copulas.

Hierarchical Archimedean copulas (HACs) are a frequently used alternative to the most popular Gaussian copulas due to their flexibility and conveniently limited number of parameters. Despite their popularity, feasible techniques for HAC estimation are addressed only in few papers. Most of them assume in the estimation process a given structure of a copula, which is motivated trough applications in economy, see [27,28]. There exists only one recently published paper [23], which addresses the estimation technique generally, i.e., the estimation also concerns the proper structure determination of the HAC.

The mentioned paper describes a multi-stage procedure, which is used both for the structure determination and the estimation of the parameters. The authors devote mainly to the estimation of the parameters using the maximum-likelihood (ML) technique and briefly mention its alternative, which uses for the parameters estimation the relationship between the copula parameter and the value of Kendall's tau computed on a bivariate margin of the copula (shortly, $\theta - \tau$ relationship). The authors present six approaches denoted as $\tau_{\Delta\tau>0}$, τ_{binary}, Chen, θ_{binary}, $\theta_{binary\ aggr.}$ and θ_{RML} to the structure determination. The first two approaches are based on the $\theta - \tau$ relationship, the third approach is based on the Chen test statistics [2] and the last three approaches are based on the ML technique. The first five approaches lead to biased estimators, what can be seen in the results of the attached simulation study, and the sixth (θ_{RML}) is used for re-estimation and thus for better approximation of the parameters of the true copula. θ_{RML} shows the best goodness-of-fit (measured by Kullback-Leibler divergence) of the resulting estimates. However, the best approximation of the true parameters with θ_{RML} is possible only in the cases, when the structure is properly determined (the estimated structure equals the true structure). But, as θ_{RML} is based on the biased $\theta_{binary\ aggr.}$, which often does not return the true structure due to the involved bias, θ_{RML} also cannot return close approximation of the true parameters in the cases, when the structure is determined improperly. Moreover, the number of those cases rapidly increases with the increasing data dimension, as we show later in Sect. 4.

In our paper, we propose the construction of the estimator for HACs that approximates the parameters of the true copula better than the previously mentioned methods, and thus also increases the ratio of properly determined structures. Avoiding the need of re-estimation, we also gain high computational efficiency. The included experiments on simulated data show that our approach outperforms all the other above mentioned methods in the sense of goodness-of-fit, the properly determined structures ratio and also in the time consumption, which is even slightly lower than the most efficient binary methods $\tau_{binary}, \theta_{binary}$.

The paper is structured as follows. The next section summarizes some necessary theoretical concepts concerning Archimedean copulas (ACs) and HACs. Section 3 presents the new approach to the HAC estimation. Section 4 describes the experiments and their results and Sect. 5 concludes this paper.

2 Preliminaries

2.1 Copulas

Definition 1. *For every $d \geq 2$, a d-dimensional copula (shortly, d-copula) is a d-variate distribution function on $\mathbb{I}^d$ ($\mathbb{I}$ is the unit interval), whose univariate margins are uniformly distributed on $\mathbb{I}$.*

Copulas establish a connection between general joint distribution functions (d.f.s) and its univariate margins (in text below we use only *margin* for term *univariate margin*), as can be seen in the following theorem.

Theorem 1. *(Sklar's Theorem) [29] Let H be a d-variate d.f. with univariate margins $F_1, ..., F_d$. Let A_j denote the range of F_j, $A_j := F_j(\overline{\mathbb{R}})(j = 1, ..., d), \overline{\mathbb{R}} := \mathbb{R} \cup \{-\infty, +\infty\}$. Then there exists a copula C such for all $(x_1, ..., x_d) \in \overline{\mathbb{R}}^d$,*

$$H(x_1, ..., x_d) = C(F_1(x_1), ..., F_d(x_d)). \tag{1}$$

Such a C is uniquely determined on $A_1 \times ... \times A_d$ and, hence, it is unique if $F_1, ..., F_d$ are all continuous. Conversely, if $F_1, ..., F_d$ are univariate d.f.s, and if C is any d-copula, then the function $H : \overline{\mathbb{R}}^d \to \mathbb{I}$ defined by (1) is a d-dimensional distribution function with margins $F_1, ..., F_d$.

Through the Sklar's theorem, one can derive for any d-variate d.f. its copula C using (1). In case that the margins $F_1, ..., F_d$ are all continuous, the copula C is given by $C(u_1, ..., u_d) = H(F_1^-(u_1), ..., F_d^-(u_d))$, where $F_i^-, i \in \{1, ..., d\}$ denotes pseudo-inverse of F_i given by $F_i^-(s) = \inf\{t| F_i(t) \geq s\}, s \in \mathbb{I}$. Many classes of copulas are derivable in this way from popular joint d.f.s, e.g., the most popular class of Gaussian copulas is derived using H corresponding to a d-variate Gaussian distribution. But, using this process often results in copulas not expressible in closed form, what can bring difficulties in some applications.

2.2 Archimedean Copulas

This drawback is overcame while using (exchangeable) Archimedean copulas, due to their different construction process. ACs are not constructed using the Sklar's theorem, but instead of it, one starts with a given functional form and asks for properties needed to obtain a proper copula. As a result of such a construction, ACs are always expressed in closed form, which is one of the main advantages of this class of copulas [10]. To construct ACs, we need the notion of an *Archimedean generator* and of *complete monotonicity*.

Definition 2. Archimedean generator *(shortly, generator) is a continuous, non-increasing function* $\psi : [0, \infty] \rightarrow [0, 1]$, *which satisfies* $\psi(0) = 1, \psi(\infty) = \lim_{t \rightarrow \infty} \psi(t) = 0$ *and is strictly decreasing on* $[0, \inf\{t : \psi(t) = 0\}]$. *We denote the set of all generators as* Ψ.

Definition 3. *A function* f *is called* completely monotone *(shortly, c.m.) on* $[a, b]$, *if* $(-1)^k f^{(k)}(x) \geq 0$ *holds for every* $k \in \mathbb{N}_0, x \in (a, b)$. *We denote the set of all completely monotonous generators as* Ψ_∞.

Definition 4. *Any d-copula* C *is called* Archimedean copula *(we denote it d-AC), if it admits the form*

$$C(\mathbf{u}) := C(\mathbf{u}; \psi) := \psi(\psi^{-1}(u_1) + ... + \psi^{-1}(u_d)), \mathbf{u} \in \mathbb{I}^d, \qquad (2)$$

where $\psi \in \Psi$ *and the* $\psi^{-1} : [0, 1] \rightarrow [0, \infty]$ *is defined* $\psi^{-1}(s) = \inf\{t : \psi(t) = s\}, s \in \mathbb{I}$.

For verifying whether function C given by (2) is a proper copula, we can use the property stated in Definition 3. A condition sufficient for C to be a copula is stated as follows.

Theorem 2. *[21] If* $\psi \in \Psi$ *is completely monotone, then the function* C *given by (2) is a copula.*

We can see from Definition 4 and from the properties of generators that having a random vector **U** distributed according to some AC, all its k-dimensional $(k < d)$ marginal copulas have the same marginal distribution. It implies that all multivariate margins of the same dimension are equal, thus, e.g., the dependence among all pairs of components is identical. This symmetry of ACs is often considered to be a rather strong restriction, especially in high dimensional applications.

Given the number of variables, to derive the explicit form of an AC to work with, we need the explicit form of generators. The reader can find many explicit forms of the generators in, e.g., [22]. In this paper, we use and present only the Clayton generator, defined $(1 + t)^{-1/\theta}$, which corresponds to the family of the Clayton copulas. Copulas based on this generator have been used, e.g., to study correlated risks, because they exhibit strong left tail dependence and relatively weak right tail dependence. The explicit parametric form of a bivariate Clayton copula is $C(u_1, u_2; \psi) = \left(u_1^{-\theta} + u_2^{-\theta} - 1\right)^{-\frac{1}{\theta}}$ [22].

2.3   Hierarchical Archimedean Copulas

To allow for asymmetries, one may consider the class of HACs (often also called *nested Archimedean copulas*), recursively defined as follows.

Definition 5. *[11] A d-dimensional copula* C *is called* hierarchical Archimedean copula *if it is an AC with arguments possibly replaced by other hierarchical Archimedean copulas. If* C *is given recursively by (2) for* $d = 2$ *and*

$$C(\mathbf{u}; \psi_1, ..., \psi_{d-1}) = \psi_1(\psi_1^{-1}(u_1) + \psi_1^{-1}(C(u_2, ..., u_d; \psi_2, ..., \psi_{d-1}))), \boldsymbol{u} \in \mathbb{I}^d, (3)$$

for $d \geq 2$, C *is called* fully-nested hierarchical Archimedean copula (FHAC)[1] *with* $d - 1$ nesting levels. *Otherwise* C *is called* partially-nested hierarchical Archimedean copula (PHAC)[2].

Remark 1. We denote a d-dimensional HAC as d-HAC, and analogously d-FHAC and d-PHAC.

From the definition, we can see that ACs are special cases of HACs. The most simple proper 3-PHAC is with two nesting levels. The copula is given by

$$C(\mathbf{u}; \psi_1, \psi_2) = C(u_1, C(u_2, u_3; \psi_2); \psi_1)$$
$$= \psi_1(\psi_1^{-1}(u_1) + \psi_1^{-1}(\psi_2(\psi_2^{-1}(u_2) + \psi_2^{-1}(u_3)))), \mathbf{u} \in \mathbb{I}^3. \quad (4)$$

As in the case of ACs, we can ask for necessary and sufficient condition for the function C given by (3) to be a proper copula. Partial answer to this question in form of sufficient condition is contained in the following theorem.

Theorem 3. *(McNeil (2008))* [20] *If* $\psi_j \in \Psi_\infty, j \in \{1, ..., d-1\}$ *such that* $\psi_k^{-1} \circ \psi_{k+1}$ *have completely monotone derivatives for all* $k \in \{1, ..., d-2\}$, *then* $C(\mathbf{u}; \psi_1, ..., \psi_{d-1})$, $\mathbf{u} \in \mathbb{I}^d$, *given by (3) is a copula.*

McNeil's theorem is stated only for fully-nested HACs, but it can be easily translated also for use with partially-nested HACs (for more see [20]). The condition for $(\psi_1^{-1} \circ \psi_2)'$ to be compete monotone is often called the *nesting condition.*

A d-HAC structure, which is given by the recursive nesting in the definition, can be expressed as a tree with $k \leq d-1$ non-leaf nodes (shortly, nodes), which correspond to the generators $\psi_1, ..., \psi_k$, and d leafs, which correspond to the variables $u_1, ..., u_d$. If the structure corresponds to a binary tree, then $k = d-1$. In other case $k < d-1$. Thus, a HAC structure is viewed as a tree in the next text. Also, for the sake of simplicity, we assume only binary HAC structures.

Let s be the structure of a d-HAC. Each 2-AC is determined just by its corresponding generator, and, if we identify each node in s with one generator, we have always nodes $\psi_1, ..., \psi_{d-1}$. For a node ψ denote as $\mathcal{D}_n(\psi)$ the set of all descendant nodes of ψ, $\mathcal{P}(\psi)$ the parent node of ψ, $\mathcal{H}_l(\psi)$ the left child of ψ and $\mathcal{H}_r(\psi)$ the right child of ψ.

For simplicity, a d-HAC structure s is denoted as a sequence of reordered indices $\{1, ..., d\}$ using parentheses to mark the variables with the same parent node. For example, the structure of the copula given by (4) is denoted as $(1(23))$. The inner parenthesis corresponds to the fact that for the variables u_2, u_3 is $\mathcal{P}(u_2) = \mathcal{P}(u_3) = \psi_2$. As u_2, u_3 are connected through their parent, we can introduce a new variable denoted as z_{23}, which represents the variables u_2, u_3 and is defined as $z_{23} = C(u_2, u_3; \psi_2)$. Then (4) turns in $\psi_1(\psi_1^{-1}(u_1) + \psi_1^{-1}(z_{23})) = C(u_1, z_{23}; \psi_1)$, and thus the outer parenthesis in the notation of the structure corresponds to the fact that for the variables u_1, z_{23} is $\mathcal{P}(u_1) = \mathcal{P}(z_{23}) = \psi_1$.

---

[1] Sometimes called *fully-nested Archimedean copula.*
[2] Sometimes called *partially-nested Archimedean copula.*

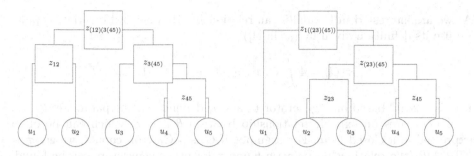

Fig. 1. On the left side is depicted the 5-PHAC structure denoted as $((12)(3(45)))$ and on the right side is depicted the 5-PHAC structure denoted as $(1((23)(45)))$.

The structure of the 4-FHAC given as in Definition 5 is denoted as $(1(2(34)))$, for 5-FHAC, it is $(1(2(3(45))))$, etc. Analogously, for PHACs, $((12)(3(45)))$ and $(1((23)(45)))$ denote the structures depicted on the left and the right side in Fig. 1.

When using HACs in applications, there exist, for example for $d = 10$, more than 280 millions of possible HAC structures (including also non-binary ones) and each 10-HAC can incorporate up to 9 parameters (using only one-parametric generators) in generators from possibly different families. If choosing the model that the best fit the data, this is much more complex situation relative to the case when using ACs, which have just one structure, one parameter and one Archimedean family.

To derive the explicit parametric form a d-HAC C, we need the explicit parametric forms of its generators $\psi_1, ..., \psi_{d-1}$, which involve the parameters $\theta_1, ..., \theta_{d-1}$ (θ_i corresponds to the generator $\psi_i, i = 1, ..., d - 1$), and its structure s. Due to this, the copula C is also denoted as $C(\psi, \theta; s)(u_1, ..., u_d)$ in the rest of the text. For example, the 3-HAC that is given by (4) and assuming both of its generators ψ_1, ψ_2 to be Clayton generators, can be denoted as $C(\psi_1, \psi_2, \theta_1, \theta_2; (1(23)))$ and its parametric form is given as

$$C(\psi_1, \psi_2, \theta_1, \theta_2; (1(23))) = \left(\left(\left(u_2^{-\theta_2} + u_3^{-\theta_2} - 1 \right)^{-\frac{1}{\theta_2}} \right)^{-\theta_1} + u_1^{-\theta_1} - 1 \right)^{-\frac{1}{\theta_1}}.$$

$$(5)$$

2.4 Kendall's Tau and Its Generalization

The standard definition of Kendall's tau for two random variables X, Y is given as follows. Let (X_1, Y_1) and (X_2, Y_2) be independent random vectors with the same distribution as (X, Y). Then the population version of Kendall's tau is defined as the probability of concordance minus the probability of discordance, i.e.,

$$\tau = \tau_{XY} = P((X_1 - X_2)(Y_1 - Y_2) > 0) - P((X_1 - X_2)(Y_1 - Y_2) < 0). \quad (6)$$

As we are interested in Kendall's tau relationship to a general bivariate copula, we use its definition given by (as in [4])

$$\tau(C) = 4 \int_{\mathbb{I}^2} C(u_1, u_2) dC(u_1, u_2) - 1. \tag{7}$$

If C is a 2-AC based on a generator ψ, and ψ depends on the parameter $\theta \in \mathbb{R}$, then (7) states an explicit relationship between θ and τ, which can be often expressed in a closed form. For example, if C is a Clayton copula, we get $\tau = \theta/(\theta + 2)$ (the relationship between θ and τ for other generators can be found, e.g., in [10]). The inversion of this relationship establish an estimator of the parameter θ, which can be based on the empirical version of τ given by (as in [4])

$$\tau_n = \frac{4}{n(n-1)} \sum_{i=1,j=1}^{n} \mathbf{1}_{\{(u_{i1}-u_{j1})(u_{i2}-u_{j2})>0\}}, \tag{8}$$

where $(u_{\bullet 1}, u_{\bullet 2})$ denotes the realizations of r.v.s $(U_1, U_2) \sim C$.

This estimation method was introduced in [5] as a method-of-moments estimator for bivariate one-parameter Archimedean copulas. The copula parameter $\theta \in \Theta \subseteq \mathbb{R}$ is estimated by $\hat{\theta}_n$ such that $\tau(\hat{\theta}_n) = \tau_n$, where $\tau(\theta)$ denotes Kendall's tau of the corresponding Archimedean family viewed as a function of the parameter $\theta \in \Theta \subseteq \mathbb{R}$, i.e., that $\hat{\theta}_n = \tau^{-1}(\tau_n)$, assuming the inverse τ^{-1} of τ exists. If the equation has no solution, this estimation method does not lead to an estimator. Unless there is an explicit form for τ^{-1}, $\hat{\theta}_n$ is computed by numerical root finding [12].

This estimation method can also be generalized for ACs when $d > 2$, see [1, 12,16,28]. The generalized method is using pairwise sample version of Kendall's tau. If τ_{ij}^n denotes the sample version of Kendall's tau between the i-th and j-th data column, then θ is estimated by

$$\hat{\theta}_n = \tau^{-1}\left(\binom{d}{2}^{-1} \sum_{1 \leq i \leq j \leq d} \tau_{ij}^n\right). \tag{9}$$

As can be seen, the parameter is chosen such that Kendall's tau equals the average over all pairwise sample versions of Kendall's tau. Properties of this estimator are not known and also not easy to derive since the average is taken over dependent data columns [12]. However, $\binom{d}{2}^{-1} \sum_{1 \leq i \leq j \leq d} \tau_{ij}^n$ is an unbiased estimator of $\tau(\theta)$. This is an important property and we transfer it later to the estimator that we use for the structure determination, which we base on appropriately selected pairwise sample versions of Kendall's tau.

To use the generalized method mentioned in the previous paragraph with HACs, we define a generalization of τ for m (possibly > 2) random variables (r.v.s). For simplification denote the set of pairs of r.v.s as $\mathbf{U}_{IJ} = \{(U_i, U_j)|(i,j) \in I \times J\}$, where $I, J \subset \{1, ..., d\}, I \neq \emptyset \neq J, (U_1, ..., U_d) \sim C$, C is a d-HAC.

Definition 6. *Let τ be the Kendall's tau, $g : [0,1]^k \to [0,1], k \in \mathbb{N}$, be an aggregation function (like, e.g., max, min or mean), which has the following properties: 1) $g(u, ..., u) = u$ for all $u \in \mathbb{I}$ and 2) $g(u_{p_1}, ..., u_{p_k}) = g(u_1, ..., u_k)$ for all $u_1, ..., u_k \in \mathbb{I}$ and all permutations p of $\{1, ..., k\}$. Then define an* aggregated *Kendall's tau τ^g as*

$$\tau^g(\mathbf{U}_{IJ}) = \begin{cases} \tau(U_i, U_j) & \text{if } I = \{i\}, J = \{j\} \\ g(\tau(U_{i_1}, U_{j_1}), \tau(U_{i_1}, U_{j_2}), ..., \tau(U_{i_l}, U_{j_q})), & \text{else,} \end{cases} \quad (10)$$

where $I = \{i_1, ..., i_l\}, J = \{j_1, ..., j_q\}$ are non-empty disjoint subsets of $\{1, ..., d\}$.

As the aggregated τ^g depends only on the pairwise τ and the aggregation function g, we can easily derive its empirical version τ_n^g just by substituting τ in τ^g by its empirical version τ_n given by (8). Then, analogously to the case of ACs, the parameter is estimated as $\hat{\theta}_n = \tau^{-1}(\tau_n^g)$. But, as all bivariate margins of a HAC are not assumed to be identical, each estimate is computed just on some appropriately selected ones. This is later explained by Remark 2.

2.5   Okhrin's Algorithm for the Structure Determination of HAC

We recall the algorithm presented in [24] for the structure determination of HAC, which returns for some unknown HAC C its structure using only the known forms of its bivariate margins. The algorithm uses the following definition.

Definition 7. *Let C be a d-HAC with generators $\psi_1, ..., \psi_{d-1}$ and $(U_1, ..., U_d) \sim C$. Then denote as $\mathcal{U}_C(\psi_k), k = 1, ..., d-1$, the set of indexes $\mathcal{U}_C(\psi_k) = \{i | (\exists U_j) (U_i, U_j) \sim C(\cdot; \psi_k) \vee (U_j, U_i) \sim C(\cdot; \psi_k), 1 \leq i < j \leq d\}, k = 1, ..., d-1$.*

Proposition 1. *[24] Defining $\mathcal{U}_C(u_i) = \{i\}$ for the leaf $i, 1 \leq i \leq d$, there is an unique disjunctive decomposition of $\mathcal{U}_C(\psi_k)$ given by*

$$\mathcal{U}_C(\psi_k) = \mathcal{U}_C(\mathcal{H}_l(\psi_k)) \cup \mathcal{U}_C(\mathcal{H}_r(\psi_k)). \quad (11)$$

For an unknown d-HAC C, knowing all its bivariate margins, its structure can be easily determined using Algorithm 1. We start from the sets $\mathcal{U}_C(u_1), ..., \mathcal{U}_C(u_d)$ joining them together through (11) until we reach the node ψ for which $\mathcal{U}_C(\psi) = \{1, ..., d\}$.

We illustrate the Algorithm 1 for a 5-HAC given by $C(C(u_1, u_2; \psi_2), C(u_3, C(u_4, u_5; \psi_4); \psi_3); \psi_1) = C(\psi_1, ..., \psi_4, ((12)(3(45))))(u_1, ..., u_5)$. The structure of this copula is depicted on the left side in Fig. 1 and its bivariate margins are:

$(U_1, U_2) \sim C(\cdot; \psi_2)$ $(U_1, U_3) \sim C(\cdot; \psi_1)$ $(U_1, U_4) \sim C(\cdot; \psi_1)$ $(U_1, U_5) \sim C(\cdot; \psi_1)$
$(U_2, U_3) \sim C(\cdot; \psi_1)$ $(U_2, U_4) \sim C(\cdot; \psi_1)$ $(U_2, U_5) \sim C(\cdot; \psi_1)$ $(U_3, U_4) \sim C(\cdot; \psi_3)$
$(U_3, U_5) \sim C(\cdot; \psi_3)$ $(U_4, U_5) \sim C(\cdot; \psi_4)$

Now assume that the structure is unknown and only the bivariate margins are known. We see that $\mathcal{U}_C(\psi_1) = \{1, 2, 3, 4, 5\}$, $\mathcal{U}_C(\psi_2) = \{1, 2\}$, $\mathcal{U}_C(\psi_3) = \{3, 4, 5\}, \mathcal{U}_C(\psi_4) = \{4, 5\}$. For leafs $u_1, ..., u_5$, it is defined $\mathcal{U}_C(u_i) = \{i\}$,

---

Algorithm 1. The HAC Structure Determination [24]

Input:
1) $\mathcal{U}_C(\psi_1), ..., \mathcal{U}_C(\psi_{d-1})$,
2) $\mathcal{I} = \{1, ..., d-1\}$

while $\mathcal{I} \neq \emptyset$ **do**
    1. $k = \text{argmin}_{i \in \mathcal{I}}(\#\mathcal{U}_C(\psi_i))$, if there are more minima, then choose as k one of them arbitrarily.
    2. Find the nodes ψ_l, ψ_r, for which $\mathcal{U}_C(\psi_k) = \mathcal{U}_C(\psi_l) \cup \mathcal{U}_C(\psi_r)$.
    3. $\mathcal{H}_l(\psi_k) := \psi_l, \mathcal{H}_r(\psi_k) := \psi_r$.
    4. Set $\mathcal{I} := \mathcal{I} \backslash \{k\}$.
end while

Output:
The structure stored in $\mathcal{H}_l(\psi_k), \mathcal{H}_r(\psi_k), k = 1, ..., d-1$

$i = 1, ..., 5$. In step 1., there are two minima: $k = 2$ and $k = 4$. We choose arbitrarily $k = 4$. As $\mathcal{U}_C(\psi_4) = \mathcal{U}_C(u_4) \cup \mathcal{U}_C(u_5)$, we set in step 3. $\mathcal{H}_l(\psi_4) := u_4$ and $\mathcal{H}_r(\psi_4) := u_5$. In step 4., we set $\mathcal{I} = \{1, 2, 3\}$. In the second loop, $k = 2$. As $\mathcal{U}_C(\psi_2) = \mathcal{U}_C(u_1) \cup \mathcal{U}_C(u_2)$, we set in step 3. $\mathcal{H}_l(\psi_2) := u_1$ and $\mathcal{H}_r(\psi_2) := u_2$. In the third loop, we have $k = 3$. As $\mathcal{U}_C(\psi_3) = \mathcal{U}_C(u_3) \cup \mathcal{U}_C(\psi_4)$, we set in step 3. $\mathcal{H}_l(\psi_3) := u_3$ and $\mathcal{H}_r(\psi_3) := \psi_4$. In the last loop, we have $k = 1$. As $\mathcal{U}_C(\psi_1) = \mathcal{U}_C(\psi_2) \cup \mathcal{U}_C(\psi_3)$, we set in step 3. $\mathcal{H}_l(\psi_1) := \psi_2$ and $\mathcal{H}_r(\psi_1) := \psi_3$. Observing the original copula form and Fig. 1, we see that we have determined the correct structure, which is stored in $\mathcal{H}_l(\psi_k), \mathcal{H}_r(\psi_k), k = 1, ..., 4$.

3 Our Approach

3.1 HAC Structure Determination

Recalling Theorem 3, the sufficient condition for C to be a proper copula is that the nesting condition must hold for each generator and its parent in a HAC structure. As this is the only known condition that assures that C is a proper copula, we deal in our work only with the copulas that fulfill this condition. The nesting condition results in constraints on the parameters θ_1, θ_2 of the involved generators ψ_1, ψ_2 (see [10,11]). As $\theta_i, i = 1, 2$ is closely related to τ through (7), there is also an important relationship between the values of τ and the HAC tree structure following from the nesting condition. This relationship is described for the fully-nested 3-HAC given by the form (4) in Remark 2.3.2 in [10]. There, it is shown that if the nesting condition holds for the parent-child pair (ψ_1, ψ_2), then $0 \leq \tau(\psi_1) \leq \tau(\psi_2)$ (as we deal only with HACs with binary structures, which are fully determined by its generator, we use as the domain of τ the set Ψ instead of the usually used set of all 2-copulas). We generalize this statement, using our notation, as follows.

Algorithm 2. The HAC Structure Determination Based on τ

Input:
  1) $\mathcal{I} = \{1, ..., d\}$,
  2) $(U_1, ..., U_d) \sim C$,
  3) τ^g ... an aggregated Kendall's tau,
  4) $z_k = u_k, \mathcal{U}_C(z_k) = \mathcal{U}_C(u_k) = \{k\}, k = 1, ..., d$

The structure determination:
for $k = 1, ..., d - 1$ do
  1. $(i, j) := \underset{i^* < j^*, i^* \in \mathcal{I}, j^* \in \mathcal{I}}{\text{argmax}} \ \tau^g(\mathbf{U}_{\mathcal{U}_C(z_{i^*}) \mathcal{U}_C(z_{j^*})})$
  2. $\mathcal{U}_C(z_{d+k}) := \mathcal{U}_C(z_i) \cup \mathcal{U}_C(z_j)$
  3. $\mathcal{I} := \mathcal{I} \cup \{d + k\} \backslash \{i, j\}$
end for

Output:
  $\mathcal{U}_C(z_{d+k}), k = 1, ..., d - 1$

Proposition 2. *Let C be a d-HAC with the structure t and the generators $\psi_1, ..., \psi_{d-1}$, where each parent-child pair satisfy the nesting condition. Then $\tau(\psi_i) \leq \tau(\psi_j)$, where $\psi_j \in \mathcal{D}_n(\psi_i)$, holds for each $\psi_i, i = 1, ..., d - 1$.*

Proof. As $\psi_j \in \mathcal{D}_n(\psi_i)$, there exists a unique sequence $\psi_{k_1}, ..., \psi_{k_l}$, where $1 \leq k_m \leq d - 1, m = 1, ..., l, l \leq d - 1, \psi_{k_1} = \psi_i, \psi_{k_l} = \psi_j$ and $\psi_{k-1} = \mathcal{P}(\psi_k)$ for $k = 2, ..., l$. Applying the above mentioned remark for each pair $(\psi_{k-1}, \psi_k), k = 2, ..., l$, we get $\tau(\psi_{k_1}) \leq ... \leq \tau(\psi_{k_l})$. $\square$

Thus, having a branch from t, all its nodes are uniquely ordered according to their value of τ assuming unequal values of τ for all parent-child pairs. This provides an alternative algorithm for the HAC structure determination. We have to assign the generators with the highest values of τ to the lowest levels of the branches in the structure and ascending to higher levels we assign the generators with lower values of τ.

Remark 2. $\tau(\psi_k) = \tau^g(\mathbf{U}_{\mathcal{U}_C(\mathcal{H}_l(\psi_k)) \mathcal{U}_C(\mathcal{H}_r(\psi_k))})$ for a d-HAC C and for each $k = 1, ..., d - 1$. This is because the bivariate margins $C_{ij}, (i, j) \in \mathcal{U}_C(\mathcal{H}_l(\psi_k)) \times \mathcal{U}_C(\mathcal{H}_r(\psi_k))$ of C are all equal and $g(u, ..., u) = u$ for all $u \in \mathbb{I}$. Thus $\tau(\psi_k)$ depends only on the population version of Kendall correlation matrix.

Computing $\tau(\psi_k), k = 1, ..., d - 1$ using Remark 2 and following Proposition 2 leads to the alternative algorithm for HAC structure determination. The algorithm is summarized in Algorithm 2 and can be used for arbitrary $d > 2$ (see [8] for more details including an example for $d = 4$). It returns the sets $\mathcal{U}_C(z_{d+k+1})$ corresponding to the sets $\mathcal{U}_C(\psi_k), k = 1, ..., d - 1$. Passing them to Algorithm 1, we avoid their computation from Definition 7 and we get the requested d-HAC structure without a need of knowing the forms of the bivariate margins. Assuming a family for each ψ_k, $\theta - \tau$ relationship for the given family can be used to obtain the parameters, i.e., $\theta_k = \tau_\theta^{-1}(\tau(\psi_k))$, $k = 1, ..., d - 1$, where τ_θ^{-1} denotes the $\theta - \tau$ relationship, e.g., for Clayton family $\tau_\theta^{-1}(\tau) = 2\tau/(1 - \tau)$. Hence we get together with the structure the whole copula.

3.2   HAC Estimation

Using τ_n^g instead of τ^g, we can easily derive the empirical version of the structure determination process represented by Algorithms 1, 2. In this way, we base the structure determination only on the values of the pairwise τ. This is an essential property of our approach. Using the $\theta - \tau$ relationship established through (7) for some selected Archimedean family, whole HAC, including its structure and its parameters, can be estimated just from Kendall correlation matrix computed for the realizations of $(U_1, ..., U_d)$, assuming all the generators to be from a selected Archimedean family.

The proposed empirical approach is summarized in Algorithm 3. The Kendall correlation matrix (τ_{ij}^n) is computed for the realizations of the pairs (U_i, U_j), $1 \leq i < j \leq d$ using (8). The algorithm returns the parameters $\hat{\theta}_1, ..., \hat{\theta}_{d-1}$ of the estimate $\hat{C}$ and the sets $\mathcal{U}_{\hat{C}}(z_{d+k})$ corresponding to the sets $\mathcal{U}_{\hat{C}}(\psi_k)$, $k = 1, ..., d - 1$. Passing the sets to Algorithm 1 we get the requested $\hat{C}$ structure.

If g is set to be the average function, and as $\tau_n^{avg}(\theta_k) = g((\tau_{ij}^n)_{(\tilde{i},\tilde{j}) \in \mathcal{U}_{\hat{C}}(z_i) \times \mathcal{U}_{\hat{C}}(z_j)})$, where i, j are the indices found in step 1. of the algorithm, then $\tau_n^{avg}(\theta_k)$ is an unbiased estimator of $\tau(\theta_k)$, and thus the structure determination is based only on unbiased estimates, what is another favourable property of the proposed method.

Due to the nesting condition, the parameter $\hat{\theta}_k$ is trimmed in step 3. in order to obtain the resulting estimate as a proper d-HAC. Note that if we allow the generators to be from different Archimedean families, the task is much more complex, and we do not concern it in the paper due to space limitations and refer the reader to [9, 10].

Note that the proposed algorithm is just a variation of another famous algorithm, namely the algorithm for agglomerative hierarchical clustering (AHC). Defining $\delta_{ij} = 1 - \tau_{ij}^n$ we establish δ_{ij} to be a standardly used distance between the random variables U_i, U_j. Setting g to be the aggregation function min, avg or max, the algorithm results in (due to $\delta_{ij} = 1 - \tau_{ij}^n$) complete-linkage, average-linkage or single-linkage AHC, respectively. As most of statistical softwares include an implementation of AHC, the implementation of the proposed algorithm is straightforward. Moreover, adding the dendrogram obtained during AHC makes the result even more understandable to the user.

Algorithm 3. The HAC Estimation

Input:

1) (τ_{ij}^n) {...Kendall correlations matrix},
2) g {...an aggregation function},
3) $\mathcal{I} = \{1, ..., d\}$,
4) $z_i = u_i, i = 1, ..., d$,
5) Archimedean family based on generator ψ and corresponding τ_θ^{-1}

Estimation:

for $k = 1, ..., d - 1$ do

   1. $(i, j) := \underset{i < j, i \in \mathcal{I}, j \in \mathcal{I}}{\mathrm{argmax}} \; g((\tau_{\tilde{i}\tilde{j}}^n)_{(\tilde{i},\tilde{j}) \in \mathcal{U}_{\hat{C}}(z_{\tilde{i}}) \times \mathcal{U}_{\hat{C}}(z_{\tilde{j}})})$

   2. $\hat{\theta}_k := \tau_\theta^{-1} \big(g((\tau_{\tilde{i}\tilde{j}}^n)_{(\tilde{i},\tilde{j}) \in \mathcal{U}_{\hat{C}}(z_i) \times \mathcal{U}_{\hat{C}}(z_j)}) \big)$

   3. $\hat{\theta}_k := \min(\hat{\theta}_k, \hat{\theta}_i, \hat{\theta}_j)$

   4. $z_{d+k} := C(u_i, u_j; \psi)$

   5. $\mathcal{U}_{\hat{C}}(z_{d+k}) := \mathcal{U}_{\hat{C}}(z_i) \cup \mathcal{U}_{\hat{C}}(z_j)$

   6. $\mathcal{I} := \mathcal{I} \cup \{d + k\} \backslash \{i, j\}$

end for

Output:

$\hat{\theta}_k, \mathcal{U}_{\hat{C}}(k), k = 1, ..., d - 1$

---

4   Experiments

We performed a large number of different experiments on simulated data involving different data dimensions, HAC structures, generators and parameters. Due to space limitations we present only one experiment, where we compare the proposed method with the other previously mentioned methods on simulated data for $d = 5, 6, 7, 9$. We simulate 100 samples of size 500, i.e., 500 rows and d columns of simulated data for each sample, according to [11] for 4 copula models based on the Clayton generator. The first considered model is $((12)_{\frac{3}{4}}(3(45)_{\frac{4}{4}})_{\frac{3}{4}})_{\frac{2}{4}}$. The natural numbers in the model notation (as in [23]) are the indexes of the copula variables, i.e., $1, ..., 5$, the parentheses correspond to each $\mathcal{U}_C(\cdot)$ of individual copulas, i.e., $\mathcal{U}_C(\psi_1) = \{1, 2\}, \mathcal{U}_C(\psi_2) = \{4, 5\}, \mathcal{U}_C(\psi_3) = \{3, 4, 5\}, \mathcal{U}_C(\psi_4) = \{1, 2, 3, 4, 5\}$, and the subscripts are the model parameters, i.e., $(\theta_1, \theta_2, \theta_3, \theta_4) = (\frac{2}{4}, \frac{3}{4}, \frac{3}{4}, \frac{4}{4})$. Note that the indexes of the 4 generators could be permuted arbitrarily and the particular selection of their ordering serves just for better illustration. The other 3 models are given with analogous notation as $(1((23)_{\frac{5}{4}}(4(56)_{\frac{6}{4}})_{\frac{5}{4}})_{\frac{4}{4}})_{\frac{2}{4}}$, $(1((23)_{\frac{5}{4}}(4(5(67)_{\frac{7}{4}})_{\frac{6}{4}})_{\frac{5}{4}})_{\frac{4}{4}})_{\frac{2}{4}}$ and $((1(2(34)_{\frac{5}{4}})_{\frac{4}{4}})_{\frac{3}{4}} \; ((56)_{\frac{4}{4}}(7(89)_{\frac{5}{4}})_{\frac{4}{4}})_{\frac{3}{4}})_{\frac{2}{4}}$. The smallest difference between the parameters is set to $\frac{1}{4}$. As we revealed, while we experimented with different parameterizations, a larger difference in the parameters could hide the impact of the bias of the concerned methods on the structure determination, and the results obtained by different methods can be similar in some of those cases. Setting it to $\frac{1}{4}$ fully reveals the impact of the bias and clearly shows the difference among the methods.

The results for each model are shown in Table 1 and are separated by the double lines. As we are interested in binary copulas, we choose for the comparison the methods θ_{binary}, θ_{RML}, τ_{binary}, which return binary copula structures as their results. The first 2 methods are based on ML estimation technique, whereas the third method is based on the $\theta - \tau$ relationship. To get the results we used their R implementation described in [25]. Our method, implemented in Matlab, is denoted as τ_{binary}^{avg}, i.e., the involved function g is selected to be the avg function due to the previously mentioned reasons. As θ_{RML} failed in most cases for $d \geq 7$, the results for the method for those dimensions are not presented.

Firstly, we assess the ability of the methods to determine the true copula structure correctly. This can be seen from the third and the fourth column. The third column shows 3 the most frequent structures obtained by the method (if the true structure was not the one of the 3 most frequent structures, then we show the 2 most frequent structures and the true structure) with average parameter values. The true structure is emphasized by bold text. The fourth column shows the frequency of the structures. τ_{binary}^{avg} clearly dominates in all four cases ($d = 5, 6, 7, 9$). The other methods show very poor ability to detect the correct structure, especially for $d \geq 7$, where, e.g., θ_{binary} did not return the correct structure for any among all 100 samples used.

Next, we assess the methods by means of goodness-of-fit. The results can be seen in the fifth and the sixth column, where the statistics $S^{(K)}, S^{(C)}$ (described in [6]) are computed on all bivariate margins and their maximum (the $S^{(K)}, S^{(C)}$ for the worst fitted bivariate margin) is shown. τ_{binary}^{avg} also dominates in all four cases. θ_{RML} shows also good results, but its time consumption for comparable results is considerably higher. The remaining methods show poor results, what is additionally illustrated by the discrepancy between the estimated average parameter values shown in the third column and the true parameter values.

The next two columns show the average Frobenius norm of the difference between the Kendall correlation matrix for the true model and the Kendall correlation matrix for the estimated model and the average Frobenius norm of the difference between the matrix of lower tail coefficients (cf. [22]) for the true model and the matrix of lower tail coefficients for the estimated model (as in [23]). The comparison results are similar to the goodness-of-fit comparison. θ_{RML} shows slightly better results than τ_{binary}^{avg} and the remaining methods show significant discrepancy between the theoretical and the empirical quantities.

The last column shows the average computing times needed for a single data sample. τ_{binary}^{avg} is slightly better that the binary methods θ_{binary}, τ_{binary}, whereas θ_{RML} shows significantly higher time consumption, particularly for $d = 6$.

Table 1. The results for the copula models for $d = 5, 6, 7, 9$. The columns contain method names; the 3 most frequent estimated structures with average parameter values; goodness-of-fit statistics $S^{(K)}$, $S^{(C)}$ (described in [6]); the Frobenius norms of the differences between estimated and true Kendall matrices and lower tail indices; the estimation time in s. The values in parenthesis are the corresponding standard deviations

d	Method	Structure(s)	%	$S_n^{(K)}$	$S_n^{(C)}$	Avg. error in τ	λ_L	time (in s)
5	θ_{binary}	(3((12)0.77(45)1.01)0.76)0.24	79	2.1478 (0.5)	0.7206 (0.3)	0.3101 (0.025)	0.6306 (0.04)	0.1517 (0.04)
		((12)0.69(3(45)1.01)0.72)0.68	18	0.4897 (0.21)	0.4089 (0.21)	0.1426 (0.024)	0.2893 (0.05)	
		((12)0.61(4(35)0.85)0.71)0.61	2	0.5546 (0.22)	0.2843 (0.04)	0.1208 (0.02)	0.2346 (0.04)	
	θ_{RML}	((12)0.71(3(45)1.00)0.77)0.54	52	0.2102 (0.08)	0.2426 (0.11)	0.0511 (0.02)	0.1016 (0.05)	0.3616 (0.06)
		((45)1.01(3(12)0.79)0.72)0.62	43	0.4959 (0.28)	0.3290 (0.14)	0.1339 (0.018)	0.2704 (0.03)	
		((12)0.80(4(35)0.93)0.81)0.52	3	0.3090 (0.12)	0.2992 (0.09)	0.0973 (0.026)	0.1743 (0.05)	
	τ_{binary}	((12)0.81(3(45)1.04)0.93)0.89	46	1.2082 (0.3)	0.5333 (0.22)	0.2751 (0.06)	0.5234 (0.11)	0.3055 (0.018)
		(1(2(3(45)1.02)0.92)0.78)0.85	23	0.9928 (0.29)	0.4469 (0.18)	0.2332 (0.07)	0.4494 (0.12)	
		2(1(3(45)0.99)0.92)0.79)0.88	21	0.9659 (0.2)	0.4022 (0.16)	0.2443 (0.04)	0.4709 (0.08)	
	τ_{binary}^{avg}	((12)0.76(3(45)1.01)0.75)0.49	92	0.1719 (0.06)	0.2372 (0.1)	0.0627 (0.028)	0.1208 (0.06)	0.1631 (0.0007)
		((12)0.68(5(34)0.95)0.87)0.52	3	0.1826 (0.05)	0.2141 (0.06)	0.0778 (0.016)	0.1362 (0.028)	
		((12)0.74(4(35)0.93)0.85)0.50	3	0.2106 (0.05)	0.3107 (0.14)	0.0829 (0.011)	0.1513 (0.019)	
6	θ_{binary}	1(4((23)1.28(56)1.53)1.28)0.55)0.18	49	2.1014 (0.4)	0.8661 (0.34)	0.4078 (0.03)	0.7367 (0.05)	0.2674 (0.08)
		1((23)1.16(4(56)1.53)1.24)1.15)0.21	25	1.1039 (0.3)	0.4969 (0.27)	0.2507 (0.04)	0.4839 (0.05)	
		(14)0.56((23)1.24(56)1.49)1.24)0.56	22	1.7606 (0.4)	0.7776 (0.27)	0.3101 (0.018)	0.5375 (0.03)	
	θ_{RML}	1((23)1.19(4(56)1.53)1.28)1.00)0.50	48	0.1965 (0.07)	0.2945 (0.12)	0.0506 (0.019)	0.0884 (0.04)	3.4299 (2.13)
		1((56)1.52(4(23)1.29)1.21)1.08)0.51	44	0.3149 (0.13)	0.3055 (0.14)	0.1026 (0.02)	0.1617 (0.04)	
		1(3(4(56)1.68)1.40)1.12)1.04)0.56	2	0.2016 (0.08)	0.3781 (0.05)	0.1006 (0.04)	0.1601 (0.08)	
	τ_{binary}	1(2(3(4(56)1.56)1.49)1.39)0.70	40	0.6187 (0.16)	0.4378 (0.16)	0.2478 (0.06)	0.3970 (0.1)	0.4983 (0.02)
		1(3(4(56)1.53)1.48)1.41)1.40)0.71	32	0.6652 (0.17)	0.4294 (0.15)	0.2541 (0.05)	0.4073 (0.07)	
		1((23)1.37(4(56)1.57)1.52)1.36)0.73	11	0.6411 (0.13)	0.4015 (0.13)	0.2474 (0.06)	0.4077 (0.1)	
	τ_{binary}^{avg}	1((23)1.27(4(56)1.54)1.25)1.00)0.51	84	0.1753 (0.06)	0.2749 (0.19)	0.0745 (0.029)	0.1263 (0.05)	0.2470 (0.06)
		1((23)1.21(5(46)1.49)1.36)1.04)0.50	4	0.1535 (0.05)	0.3090 (0.12)	0.1017 (0.04)	0.1640 (0.08)	
		1(3(2(4(56)1.62)1.38)1.20)1.06)0.54	3	0.1657 (0.01)	0.1743 (0.05)	0.1174 (0.029)	0.1738 (0.04)	
7	θ_{binary}	(14)0.52((23)1.24(5(67)1.74)1.41)1.24)0.52	48	2.3349 (0.5)	1.0978 (0.6)	0.3810 (0.03)	0.6637 (0.06)	0.3827 (0.03)
		1(4((23)1.25(5(67)1.77)1.43)1.24)0.48)0.14	18	2.7023 (0.4)	1.2764 (0.6)	0.5236 (0.04)	0.9294 (0.06)	
		1((45)1.17((23)1.35(67)1.77)1.34)1.16)0.19	16	1.3054 (0.4)	0.5234 (0.2)	0.3388 (0.03)	0.6068 (0.03)	
	τ_{binary}	1(2(3(4(5(67)1.79)1.73)1.63)1.46)1.45)0.70	45	0.8215 (0.19)	0.4797 (0.17)	0.3173 (0.07)	0.4776 (0.11)	0.7435 (0.021)
		1(3(2(4(5(67)1.81)1.76)1.66)1.47)1.46)0.72	32	0.8420 (0.2)	0.5341 (0.19)	0.3333 (0.07)	0.5047 (0.1)	
		1((23)1.48(4(5(67)1.85)1.67)1.48)0.67	3	0.8633 (0.1)	0.4852 (0.11)	0.3373 (0.14)	0.5019 (0.2)	
	τ_{binary}^{avg}	1((23)1.27(4(5(67)1.80)1.52)1.25)1.00)0.50	77	0.1877 (0.05)	0.3065 (0.15)	0.0895 (0.04)	0.1472 (0.07)	0.3255 (0.07)
		1((23)1.26(4(7(56)1.65)1.55)1.28)1.02)0.49	6	0.1854 (0.05)	0.3338 (0.2)	0.0902 (0.018)	0.1394 (0.04)	
		1((23)1.25(4(6(57)1.55)1.42)1.25)1.02)0.50	5	0.2094 (0.08)	0.4709 (0.26)	0.0951 (0.027)	0.1514 (0.06)	
9	θ_{binary}	(17)0.51((2(34)1.25)0.90((56)1.02(89)1.26)1.02)0.89)0.51	58	1.6487 (0.4)	0.7410 (0.26)	0.4771 (0.04)	0.9144 (0.08)	0.7862 (0.06)
		1((2(34)1.25)0.86((56)0.96(7(89)1.33)1.01)0.96)0.86)0.13	11	3.5263 (0.6)	0.9699 (0.3)	0.6364 (0.03)	1.1800 (0.05)	
		1((56)0.91((2(34)1.32)0.96(7(89)1.30)0.99)0.94)0.72)0.13	10	4.1839 (0.5)	1.2621 (0.4)	0.6296 (0.024)	1.1628 (0.05)	
	τ_{binary}	(1(2(34)1.34)1.22)1.06(6(5(7(89)1.28)1.06)1.06)1.11	15	2.6079 (0.4)	0.9986 (0.26)	0.7463 (0.09)	1.3381 (0.13)	1.4654 (0.02)
		(1(2(34)1.31)1.24)1.12(5(6(7(89)1.29)1.23)1.12)1.11)1.12	13	2.3948 (0.3)	0.9770 (0.28)	0.7620 (0.11)	1.3583 (0.15)	
		(1(2(34)1.21)1.17)1.04((56)1.06(7(89)1.13)1.10)1.05	4	2.3784 (0.29)	0.8742 (0.3)	0.6753 (0.15)	1.2305 (0.23)	
	τ_{binary}^{avg}	(1(2(34)1.27)0.99(6(7(89)1.28)1.01)0.75)0.50	81	0.2491 (0.07)	0.3328 (0.12)	0.1134 (0.04)	0.2096 (0.09)	0.4851 (0.0019)
		(1(3(24)1.17)1.07)0.72((56)0.97(7(89)1.27)0.99)0.76)0.49	4	0.2264 (0.06)	0.1860 (0.04)	0.1264 (0.06)	0.2400 (0.14)	
		(1(2(34)1.41)1.07)0.84((56)1.05(9(78)1.26)1.12)0.83)0.56	3	0.1921 (0.03)	0.3401 (0.21)	0.1444 (0.022)	0.2576 (0.04)	

5   Conclusion

Copulas are a feasible tool for the modeling of complex patters. A popular alternative to Gaussian copulas, the hierarchical Archimedean copulas, are convenient copula models even in high dimensions due to their flexibility and rather limited number of parameters. Despite their popularity, a general approach for their estimation has been addressed only in one recently published paper [23], which proposes several methods for the estimation task.

We propose an alternative approach to structure determination and estimation of a hierarchical Archimedean copula, which combines the advantages and avoids the disadvantages of the previously mentioned methods in the terms of the correctly determined structures ratio, the goodness-of-fit of the estimates, and computation time. This is confirmed in the experiments on simulated data performed for different dimensions and copula models. The proposed method should be preferred to the other mentioned methods and is particularly attractive in applications, where a good approximation and computational efficiency are both crucial issues.

Acknowledgment. The research reported in this paper has been supported by the Czech Science Foundation (GA ČR) grant 13-17187S.

References

1. Berg, D.: Copula goodness-of-fit testing: an overview and power comparison. Eur. J. Finance **15**(7–8), 675–701 (2009)
2. Chen, X., Fan, Y., Patton, A.J.: Simple tests for models of dependence between multiple financial time series, with applications to us equity returns and exchange rates. Discussion paper 483, Financial Markets Group, London School of Economics (2004)
3. Cuvelier, E., Noirhomme-Fraitur, M.: Clayton copula and mixture decomposition. In: Janssen, J., Lenca, P. (eds.) Applied Stochastic Models and Data Analysis, ASMDA'05, Brest (2005)
4. Genest, C., Favre, A.: Everything you always wanted to know about copula modeling but were afraid to ask. Hydrol. Eng. **12**, 347–368 (2007)
5. Genest, C., Rivest, L.-P.: Statistical inference procedures for bivariate Archimedean copulas. J. Am. Stat. Assoc. **88**(423), 1034–1043 (1993)
6. Genest, C., Rémillard, B., Beaudoin, D.: Goodness-of-fit tests for copulas: a review and a power study. Insur. Math. Econ. **44**(2), 199–213 (2009)
7. González-Fernández, Y., Soto, M.: Copulaedas: an R package for estimation of distribution algorithms based on copulas. CoRR, abs/1209.5429 (2012)
8. Górecki, J., Holeňa, M.: An alternative approach to the structure determination of hierarchical Archimedean copulas. 31st International Conference on Mathematical Methods in Economics 2013, pp. 201–206, Jihlava (2013)
9. Hofert, M.: Construction and sampling of nested Archimedean copulas. In: Jaworski, P., Durante, F., Hardle, W.K., Rychlik, T. (eds.) Copula Theory and Its Applications. Lecture Notes in Statistics, vol. 198, pp. 147–160. Springer, Berlin (2010)

10. Hofert, M.: Sampling nested Archimedean copulas with applications to CDO pricing. Ph.D. thesis, Ulm University (2010)
11. Hofert, M.: Efficiently sampling nested Archimedean copulas. Comput. Stat. Data Anal. **55**(1), 57–70 (2011)
12. Hofert, M., Mächler, M., McNeil, A.J.: Estimators for Archimedean copulas in high dimensions: a comparison. arXiv preprint arXiv:1207.1708 (2012)
13. Kao, S.-C., Ganguly, A.R., Steinhaeuser, K.: Motivating complex dependence structures in data mining: a case study with anomaly detection in climate. International Conference on Data Mining Workshops, vol. 0, pp. 223–230 (2009)
14. Kao, S.-C., Govindaraju, R.S.: Trivariate statistical analysis of extreme rainfall events via Plackett family of copulas. Water Resour. Res. **44**, 1–19 (2008)
15. Kojadinovic, I.: Hierarchical clustering of continuous variables based on the empirical copula process and permutation linkages. Comput. Stat. Data Anal. **54**(1), 90–108 (2010)
16. Kojadinovic, I., Yan, J.: Modeling multivariate distributions with continuous margins using the copula R package. J. Stat. Softw. **34**(9), 1–20 (2010)
17. Kuhn, G., Khan, S., Ganguly, A.R., Branstetter, M.L.: Geospatial-temporal dependence among weekly precipitation extremes with applications to observations and climate model simulations in South America. Adv. Water Resour. **30**(12), 2401–2423 (2007)
18. Lascio, F., Giannerini, S.: A copula-based algorithm for discovering patterns of dependent observations. J. Classif. **29**, 50–75 (2012)
19. Maity, R., Kumar, D.N.: Probabilistic prediction of hydroclimatic variables with nonparametric quantification of uncertainty. J. Geophys. Res. **113**, D14105 (2008)
20. McNeil, A.J.: Sampling nested Archimedean copulas. J. Stat. Comput. Simul. **78**(6), 567–581 (2008)
21. McNeil, A.J., Nešlehová, J.: Multivariate Archimedean copulas, d-monotone functions and l1-norm symmetric distributions. Ann. Stat. **37**, 3059–3097 (2009)
22. Nelsen, R.: An Introduction to Copulas, 2nd edn. Springer, New York (2006)
23. Okhrin, O., Okhrin, Y., Schmid, W.: On the structure and estimation of hierarchical Archimedean copulas. J. Econom. **173**(2), 189–204 (2013)
24. Okhrin, O., Okhrin, Y., Schmid, W.: Properties of hierarchical Archimedean copulas. Stat. Risk Model. **30**(1), 21–54 (2013)
25. Okhrin, O., Ristig, A.: Hierarchical Archimedean copulae: the HAC package. Discussion paper 2012, 036, CRC 649, Economic Risk (2012)
26. Rey, M., Roth, V.: Copula mixture model for dependency-seeking clustering. In: Proceedings of the 29th International Conference on Machine Learning (ICML 2012), Edinburgh, Scotland, UK (2012)
27. Savu, C., Trede, M.: Goodness-of-fit tests for parametric families of Archimedean copulas. Quant. Finance **8**(2), 109–116 (2008)
28. Savu, C., Trede, M.: Hierarchies of Archimedean copulas. Quant. Finance **10**, 295–304 (2010)
29. Sklar, A.: Fonctions de répartition a n dimensions et leurs marges. Publ. Inst. Stat. Univ. Paris **8**, 229–231 (1959)
30. Wang, L., Guo, X., Zeng, J., Hong, Y.: Copula estimation of distribution algorithms based on exchangeable Archimedean copula. Int. J. Comput. Appl. Technol. **43**, 13–20 (2012)
31. Yuan, A., Chen, G., Zhou, Z.-C., Bonney, G., Rotimi, C.: Gene copy number analysis for family data using semiparametric copula model. Bioinform. Biol. Insights **2**, 343–355 (2008)

ReliefF for Hierarchical Multi-label Classification

Ivica Slavkov[1][(✉)], Jana Karcheska[2], Dragi Kocev[1], Slobodan Kalajdziski[2], and Sašo Džeroski[1]

[1] Department of Knowledge Technologies, Jožef Stefan Institute, Ljubljana, Slovenia
{ivica.slavkov,dragi.kocev,saso}@ijs.si
[2] Faculty of Computer Science and Engineering,
Ss. Cyril and Methodius University, Skopje, Macedonia
j.karcheska@gmail.com,
slobodan.kalajdziski@inki.ukim.mk

Abstract. In machine learning, the data available for analysis is becoming more complex both in terms of high-dimensionality and the way it is structured. This emphasises the need for developing machine learning algorithms that are able to tackle both the high-dimensionality and the complex structure of the data. Our work in this paper, focuses on extending a feature ranking algorithm that can be used as a filter method for a specific type of structured data. More specifically, we adapt the RReliefF algorithm for regression, for the task of hierarchical multi-label classification (HMC). We evaluate this algorithm experimentally in a filter-like setting by employing ensembles of predictive clustering trees for HMC as a classifier. In the experimental evaluation, we consider datasets from two prominent domains for HMC - functional genomics and image annotation. The results show that HMC-ReliefF can identify the relevant features present in the data and produces a ranking where they are placed among the top ranked ones.

Keywords: Feature selection · Feature ranking · Feature relevance · Structured data · Hierarchical multi-label classification · Multi-label classification · ReliefF

1 Introduction

The current trend in machine learning is that the data available for analysis is becoming increasingly more complex. The complexity arises both from the data being high-dimensional and from the data being more structured. On one hand, high-dimensional data presents specific challenges for many machine learning algorithms, especially with the stability of the produced results [11]. On the other, mining complex data and extracting knowledge from it has been identified as one of the most challenging problems in machine learning [6,17].

Various feature selection methods exist for dealing with the high-dimensionality of the data. They usually precede the induction of predictive models and can be

A. Appice et al. (Eds.): NFMCP 2013, LNAI 8399, pp. 148–161, 2014.
DOI: 10.1007/978-3-319-08407-7_10, © Springer International Publishing Switzerland 2014

classified as filter, wrapper and embedded methods [10]. Filter methods [3] are the simplest ones and they usually involve a feature ranking algorithm that produces a list of relevant features. Wrapper methods [15] rely on classification algorithms to perform feature selection and are computationally expensive. Embedded methods [10] are basically classification algorithms that have the feature selection embedded in the model induction phase.

Learning in a supervised context, where the target is structured, has also attracted much attention. Several algorithms that were previously employed only for classification or regression purposes, have been extended to also work with structured targets. These include decision trees for hierarchical targets [23], SVMs for multi-label and hierarchical multi-label problems [9], as well as tree ensembles that can be additionally employed for vectors of multiple targets [14].

Our work in this paper focuses on tackling the feature selection problem in the context of structured targets. We consider this a relevant problem in machine learning that relates to both of the previously discussed trends. So far, structured prediction has not been extensively researched in the context of feature ranking methods and we consider this a novel and interesting line of work to pursue.

More specifically, we focus on the ReliefF [20] algorithm for feature ranking. This algorithm is an intuitive, instance based algorithm and its theoretical properties have been extensively explored [20]. We extend ReliefF for a specific type of structured prediction problems, namely those from the Hierarchical Multi-Label Classification (HMC) domain [21]. The target that is predicted for these problems is defined with a hierarchy of classes and each instance in the dataset can be labelled with more than one class at a time. By definition, when an instance is labelled with one class it is also labelled with all of its parent classes according to the given hierarchy.

In practice, this type of problems appear in different domains, for example in biology for the task of gene function prediction or in image retrieval for the task of image annotation. For the task of gene function prediction, each gene can be annotated by multiple functions and the functions are organised into a tree-shaped hierarchy or a directed acyclic graph such as the Gene Ontology [2]. Thus, predicting the function of a gene from certain gene properties would have to take into account the multi-label annotation of each gene and also the hierarchical connections of these labels.

In the remainder of this paper, we present the details of our work organised as follows. In Sect. 2, we define more formally the HMC setting and present the distance measures appropriate for this setting. Next, in Sect. 3, we discuss in depth the original RReliefF algorithm for regression and explain our HMC-ReliefF extension of the algorithm. We present our experimental evaluation of the proposed HMC-ReliefF algorithm in Sect. 4. Finally, in Sect. 5, we present our conclusions and discuss directions of possible further work.

2   Hierarchical Multi-label Classification

In our work we extend the ReliefF algorithm for the task of hierarchical multi-label classification (HMC). Hierarchical classification is a specific type of a classification task in which the classes are organised in a hierarchy. An example that belongs to a given class automatically belongs to all its super-classes (this is known as the *hierarchy constraint*). Furthermore, if an example can belong simultaneously to multiple classes that can follow multiple paths from the root class, then the task is called hierarchical multi-label classification (HMC) [21,23].

We formally define the hierarchical multi-label classification setting as follows:

- A description space X that consists of tuples of values of primitive data types (discrete or continuous), i.e., $\forall X_i \in X, X_i = (x_{i_1}, x_{i_2}, ..., x_{i_D})$, where D is the size of the tuple (or number of descriptive variables),
- a target space S, defined with a class hierarchy $(C, \leq_h)$, where C is a set of classes and $\leq_h$ is a partial order (e.g., structured as a rooted tree) representing the superclass relationship ($\forall\ c_1, c_2 \in C : c_1 \leq_h c_2$ if and only if c_1 is a superclass of c_2),
- a set E, where each example is a pair of a tuple and a set, from the descriptive and target space respectively, and each set satisfies the hierarchy constraint, i.e., $E = \{(X_i, S_i)|X_i \in X, S_i \subseteq C, c \in S_i \Rightarrow \forall c' \leq_h c : c' \in S_i, 1 \leq i \leq N\}$ and N is the number of examples in E ($N = |E|$)

Two toy examples of classes organised in hierarchies can be seen in Fig. 1. The first hierarchy in Fig. 1(a) consists of five classes $\{c_1, c_2, c_3, c_{2.1}, c_{2.2}\}$, organised in a tree-like structure. The other hierarchy in Fig. 1(c), contains six classes ($c_1 - c_6$) and they are organised in a directed acyclic graph (DAG), where each class can have multiple parents.

Calculating the distance between two different instances of the target space S_1 and S_2, can be done in different ways. These distances include: a weighted Euclidean distance for HMC [23], Jaccard distance (also known as Union-intersection distance/score) [12], simGIC (Similarity for Graph Information Content) [18] and ImageCLEF (evaluation score of the ImageCLEF image annotation task) [5]. An experimental evaluation comparing these distances in the context of HMC [1] has shown that learning predictive models that use the different distances, does not produce statistically significant differences in predictive performance.

In our work, we chose to extend the RReliefF algorithm by using a weighted Euclidean distance for HMC [23]. With this weighted Euclidean distance, the hierarchical aspect is incorporated by relating the class weight with the depth of the class within the hierarchy. Extending RReliefF with this distance is the most straightforward choice, considering that the original algorithm uses the Euclidean distance for calculating the distance for the target variable.

Before calculating the distance between two instances of the hierarchy, they are first represented as a vector of binary values [23]. The vector is created by traversing the tree or DAG that is representing the hierarchy in pre-order and assigning a 1 or 0 sequentially in the vector for a present or absent label respectively. For example, consider an instance of the toy class hierarchy S_1, given in boldface in Fig. 1(b). This particular instance consists of three classes,

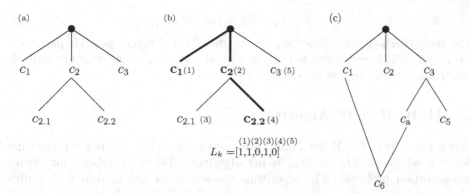

Fig. 1. Toy examples of hierarchies structured as a tree and a DAG. (a) Class label names contain information about the position in the hierarchy, e.g., $c_{2.1}$ is a subclass of c_2. (b) The set of classes $S_1 = \{c_1, c_2, c_{2.2}\}$, shown in bold in the hierarchy, represented as a vector (L_k). (c) A class hierarchy structured as a DAG. The class c_6 has two parents: c_1 and c_4.

namely $\{c_1, c_2, c_{2.2}\}$ and its corresponding vector representation would be $L_1 = [1, 1, 0, 1, 0]$.

If we additionally consider another instance S_2, labelled just with class $\{c_2\}$, with a vector representation $L_2 = [0, 1, 0, 0, 0]$, then the distance between S_1 and S_2 would be obtained by simply comparing the two binary vectors. In our HMC-ReliefF algorithm we use a weighted Euclidean distance measure given with the following equation:

$$d(L_1, L_2) = \sqrt{\sum_i w(c_i)(L_{1,i} - L_{2,i})^2}, \qquad (1)$$

The weighting function $w(c)$ allows for the hierarchical structure of the classes to be taken into account by making the value dependent on the depth of the hierarchy:

$$w(c) = w_0^{depth(c)}, 0 < w_0 < 1. \qquad (2)$$

This scheme ensures that the differences higher in the hierarchy have larger influence on the total distance.

For the specific case of comparing S_1 and S_2, the distance is calculated as follows:

$$d(S_1, S_2) = d([1, 1, 0, 1, 0], [0, 1, 0, 0, 0]) = \sqrt{w_0 + w_0^2}.$$

where $w(c_1) = w_0$ and $w(c_3) = w_0^2$.

If the hierarchy is represented with a DAG, this scheme needs to be modified. In this case, more than one path from the root to a given class may exist and thus a node can have different depths. This problem is solved with the following recursive equation:

$$w(c) = w_0 \cdot avg(w(parent_j(c))). \tag{3}$$

By using this weighting function, the weight of the different possible parents is averaged. This is recommended [23] as a good way to take into account multiple inheritance which occurs in DAGs.

3   HMC-ReliefF Algorithm

Algorithms from the Relief family are instance-based methods for estimating feature relevance. The original Relief algorithm [13] is formulated for binary classification problems. The algorithm was extended [16] to deal with multi-class problems and the extension was named ReliefF. Later, it was also adapted for regression problems [19] and named RReliefF.

In general, the feature relevance value assigned by the Relief algorithm to a feature F is an approximation of the following difference of probabilities [16]:

$$W[F] = P(\text{diff. value of } F | \text{nearest inst. from diff. class}) - \tag{4}$$
$$P(\text{diff. value of } F | \text{nearest inst. from same class})$$

In the case of classification, the basic intuition behind the ReliefF algorithm is to estimate the relevance of a feature according to how well it distinguishes between neighbouring instances. If the feature has different values for neighbouring instances that are of different class (nearest miss), then it is awarded a higher relevance values. However, if the values of the class for the neighbouring instances are the same (nearest hit), then the relevance value is decreased.

Although the hierarchical multi-label setting is a classification one, extending the ReliefF algorithm is not a good idea. Namely, if we simply treat two instances annotated by different parts of the hierarchy in a simple hit/miss scenario, we would simply translate the HMC problem to a multi-class one, therefore ignoring both the hierarchical and the multi-label aspect. Having in mind that the definition of the HMC distance in Sect. 2 is actually weighted Euclidean, it is more suited to be included in the RReliefF algorithm, originally designed for regression.

In a regression setting, the target space is continuous and the concept of nearest hit/miss does not apply. Therefore, the feature relevance $W[F]$ is reformulated as the difference between the following probabilities:

$$W[F] = P(\text{diff. value of } F | \text{nearest inst. with diff. prediction}) - \tag{5}$$
$$P(\text{diff. value of } F | \text{nearest inst. with same prediction})$$

Additionally, if we introduce the following probabilities:

$$P_{diffF}(\text{diff. value of F} | \text{nearest instance})$$

and

$$P_{diffC}(\text{diff. prediction} | \text{nearest instance}),$$

as well as the conditional probability:

$P_{diffC|diffF}$(diff. prediction|diff. value of F and nearest instances).

Finally, by using the Bayes rule, we obtain:

$$W[F] = \frac{P_{diffC|diffF}P_{diffF}}{P_{diffC}} - \frac{(1 - P_{diffC|diffF})P_{diffF}}{1 - P_{diffC}} \qquad (6)$$

The details of the RReliefF algorithm are given in pseudocode form in Algorithm 1. The algorithm begins by selecting a random instance (R_i) and finding the k nearest instances I_j to it. From these instances, it then approximates the relevance $W[F]$ from Eq. 6 of each feature by calculating N_{dC}, $N_{dF}[F]$ and $N_{dC\&dF}[F]$, described in lines 6,8 and 9 of Algorithm 1. The estimations of these values is based on the distance calculation in the feature space, $diff(F, R_i, I_j)$, (lines 8 and 9) and in the target space, $diff(\tau(\cdot), R_i, I_j)$, (lines 6 and 9).

Algorithm 1. Pseudocode for the RReliefF algorithm, taken from [20].

Input: for each training instance a vector of feature values **x** and predicted value $\tau(\mathbf{x})$
Output: the vector W of estimations of the relevance of features
1: set all N_{dC}, $N_{dF}[F]$, $N_{dC\&dF}[F]$, $W[F]$ to 0
2: **for** $i = 1$ to m **do**
3:     randomly select an instance R_i
4:     select k instances I_j nearest to R_i
5:     **for** $j = 1$ to m **do**
6:         $N_{dC} = N_{dC} + diff(\tau(\cdot), R_i, I_j) \cdot d(i,j)$
7:         **for** $F = 1$ to f **do**
8:             $N_{dF}[F] = N_{dF}[F] + diff(F, R_i, I_j) \cdot d(i,j)$
9:             $N_{dC\&dF}[F] = N_{dC\&dF}[F] + diff(\tau(\cdot), R_i, I_j) \cdot diff(F, R_i, I_j) \cdot d(i,j)$
10:         **end for**
11:     **end for**
12: **end for**
13: **for** $F = 1$ to f **do**
14:     $W[F] = N_{dC\&dF}[F]/N_{dC} - (N_{dF}[F] - N_{dC\&dF}[F])/(m - N_{dC})$
15: **end for**

Our original purpose is to extend the RReliefF algorithm for hierarchical multi-label classification problems. Considering that the HMC refers to the target space, we extend the RReliefF algorithm by changing the way that $diff(\tau(\cdot), R_i, I_j)$, from lines 6 and 9, is calculated. From Sect. 2 and Eq. 1 we obtain:

$$diff(\tau(\cdot), R_i, I_j) = diff(S_i, S_j) = \sqrt{\sum_k w(c_k)(L_{i,k} - L_{j,k})^2} \qquad (7)$$

where S_i and S_j are the target descriptions of R_i and I_j correspondingly, while $L_{i,k}$ and $L_{j,k}$ are their binary representations. In this way, by changing the way the distance is calculated, the original RReliefF algorithm is extended to work for HMC problems and we name this extension HMC-ReliefF.

4   Experiments

Our experimental evaluation of the HMC-ReliefF is based on the intuition of what is the expected output of a good feature ranking algorithm. Namely, a good feature ranking algorithm would output the relevant features on top of the ranked list of features. A bad ranking algorithm would not necessarily be the one that gives an inverse ranking according to relevance, but the one that outputs a random ranking. In the random ranking, the distribution of the relevant features is expected to be uniform throughout the list.

Having this in mind, we employ a stepwise filter-like procedure [22] to evaluate our HMC-ReliefF algorithm. The idea is that starting from the ranked list of features, we construct classifiers for different numbers of top-k ranked features. If there are relevant features on top of the feature ranking, then we can construct a classifier that has a good predictive performance. If the ranking is random then the number of relevant features in the top-k ranked features is expected to be smaller.

Formally, if we have a feature ranking algorithm r that we use on a dataset $\mathscr{D}$, then the output would be a feature ranking $\mathbf{R}$, namely:

$$r(\mathscr{D}) \to \mathbf{R}.$$

The feature ranking $\mathbf{R}$ is defined as an ordered list of features F, more specifically:

$$\mathbf{R} = (F_{r1}, \ldots, F_{rj}, \ldots, F_{rk})$$

where:

$$rank(F_{r1}) \le \cdots \le rank(F_{rj}) \le \cdots \le rank(F_{rk})$$

If we assume that we can induce and evaluate a predictive model $\mathscr{M}(R_i, F_t)$, where $R_i \subseteq \mathbf{R}$ and F_t is a target feature, then our whole evaluation procedure can be described as in Algorithm 2.

Algorithm 2. Stepwise evaluation of the top-k ranked features

---

Input: Feature Ranking, $\mathbf{R} = \{F_{r1}, \ldots, F_{rn}\}$; Target Feature, F_t
Output: FFA Curve, FFA, where $|FFA| = n$
　$\mathbf{R}_S \Leftarrow \emptyset$
　for $k = 1$ to n **do**
　　$\mathbf{R}_S \Leftarrow \mathbf{R}_S \cup feature(\mathbf{R}, i)$
　　$FFA[i] = qual(\mathscr{M}(\mathbf{R}_S, F_t))$
　end for
　　return FFA

For each step k of the filtering, i.e., for each subset of top-k ranked feature subsets, we induce a classification model and evaluate its performance. This process of generating feature sets from the feature ranking is performed in a forward manner, by adding more and more of the top ranked features, which we name *forward feature addition* (FFA). At the end, we obtain a vector of model

Table 1. Properties of the datasets with hierarchical targets; N_{tr} is the number of instances in the training dataset, D/C is the number of descriptive attributes (discrete/continuous), $|\mathcal{H}|$ is the number of classes in the hierarchy, $\mathcal{H}_d$ is the maximal depth of the classes in the hierarchy, $\overline{\mathcal{L}}$ is the average number of labels per example, and $\overline{\mathcal{L}}_L$ is the average number of leaf labels per example. Note that the values for $\mathcal{H}_d$ are not always a natural number because the hierarchy has a form of a DAG and the maximal depth of a node is calculated as the average of the depths of its parents.

Domain	N_{tr}	$|D|/|C|$	$|\mathcal{H}|$	$\mathcal{H}_d$	$\overline{\mathcal{L}}$	$\overline{\mathcal{L}}_L$
Diatoms	1098	0/200	107	2.0	1.98	0.98
ImCLEF07D	10006	0/80	46	3.0	3.0	1.0
ImCLEF07A	10006	0/80	96	3.0	3.0	1.0
SCOP-GO	9843	0/2003	572	5.5	6.26	0.95
SCOP-FUN	3097	0/2003	250	4.0	3.41	0.95
Yeast-GO	2310	5588/342	133	6.33	5.63	0.64

quality estimates that we can plot as a curve, thus obtaining a *FFA curve* that we use to estimate the performance of the feature ranking algorithm. In order to say that the FFA curve of a certain feature ranking algorithm is better than that of a random ranking, the model quality estimates of the ranking must be larger than those of the models from the random ranking. Visually, this would mean that the FFA curve of the algorithm would be above the FFA curve of the random ranking.

4.1 Experimental Setup

In the HMC-ReliefF algorithm, given in Algorithm 1, there are two basic parameters that can be specified by users and which influence the relevance estimation. These are the number of random instances m that are chosen and the number of nearest neighbours k that are used to calculate the feature relevance values. Therefore, in our experiments, we decided to explore a reasonable set of values of these parameters in order to evaluate the algorithm performance.

For the number of random instances m, instead of considering an absolute number, we consider sampling a percentage of the datasets instance space, while for the number of nearest neighbours k we consider absolute values. More specifically, we consider the following parameters:

- $m = \{1\,\%, 5\,\%, 10\,\%, 20\,\%, 25\,\%\}$
- $k = \{5, 10, 25, 50\}$.

As a baseline for our comparisons, we use a set of 50 random rankings for each different dataset. For each of these rankings, we perform the previously described procedure in Sect. 4 and generate a separate FFA curve. For the random rankings, we average the results of the 50 individual FFA curves, thus generating an expected FFA curve for a given dataset.

As a predictive model which we induce and evaluate, we use random forests of so-called predictive clustering trees for hierarchical multi-label classification

(PCT-HMCs) [14,23]. The specific parameters that we used for the random forests of PCTs were 100 trees and a feature subset size of 10 % of the all features in the dataset. For estimating the PCT-HMCs performance, we use ten-fold cross validation.

In the HMC context, there are various error measures that can be considered. We use the area of a variant of a precision-recall curve, namely the Pooled Area Under the Precision-Recall Curve ($AU(\overline{PRC})$), details discussed in [23]. For this measure, the precision and recall are micro averaged for all classes from the hierarchy. In the datasets domains that we consider, the positive examples for a given class are only few as compared to the negative ones. The Precision-Recall evaluation of these algorithms is most suitable in this context, because we are more interested in correctly predicting the positive examples (i.e., that an example belongs to a given class), rather than correctly predicting negative instances.

For the experiments, we use datasets from two domains which have classes organised in a hierarchy. We use 6 datasets from 2 domains, more specifically: biology (*Yeast-GO* [4], *SCOP-GO* [4] and *SCOP-FUN* [4]) and image annotation/classification (*Diatoms* [8], *ImCLEF07D* [7] and *ImCLEF07A* [7]). The relevant properties that characterize each dataset are given in Table 1. Note that the Yeast-GO and the SCOP-GO datasets have a hierarchy organised as a DAG, while the remaining datasets have tree-shaped hierarchies. For more details on the datasets, we refer the reader to the referenced literature.

4.2   Results and Discussion

In this section, we present the results from our experimental evaluation. In Fig. 2, we give the FFA curves for the datasets from the image annotation domain, while in Fig. 3, we present the FFA curves for datasets from the functional genomics domain. The graphs on the left-hand side of Figs. 2 and 3 represent the FFA curves for a fixed value of m, while the value of k is varied. Correspondingly, the graphs on the right-hand side contain FFA curves for a fixed value of k, while the value of m is varied. The fixed values of m and k are chosen for the best FFA curves.

Overall, it can be observed that all of the FFA curves of the HMC-ReliefF algorithm are most of the time above the FFA curves of the random rankings. This means that at the top of the rankings produced by HMC-ReliefF, for different settings of m and k, relevant features can be found. It also means that this is not by chance, as the $AU(\overline{PRC})$ of the produced models is larger than the expected value of a random ranking. However, there are differences in the obtained curves for the different datasets, which we will discuss in detail.

We first consider the datasets from the image annotation domain, given in Fig. 2. It can be noticed that all of the FFA curves produced by HMC-ReliefF, are only slightly higher, i.e., are only slightly better, than the expected FFA curves of the random rankings. Also, there is no great variability of the FFA curves with respect to the different number of m and k. This is expected if we take into account this specific domain and the way the features are produced. Namely, most of the features are image descriptors, which are informative about

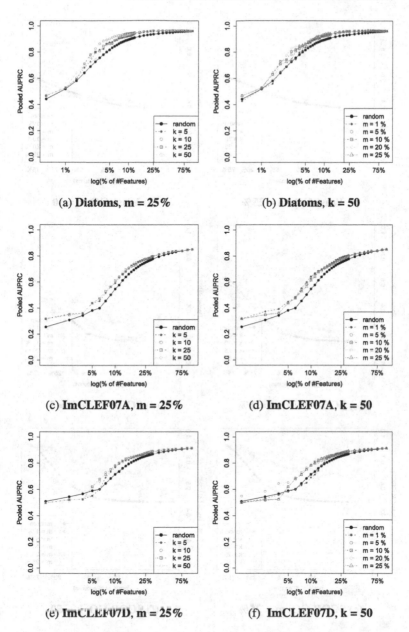

Fig. 2. Comparison of different FFA curves obtained by varying the number of m and k for datasets from the image annotation domain

the image and most of them are relevant. This can also be concluded if we observe just the expected FFA curve of the random rankings.

Next, if we consider the results from the functional genomics domain in Fig. 3, a more complex interpretation is necessary. First, the FFA curves of the

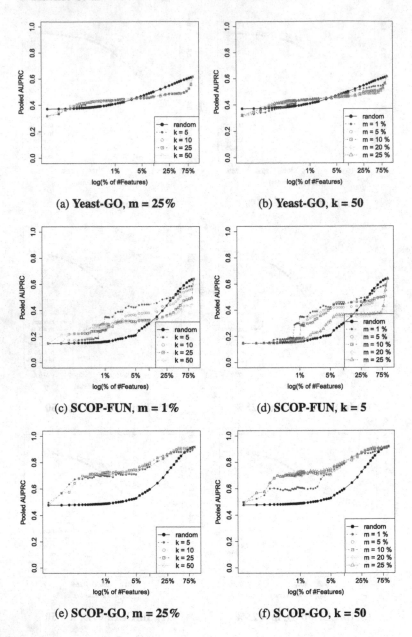

Fig. 3. Comparison of different FFA curves obtained by varying the number of m and k for datasets from the functional genomics domain

Yeast-GO dataset in Fig. 3a and b, show only slight improvement over the random FFA curves at the beginning of the ranking (top 1 % of the features). After that, seemingly irrelevant or redundant features are added, up to 75 % of the features. After this point there is a jump in the number of relevant features that are added, as the $AU(\overline{PRC})$ values become larger. For a fixed k in Fig. 3b, this effect is more pronounced as the percent of sampled instances m increases.

Upon closer inspection of the produced rankings of the *Yeast-GO* dataset, all of the numerical features were located among the top-ranked 1 % of the features and the bottom 25 % of the features, while the binary features were in the remaining part of the ranking. Although most of the numerical features were relevant, the corresponding relevance values for part of them seemed to be underestimated. This problem of underestimation of numerical attributes was also noted by Robnik-Šikonja and Kononenko [20], especially in the domains with both numeric and nominal features. To alleviate this issue, the use of a ramp function was proposed when calculating the distance between the numerical attributes. In our implementation a ramp function was also used, however different threshold parameters of this function were not explored. Robnik-Šikonja and Kononenko in [20], noted that for different domains, different thresholds might be appropriate and we believe that this is the probable cause of the underestimation of the relevance for part of the numeric features.

The FFA curves of the *SCOP-FUN* dataset, in Fig. 3c and d are the only ones that show variability of the curves with respect to m and k. Unlike the other datasets, the best FFA curves were obtained for a small number of m and of k. This is consistent with the analysis of ReliefF in [20] where it is stated that the values of m and k are often problem dependent and often smaller values might be better in order to preserve "locality" of the relevance estimations.

The best results were obtained for the *SCOP-GO* dataset, which we present in Fig. 3e and f. Both for a fixed m and k, the values of the FFA curves produced by HMC-ReliefF are much higher than those of the random rankings. For a fixed m varying the values of k does not influence the results (Fig. 3e). For a large fixed k, there is only a difference for the FFA curve produced for $m = 1$ % of the instance space, which produces lower $AU(\overline{PRC})$ values than the other values of the parameter m.

5   Conclusions and Further Work

In this paper, we presented the HMC-ReliefF algorithm, which is an extension of the RReliefF algorithm for the task of Hierarchical Multi-label Classification. We believe that this is both an interesting and novel line of work, in the context of feature ranking algorithms. To the best of our knowledge, there has not been any work for feature ranking within the context of structured data. We specifically focused on the ReliefF algorithm, due to its success in both classification and regression settings. The specific type of structured problems that we considered (HMC), was motivated by the fact that this kind of data can be found in various domains including biology and image annotation.

We evaluated the HMC-ReliefF algorithm on datasets from different domains and with different properties of the hierarchies. We first investigated if our algorithm was able to detect relevant features in a dataset and put them on top of the ranking. We consider this to be a minimum requirement of any feature ranking algorithm. Additionally, we also explored a reasonable set of parameter settings of HMC-ReliefF, which have influence on the feature relevance estimations.

The results of our experiments showed that, for various datasets, the HMC-ReliefF algorithm performed well, as evaluated by a stepwise filter like approach of constructing FFA curves. This performance was compared to an expected FFA curve, obtained from a set of random rankings. The exploration of the various parameters of HMC-ReliefF showed the following. For the image annotation datasets, large values of m and k were preferred and the FFA curves did not show much variability with respect to the parameters. The FFA curves produced by HMC-ReliefF were above the expected FFA curves with small differences. This was due to the nature of the domain and due to the fact that most of the features in the image annotation datasets were relevant.

For the functional genomics datasets, the results were more complex. The effect of underestimation of relevance of numeric features with respect to binary ones was observed, which has also been noted in the original ReliefF. The FFA curves of one of the datasets, were sensitive to the change of m and k, producing better FFA curves for smaller values. Finally, the last investigated dataset from this domain provided the best FFA curves, with values significantly larger than those of the expected FFA curves.

With this paper and the results presented we performed an initial investigation of the HMC-ReliefF algorithm. The directions for further work regarding our HMC-ReliefF algorithm are numerous. One major direction would be to define an artificial, controlled setting for investigating HMC problems in the context of feature ranking. Different types of hierarchies should be considered, which are also differently structured (balanced vs. unbalanced, different width, different depth), or differently populated by instances (sparse vs. non-sparse). Within this setting, the effects of the various parameters of HMC-ReliefF can be investigated and the advantages and limitations of the algorithm can be explored. Another major direction is to consider different types of structured outputs, such as multi-label or multi-target classification.

Acknowledgements. We would like to acknowledge the support of the European Commission through the project MAESTRA - Learning from Massive, Incompletely annotated, and Structured Data (Grant number ICT-2013-612944).

References

1. Aleksovski, D., Kocev, D., Džeroski, S.: Evaluation of distance measures for hierarchical multi-label classification in functional genomics. In: ECML/PKDD 2009 Workshop on Learning from Multi-Label Data, pp. 5–16 (2009)
2. Ashburner, M., Ball, C.A., Blake, J.A., Botstein, D., Butler, H., Cherry, J.M., Davis, A.P., Dolinski, K., Dwight, S.S., Eppig, J.T., Harris, M.A., Hill, D.P., Issel-Tarver, L., Kasarskis, A., Lewis, S., Matese, J.C., Richardson, J.E., Ringwald, M., Rubin, G.M., Sherlock, G.: Gene ontology: tool for the unification of biology. The gene ontology consortium. Nat. Genet. **25**(1), 25–29 (2000). http://dx.doi.org/10.1038/75556

3. Blum, A.L., Langley, P.: Selection of relevant features and examples in machine learning. Artif. Intell. **97**, 245–271 (1997)
4. Clare, A.: Machine learning and data mining for yeast functional genomics. Ph.D. thesis, University of Wales Aberystwyth, Aberystwyth, Wales, UK (2003)
5. Deselaers, T., Deserno, T.M., Mller, H.: Automatic medical image annotation in ImageCLEF 2007: overview, results, and discussion. Pattern Recogn. Lett. **29**(15), 1988–1995 (2008)
6. Dietterich, T.G., Domingos, P., Getoor, L., Muggleton, S., Tadepalli, P.: Structured machine learning: the next ten years. Mach. Learn. **73**(1), 3–23 (2008)
7. Dimitrovski, I., Kocev, D., Loskovska, S., Džeroski, S.: Hierchical annotation of medical images. In: Proceedings of the 11th International Multiconference - Information Society IS 2008, pp. 174–181. IJS, Ljubljana (2008)
8. Dimitrovski, I., Kocev, D., Loskovska, S., Džeroski, S.: Hierarchical classification of diatom images using ensembles of predictive clustering trees. Ecol. Inform. **7**(1), 19–29 (2012)
9. Gärtner, T., Vembu, S.: On structured output training: hard cases and an efficient alternative. Mach. Learn. **76**, 227–242 (2009)
10. Guyon, I., Elisseeff, A.: An introduction to variable and feature selection. J. Mach. Learn. Res. **3**, 1157–1182 (2003)
11. He, Z., Yu, W.: Review article: stable feature selection for biomarker discovery. Comput. Biol. Chem. **34**, 215–225 (2010)
12. Jaccard, P.: Étude comparative de la distribution florale dans une portion des alpes et des jura. Bulletin de la Société Vaudoise des Sciences Naturelles **37**, 547–579 (1901)
13. Kira, K., Rendell, L.A.: A practical approach to feature selection. In: ML92: Proceedings of the Ninth International Workshop on Machine Learning, pp. 249–256. Morgan Kaufmann Publishers Inc., San Francisco, CA, USA (1992)
14. Kocev, D., Vens, C., Struyf, J., Džeroski, S.: Tree ensembles for predicting structured outputs. Pattern Recogn. **46**(3), 817–833 (2013)
15. Kohavi, R., John, G.H.: Wrappers for feature subset selection. Artif. Intell. **97**, 273–324 (1997)
16. Kononenko, I.: Estimating attributes: analysis and extensions of RELIEF. In: Bergadano, F., De Raedt, L. (eds.) ECML 1994. LNCS, vol. 784. Springer, Heidelberg (1994)
17. Kriegel, H.P., Borgwardt, K., Kröger, P., Pryakhin, A., Schubert, M., Zimek, A.: Future trends in data mining. Data Min. Knowl. Discov. **15**, 87–97 (2007)
18. Pesquita, C., Faria, D., Bastos, H., Falcao, A.O., Couto, F.: Evaluating go-based semantic similarity measures. In: BioOntologies SIG at ISMB/ECCB - 15th Annual International Conference on Intelligent Systems for Molecular Biology (ISMB) (2007)
19. Robnik-Šikonja, M., Kononenko, I.: An adaptation of relief for attribute estimation in regression. In: Fisher, D.H. (ed.) ICML, pp. 296–304. Morgan Kaufmann, San Francisco (1997)
20. Robnik-Šikonja, M., Kononenko, I.: Theoretical and empirical analysis of ReliefF and RReliefF. Mach. Learn. **53**, 23–69 (2003)
21. Silla, C., Freitas, A.: A survey of hierarchical classification across different application domains. Data Min. Knowl. Discov. **22**(1–2), 31–72 (2011)
22. Slavkov, I.: An evaluation method for feature rankings. Ph.D. thesis, IPS Jožef Stefan, Ljubljana, Slovenia (2012)
23. Vens, C., Struyf, J., Schietgat, L., Džeroski, S., Blockeel, H.: Decision trees for hierarchical multi-label classification. Mach. Learn. **73**(2), 185–214 (2008)

The Use of the Label Hierarchy in Hierarchical Multi-label Classification Improves Performance

Jurica Levatić[1,2]([✉]), Dragi Kocev[1], and Sašo Džeroski[1,2]

[1] Department of Knowledge Technologies, Jožef Stefan Institute, Ljubljana, Slovenia
{Jurica.Levatic,Dragi.Kocev,Saso.Dzeroski}@ijs.si
[2] Jožef Stefan International Postgraduate School, Ljubljana, Slovenia

Abstract. We address the task of learning models for predicting structured outputs. We consider both global and local approaches to the prediction of structured outputs, the former based on a single model that predicts the entire output structure and the latter based on a collection of models, each predicting a component of the output structure. More specifically, we compare local and global approaches in terms of predictive performance, learning time and model complexity. Moreover, we discuss the interpretability of the obtained models. We evaluate the predictive performance of the considered approaches on six case studies from three domains: ecological modelling, text classification and image classification. Finally, we identify the properties of the tasks at hand that lead to the differences in performance.

Keywords: Predictive clustering trees · Hierarchical multi-label classification · Multi-label classification · Habitat modelling · Text classification · Image classification

1 Introduction

Supervised learning is one of the most widely researched and investigated areas of machine learning. The goal in supervised learning is to learn, from a set of examples with known class, a function that outputs a prediction for the class of a previously unseen example. If the examples belong to two classes (e.g., the example has some property or not) the task is called binary classification. The task where the examples can belong to a single class from a given set of m classes ($m \geq 3$) is known as multi-class classification. The case where the output is a real value is called regression.

However, in many real life problems of predictive modelling the output (i.e., the target) is structured, meaning that there can be dependencies between classes (e.g., classes are organized into a tree-shaped hierarchy or a directed acyclic graph) or some internal relations between the classes (e.g., sequences). These types of problems occur very often in various domains, such as life sciences (predicting gene function, finding the most important genes for a given disease, predicting toxicity of molecules, etc.), ecology (analysis of remotely sensed data,

A. Appice et al. (Eds.): NFMCP 2013, LNAI 8399, pp. 162–177, 2014.
DOI: 10.1007/978-3-319-08407-7_11, © Springer International Publishing Switzerland 2014

habitat modelling), multimedia (annotation and retrieval of images and videos) and the semantic web (categorization and analysis of text and web pages). Having in mind the needs of these application domains and the increasing quantities of structured data, Kriegel et al. [1] and Dietterich et al. [2] listed the task of "mining complex knowledge from complex data" as one of the most challenging problems in machine learning.

A variety of methods, specialized in predicting a given type of structured output (e.g., a hierarchy of classes [3]), have been proposed [4]. These methods can be categorized into two groups of methods for solving the problem of predicting structured outputs [3,4]. Local methods construct models for predicting component(s) of the output and then combine the individual models to get the overall model (i.e., they construct an architecture of several simple(r) models). Global methods that construct models for predicting the complete structure as a whole (also known as 'big-bang' approaches).

The global methods have several advantages over the local methods. First, they exploit and use the dependencies that may exist between the components of the structured output in the model learning phase, which can result in better predictive performance of the learned models. Next, they are typically more efficient: it can easily happen that the number of components in the output is very large (e.g., hierarchies in functional genomics can have several thousands of components), in which case learning a model for each component is not feasible. Furthermore, they produce models that are typically smaller than the sum of the sizes of the models built for each of the components.

Despite the many developed methods and their interesting applications, it is not clear when it is favorable (performance wise) to apply global and when local approaches. In this work, we focus on clarifying this important issue for the task of hierarchical multi-label classification (HMC). HMC is a variant of classification, where a single example may belong to multiple classes at the same time and the classes are organized in the form of a hierarchy. An example that belongs to some class c automatically belongs to all super-classes of c: This is called the hierarchical constraint. Problems of this kind can be found in many domains including text classification, functional genomics, and object/scene classification. Silla and Freitas [3] give a detailed overview of the possible application areas and the different approaches to HMC.

More specifically, we construct four types of predictive models that exploit different amounts of the information provided by the output structure, i.e., the hierarchical organization of the classes. This corresponds to four different machine learning tasks that can be formulated to solving the task of HMC: binary classification, hierarchical single-label classification, multi-label classification and hierarchical multi-label classification. The first two tasks construct (an architecture of) local predictive models, while the last two tasks construct global models.

To properly evaluate the predictive performance of the different models one needs to select predictive models from the same type that can solve the four tasks enumerated above. To this end, we consider predictive clustering trees (PCTs) as predictive models. PCTs can be viewed as a generalization of standard decision trees towards predicting structured outputs. PCTs offer a unifying approach for dealing with different types of structured outputs and construct the predictive models very efficiently. They are able to make predictions for several types of structured outputs: tuples of continuous/discrete variables, hierarchies of classes, and time series [5–7].

We perform the evaluation of the predictive models on six practically relevant HMC datasets. The datasets come from three different domains: habitat modelling, image classification and text classification. We consider habitat models for Collembola communities in the soils of Denmark [8] and communities of organisms living in Slovenian rivers [9]. Next, we use two datasets from the 2007 CLEF cross-language image retrieval campaign [10], where the goal is to annotate medical X-ray images. From the domain of text classification, we use two well known datasets: categorization of e-mails from officials of the Enron corporation [11] and categorization of Reuters newswire stories [12].

The remainder of this paper is organized as follows. Section 2 explains the predictive clustering trees framework and the extensions for the different tasks considered here. The experimental setup is presented in Sect. 3. Section 4 presents the obtained results. Finally, the conclusions are stated in Sect. 5.

2   Predictive Modelling for HMC

In this section, we present in more detail methodology used to construct the predictive models. We first present global approaches that predict the complete output (i.e., a single model for all of the possible labels in the dataset) with a single model. We then briefly describe local approaches that construct several models - each one predicting a part of the output (i.e., a model for each label separately).

2.1   Global Predictive Models

The Predictive Clustering Trees (PCTs) framework views a decision tree as a hierarchy of clusters: the top-node corresponds to one cluster containing all data, which is recursively partitioned into smaller clusters while moving down the tree. The PCT framework is implemented in the CLUS system [13], which is available for download at http://clus.sourceforge.net.

PCTs are induced with a standard *top-down induction of decision trees* (TDIDT) algorithm [14]. The algorithm is presented in Table 1. It takes as input a set of examples (E) and outputs a tree. The heuristic (h) that is used for selecting the tests (t) is the reduction in variance caused by the partitioning ($\mathcal{P}$) of the instances corresponding to the tests (t) (see line 4 of the BestTest procedure in Table 1). By maximizing the variance reduction, the cluster homogeneity is maximized and the predictive performance is improved.

The main difference between the algorithm for learning PCTs and a standard decision tree learner is that the former considers the variance function and the

Table 1. The top-down induction algorithm for PCTs.

procedure PCT	**procedure** BestTest
Input: A dataset E	**Input:** A dataset E
Output: A predictive clustering tree	**Output:** the best test (t^*), its heuristic score (h^*) and the partition $(\mathcal{P}^*)$ it induces on the dataset (E)
1: $(t^*, h^*, \mathcal{P}^*) = \text{BestTest}(E)$	1: $(t^*, h^*, \mathcal{P}^*) = (none, 0, \emptyset)$
2: **if** $t^* \neq none$ **then**	2: **for each** possible test t **do**
3: **for each** $E_i \in \mathcal{P}^*$ **do**	3: $\mathcal{P}$ = partition induced by t on E
4: $tree_i = \text{PCT}(E_i)$	4: $h = Var(E) - \sum_{E_i \in \mathcal{P}} \frac{|E_i|}{|E|} Var(E_i)$
5: **return** node$(t^*, \bigcup_i \{tree_i\})$	5: **if** $(h > h^*) \wedge \text{Acceptable}(t, \mathcal{P})$ **then**
6: **else**	6: $(t^*, h^*, \mathcal{P}^*) = (t, h, \mathcal{P})$
7: **return** leaf(Prototype(E))	7: **return** $(t^*, h^*, \mathcal{P}^*)$

prototype function (that computes a label for each leaf) as *parameters* that can be instantiated for a given learning task. So far, PCTs have been instantiated for the following tasks: multi-target prediction (which includes multi-label classification) [6], hierarchical multi-label classification [7] and prediction of time-series [15]. In this article, we focus on the first two tasks.

PCTs for Multi-label Classification. PCTs for multi-label classification can be considered as PCTs that are able to predict multiple binary (and thus discrete) targets simultaneously. Therefore, the variance function for the PCTs for MLC is computed as the sum of the Gini indices of the target variables, i.e., $Var(E) = \sum_{i=1}^{T} Gini(E, Y_i)$. Alternatively, one can also use the sum of the entropies of class variables as a variance function, i.e., $Var(E) = \sum_{i=1}^{T} Entropy$ (E, Y_i) (this definition has also been used in the context of multi-label prediction [16]). The CLUS system also implements other variance functions, such as reduced error, gain ratio and the m-estimate. The prototype function returns a vector of probabilities that an instance belongs to a given class for each target variable. Using these probabilities, the most probable (majority) class value for each target can be calculated.

PCTs for Hierarchical Multi-label Classification. CLUS-HMC is the instantiation (with the distances and prototypes as defined below) of the PCT algorithm for hierarchical classification implemented in the CLUS system [7]. The variance and prototype are defined as follows. First, the set of labels of each example is represented as a vector with binary components; the ith component of the vector is 1 if the example belongs to class c_i and 0 otherwise. It is easily checked that the arithmetic mean of a set of such vectors contains as i^{th} component the proportion of examples of the set belonging to class c_i. The variance of a set of examples E is defined as the average squared distance between each

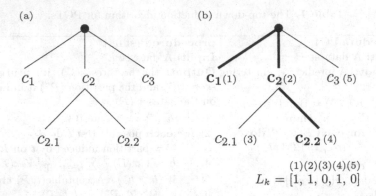

Fig. 1. Toy examples of a hierarchy structured as a tree. (a) Class label names contain information about the position in the hierarchy, e.g., $c_{2.1}$ is a subclass of c_2. (b) The set of classes $S_1 = \{c_1, c_2, c_{2.2}\}$, shown in bold, are represented as a vector (L_k).

example's class vector (L_i) and the set's mean class vector $(\overline{L})$, i.e.,

$$Var(E) = \frac{1}{|E|} \cdot \sum_{E_i \in E} d(L_i, \overline{L})^2.$$

In the HMC context, the similarity at higher levels of the hierarchy is more important than the similarity at lower levels. This is reflected in the distance measure used in the above formula, which is a weighted Euclidean distance:

$$d(L_1, L_2) = \sqrt{\sum_{l=1}^{|L|} w(c_l) \cdot (L_{1,l} - L_{2,l})^2},$$

where $L_{i,l}$ is the lth component of the class vector L_i of an instance E_i, $|L|$ is the size of the class vector, and the class weights $w(c)$ decrease with the depth of the class in the hierarchy. More precisely, $w(c) = w_0 \cdot w(p(c))$, where $p(c)$ denotes the parent of class c and $0 < w_0 < 1$).

For example, consider the toy class hierarchy shown in Fig. 1(a,b), and two data examples: (X_1, S_1) and (X_2, S_2) that belong to the classes $S_1 = \{c_1, c_2, c_{2.2}\}$ (boldface in Fig. 1(b)) and $S_2 = \{c_2\}$, respectively. We use a vector representation with consecutive components representing membership in the classes c_1, c_2, $c_{2.1}$, $c_{2.2}$ and c_3, in that order (preorder traversal of the tree of class labels). The distance is then calculated as follows:

$$d(S_1, S_2) = d([1, 1, 0, 1, 0], [0, 1, 0, 0, 0]) = \sqrt{w_0 + w_0^2}.$$

Recall that the instantiation of PCTs for a given task requires a proper instantiation of the variance and prototype functions. The variance function for the HMC task is instantiated by using the weighted Euclidean distance measure (as given above), which is further used to select the best test for a given node by calculating the heuristic score (line 4 from the algorithm in Table 1). We now discuss the instantiation of the prototype function for the HMC task.

A classification tree stores in a leaf the majority class for that leaf, which will be the tree's prediction for all examples that will arrive in the leaf. In the case of HMC, an example may have multiple classes, thus the notion of *majority class* does not apply in a straightforward manner. Instead, the mean $\bar{L}$ of the class vectors of the examples in the leaf is stored as a prediction. Note that the value for the ith component of $\bar{L}$ can be interpreted as the probability that an example arriving at the given leaf belongs to class c_i.

The prediction for an example that arrives at the leaf can be obtained by applying a user defined threshold τ to the probability; if the ith component of $\bar{L}$ is above τ then the examples belong to class c_i. When a PCT is making a prediction, it preserves the hierarchy constraint (the predictions comply with the parent-child relationships from the hierarchy) if the values for the thresholds τ are chosen as follows: $\tau_i \leq \tau_j$ whenever $c_i \leq_h c_j$ (c_i is ancestor of c_j). The threshold τ is selected depending on the context. The user may set the threshold such that the resulting classifier has high precision at the cost of lower recall or vice versa, to maximize the F-score, to maximize the interpretability or plausibility of the resulting model etc. In this work, we use a threshold-independent measure (precision-recall curves) to evaluate the performance of the models.

2.2   Local Predictive Models

Local models for predicting structured outputs use a collection of predictive models, each predicting a component of the overall structure that needs to be predicted. For the task of predicting multiple targets, local predictive models are constructed by learning a predictive model for each of the targets separately. In the task of hierarchical multi-label classification, however, there are four different approaches that can be used: flat classification, local classifiers per level, local classifiers per node, and local classifiers per parent node (see [3] for details).

Vens et al. [7] investigated the performance of the last two approaches with local classifiers over a large collection of datasets from functional genomics. The conclusion of the study was that the last approach (called hierarchical single-label classification - HSC) performs better in terms of predictive performance, smaller total model size and faster induction times.

In particular, the CLUS-HSC algorithm by Vens et al. [7] constructs a decision tree classifier for each edge (connecting a class c with a parent class $par(c)$) in the hierarchy, thus creating an architecture of classifiers. The tree that predicts membership to class c is learnt using the instances that belong to $par(c)$. The construction of this type of trees uses few instances, as only instances labeled with $par(c)$ are used for training. The instances labeled with class c are positive while the ones labeled with $par(c)$, but not with c are negative.

The resulting HSC tree architecture predicts the conditional probability $P(c|par(c))$. A new instance is predicted by recursive application of the product rule $P(c) = P(c|par(c)) \cdot P(par(c))$, starting from the tree for the top-level class. Again, the probabilities are thresholded to obtain the set of predicted classes.

To satisfy the hierarchy constraint, the threshold τ should be chosen as in the case of CLUS-HMC.

In this work, we also consider the task of single-label classification. We consider this to be a special case of multi-label classification where the number of labels is 1. To this end, we use the same algorithm as for the multi-label classification trees. We call these models single-label classification trees.

3   Experimental Design

In this section, we present the design of the experimental evaluation of the predictive models built for the four machine learning tasks considered. We begin by describing the data used. We then outline the specific experimental setup for constructing the predictive models. Finally, we present the evaluation measure for assessing the predictive performance of the models.

3.1   Data Description

We use six datasets, which come from three domains: habitat modeling, image classification and text classification. The main statistics of the datasets are given in Table 2. We can observe that the datasets vary in the size, number of attributes and characteristics of the label hierarchy.

Habitat modelling [17] focuses on spatial aspects of the distribution and abundance of plants and animals. It studies the relationships between environmental variables and the presence/abundance of plants and animals. This is typically done under the implicit assumption that both are observed at a single point in time for a given spatial unit (i.e., sampling site). We investigate the effect of environmental conditions on communities of organisms in two different ecosystems: river and soil. Namely, we construct habitat models for river water organisms living in Slovenian rivers [9] and for soil microarthropods from Danish farms [8]. The data about the organisms that live in the water of Slovenian rivers was collected during six years (1990 to 1995) of monitoring of water quality performed by the Hydro-meteorological Institute of Slovenia (now Environmental Agency of Slovenia). The data for the soil microarthropods from Danish farms describes four experimental farming systems (observed during the period 1989–1993) and a number of organic farms (observed during the period 2002–2003). The structured output space in these case studies is the taxonomic hierarchy of the species. Since different species are considered in the two domains, their respective output spaces will be different.

In image classification, the goal is to automatically annotate the image content with labels. The labels typically represent visual concepts that are present in the images. In this work, we are concerned with the annotation of medical X-ray images. We use two datasets from the 2007 CLEF cross-language image retrieval campaign [10]: ImCLEF07A and ImCLEF07D. The goal in these datasets is to recognize which part of the human anatomy is present in the image or the orientation of the body part, respectively. Images are represented by using edge

Table 2. Characteristics of the datasets: N is the number of instances, D/C is the number of descriptive attributes (discrete/continuous), $\mathcal{L}$ is the number of labels (leafs in the hierarchy), $|\mathcal{H}|$ is the number of nodes in the hierarchy, $\mathcal{H}_d$ is the maximal depth of the hierarchy, $\overline{\mathcal{L}_L}$ is the average number of labels per example.

Domain	N	D/C	$\mathcal{L}$	$|\mathcal{H}|$	$\mathcal{H}_d$	$\overline{\mathcal{L}_L}$
Slovenian rivers [9]	1060	0/16	491	724	4	25
Danish farms [8]	1944	132/5	35	72	3	7
ImCLEF07A [10]	11006	0/80	63	96	3	1
ImCLEF07D [10]	11006	0/80	26	46	3	1
Enron [11]	1648	0/1001	50	54	3	2.84
Reuters [12]	6000	0/47236	77	100	4	1.2

histograms. An edge histogram represents the frequency and the directionality of the brightness changes in the image. The structured output space consists of labels organized in hierarchy. They correspond to the anatomical (ImCLEF07A) and directional (ImCLEF07D) axis of the IRMA (Image Retrieval in Medical Applications) code [18].

Text classification is the problem of automatic annotation of textual documents to one or more categories. We used two datasets from this domain: Enron and Reuters. Enron is a labeled subset of the Enron corpus [11], prepared and annotated by the UCBerkeley Enron Email Analysis Project[1]. The e-mails are categorized into several hierarchically organized categories concerning the characteristics of the e-mail, such as genre, emotional tone or topic. Reuters is a subset of the 'Topics' category of the Reuters Corpus Volume I (RCV1) [12]. RCV1 is a collection of English language stories published by the Reuters agency between August 20, 1996, and August 19, 1997. Stories are categorized into hierarchical groups according to the major subjects of a story, such as Economics, Industrial or Government. In both domains, the text documents are described with their respective bag-of-words representation.

3.2   Experimental Design

We constructed four types of predictive models, as described in the previous section, for each of the case studies. First, we constructed single-label classification trees for each label (i.e., leaf in the label hierarchy) separately. Next, we constructed hierarchical single-label classification tree architecture. Furthermore, we constructed a multi-label classification tree for all of the leaf labels, without using the hierarchy. Finally, we constructed a hierarchical multi-label classification tree for all of the labels by using the hierarchy.

We used F-test pruning to ensure that the produced models are not overfitted and have better predictive performance [7]. The exact Fisher test is used to check whether a given split/test in an internal node of the tree results in a

---

[1] http://bailando.sims.berkeley.edu/enron_email.html

statistically significant reduction in variance. If there is no such split/test, the node is converted to a leaf. A significance level is selected from the values 0.125, 0.1, 0.05, 0.01, 0.005 and 0.001 to optimize predictive performance by using internal 3-fold cross validation.

We evaluate the predictive performance of the models on the classes/labels that are leafs in the target hierarchy. We made this choice in order to ensure a fair comparison across the different tasks. Namely, if we consider all labels (the leaf labels and the inner nodes labels), the single-label classification task will be very close to the task of hierarchical single-label classification; similarly, the task of multi-label classification becomes very close to the task of hierarchical multi-label classification. Moreover, by evaluating only the performance on leaf labels, we are measuring more precisely the influence of the inclusion of the different kinds of information in the learning process on the predictive performance of the models. To further ensure this, we set the w_0 parameter for the weighted Euclidean distance for HMC to the value of 1: all labels in the hierarchy contribute equally. By doing this, we measure only the effect of including the multi-label information (considering the multiple labels simultaneously) and the hierarchy information.

3.3 Evaluation Measures

We evaluate the algorithms by using the Area Under the Precision-Recall Curve (AUPRC), and in particular, the Area Under the Average Precision-Recall Curve ($\overline{AUPRC}$) as suggested by Vens et al. [7]. The points in the PR space are obtained by varying the value for the threshold τ from 0 to 1 with step 0.02. For each value of the threshold τ, precision and recall are micro-averaged as follows:

$$\overline{Prec} = \frac{\sum_i TP_i}{\sum_i TP_i + \sum_i FP_i}, \quad \text{and} \quad \overline{Rec} = \frac{\sum_i TP_i}{\sum_i TP_i + \sum_i FN_i}$$

where i ranges over all classes that are leafs in the output hierarchies.

We measure the performance of the predictive models along several dimensions. First, we estimate the predictive performance of the models using 10-fold cross-validation. Second, we assess the descriptive power of the models by evaluating them on the training set. Next, we measure how much the different models tend to over-fit on the training data. To this end, we use the relative decrease of the performance from the training set to the one obtained with 10-fold cross-validation. We define this as over-fit score ($OS = \frac{\overline{AUPRC}_{train} - \overline{AUPRC}_{test}}{\overline{AUPRC}_{train}}$). The smaller values of this score mean that the overfitting of the models is smaller. Finally, we measure the model complexity and the time efficiency of the predictive models. The model complexity for the global models is the number of nodes in a given tree, while the model complexity for the local models is the sum of all nodes from all trees. Similarly, the running time of the global models is the time needed to construct the model, while the running time for the local models is the time needed to construct all of the models.

We adopt the recommendations by Demšar [19] for the statistical evaluation of the results. We use the corrected non-parametric Friedman test for statistical significance on the per-fold-data for the folds of 10-fold cross validation for each dataset separately. Afterwards, to check where the statistically significant differences appear (between which methods), we use the Nemenyi post-hoc test (Nemenyi, 1963). We present the result from the Nemenyi post hoc test with an average ranks diagram as suggested by Demšar [8]. The ranks are depicted on the axis, in such a manner that the best ranking algorithms are at the right-most side of the diagram. The algorithms that do not differ significantly (in performance) are connected with a line.

4   Results and Discussion

In this section, we present the results from the experimental evaluation. We discuss the obtained models first in terms of their performance (predictive and efficiency) and then in terms of their interpretability.

The results from the evaluation of the predictive models are given in Table 3. A quick inspection of the performance reveals that the best results are obtained by models that exploit the information about the underlying output hierarchy. Next, the models that include the hierarchy information tend to over fit less as compared to the other models. Moreover, the results indicate that the HMC trees over-fit the least on these datasets. Finally, the global models (especially HMC) are more efficient than their local counterparts, in terms of both running time and model complexity.

We further examine the results by performing a statistical significance test. In particular, we performed the Friedman test to check whether the observed differences in performance are statistically significant for each dataset separately. The results from this analysis show that the difference in performance is statistically significant for each dataset with *p-value* smaller than $3 \cdot 10^{-5}$.

Figure 2 presents the average ranks from the Nemenyi post-hoc test for all types of models. The diagrams show that the HMC models are best performing on three domains (Slovenian rivers, Danish farms and Enron), while on the other three domains (ImCLEF07A, ImCLEF07D and Reuters) the best performing type of model is the HSC architecture. We next discuss the statistically significant differences in the datasets in more detail.

When HMC trees are the best performing method, they are statistically significantly better than the single-label trees. In the remaining cases, the differences are not statistically significant (although HMC trees are better than single-label trees also on ImCLEF07A and ImCLEF07D). HMC trees are statistically significantly better than HSC tree architecture only on the Slovenian rivers dataset, and HSC tree architecture is statistically significantly better than HMC trees on the Reuters dataset.

We further complement the information on the performance with the dataset properties from Table 2. HMC trees perform best on datasets with a large number of labels per example (25, 7 and 2.84 labels per example for the Slovenian rivers, Danish farms and Enron datasets, respectively). Conversely, HSC tree architectures perform better on datasets with a small number of labels per example (1.2, 1 and 1 for Reuters, ImCLEF07A and ImCLEF07D datasets, respectively).

Table 3. Performance of the methods in terms of AUPRC, decrease of training set performance relative to test set performance. (*OS*), Learning time (in seconds) and model complexity (the number of nodes in the decision trees). The best predictive performance for each dataset is shown in bold.

Dataset	Method	AUPRC	OS	Learning time	Complexity
Slovenian rivers	Single-label	0.239	0.692	23.3	15336
	HSC	0.309	0.591	10.2	25035
	Multi-label	0.322	0.007	9.4	1
	HMC	**0.374**	0.132	0.6	37
Danish farms	Single-label	0.790	0.099	3.7	2605
	HSC	0.808	0.083	1.3	2873
	Multi-label	0.801	0.112	0.7	265
	HMC	**0.815**	0.065	0.4	259
ImCLEF07A	Single-label	0.571	0.375	74.4	3957
	HSC	**0.665**	0.324	27.3	10054
	Multi-label	0.530	0.462	13.5	3553
	HMC	0.592	0.182	3.4	635
ImCLEF07D	Single-label	0.515	0.483	35.4	7418
	HSC	**0.631**	0.361	20.1	9764
	Multi-label	0.511	0.484	7.78	3675
	HMC	0.615	0.198	3.0	685
Enron	Single-label	0.398	0.495	114.7	1740
	HSC	0.466	0.434	25.1	3168
	Multi-label	0.385	0.584	13.8	1259
	HMC	**0.488**	0.110	3.3	55
Reuters	Single-label	0.431	0.546	970.8	3591
	HSC	**0.481**	0.510	781.4	7004
	Multi-label	0.332	0.654	191.8	2949
	HMC	0.373	0.365	42.5	593

The output hierarchy is much more populated in the former case, thus, allowing the learning of HMC trees to fully exploit the dependencies between the labels. This in turn provides predictive models with better predictive power. Similar behavior can be observed for the models that do not exploit the output hierarchy: the multi-label trees are better on datasets with more labels per example, while the single-label tree are better on datasets with fewer labels per example.

We next discuss the poor performance of the global models on the Reuters dataset. This is the only dataset where HMC trees have worse predictive performance than single-label trees. The poor predictive performance is mainly due to two reasons: (1) the dataset has a small number of labels per examples and (2) the dataset is extremely high-dimensional and sparse. However, this prompts for additional investigation and analysis using more benchmark datasets that exhibit similar properties.

Besides the predictive power of the models, their interpretability is often a highly desired property, especially in domains such as habitat modelling.

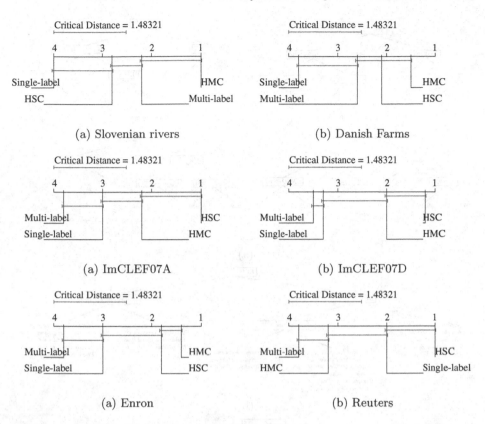

Fig. 2. Average ranks diagrams for the performance of the four methods in terms of AU$\overline{PRC}$ for each of the six datasets. Better algorithms are positioned on the right-hand side, the ones that differ by less than the critical distance for a *p-value* = 0.05 are connected with a line.

We discuss the interpretability of the models from the perspective of this domain. The predictive models that we consider here (PCTs) are readily interpretable. However, the difference in the interpretability of the local and global models is easy to notice. Firstly, global models, especially HMC trees , have considerably smaller complexity than the (collections of) local models (Table 3). In Fig. 3, we present illustrative examples of the predictive models for the Slovenian rivers dataset. We show several PCTs for single-label classification, a tree for multi-label classification and a tree for hierarchical multi-label classification.

We can immediately notice the differences between the local and global predictive models. The local models[2] offer information only for a part for the output

---

[2] Note that the hierarchical single-label classification models will be similar to the single-label classification models, with the difference that the predictive models are organized into a hierarchical architecture. This makes the interpretation of the HSC models an even more difficult task.

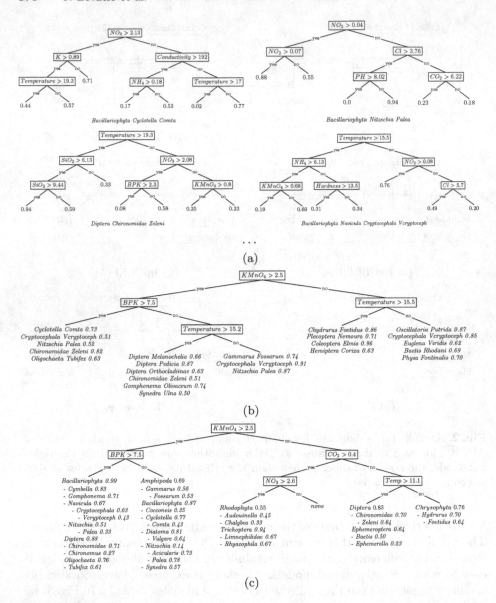

Fig. 3. Illustrative examples of decision trees (PCTs) learnt for the Slovenian rivers dataset. Single-label classification (a) produces a separate model for each of the species, whereas multi-label classification (b) and hierarchical multi-label classification (c) consider all of the species in a single tree.

space, i.e., they are valid just for a single species. In order to reconstruct the complete community model, one needs to look at the separate models and then try to make some overall conclusions. However, this could be very tedious or even impossible in domains with high biodiversity where there are hundreds of species present, such as the domain we consider here - Slovenian rivers.

On the other hand, the global models are much easier to interpret. The single global model is valid for the complete structured output, i.e., for the whole community of species present in the ecosystem. The global models are able to capture the interactions present between the species, i.e., which species can co-exist at a locations with given physico-chemical properties. Moreover, the HMC models, as compared to the multi-label models, offer additional information about the higher taxonomic ranks. For example, the HMC model could state that there is a low probability (0.27) that the species *Diptera chironomus* is present under the given environmental conditions, while the is a high probability (0.88) that the genus *Diptera* is present (left-most leaf of the HMC tree in Fig. 3).

5   Conclusions

We address the task of learning predictive models for hierarchical multi-label classification, which take as input a tuple of attribute values and predict a set of classes organized into a hierarchy. We consider both global and local approaches for prediction of structured outputs. The former are based on a single model that predicts the entire output structure, while the latter are based on a collection of models, each predicting a component of the output structure.

We investigate the differences in performance and interpretability of the local and global models. More specifically, we examine whether including information in the form of hierarchical relationships among the labels and considering the multiple labels simultaneously helps to improve the performance of the predictive models. To this end, we consider four machine learning tasks: single-label classification, hierarchical single-label classification, multi-label classification and hierarchical multi-label classification.

We use predictive clustering trees as predictive models, since they can be used for solving all of the four tasks considered here. We construct and evaluate four types of trees: single-label trees, hierarchical single-label trees, multi-label trees and hierarchical multi-label trees.

We compare the performance of local and global predictive models on six datasets from three practically relevant tasks: habitat modelling, image classification and text classification. The results show that the inclusion of the hierarchical information in the model construction phase, i.e., for HMC trees and for HSC tree architecture, improves the predictive performance. The improvement in performance for HMC trees is more pronounced on domains that have a more populated hierarchy, i.e., on datasets with a larger number of labels per example. On the other hand, HSC tree architecture perform better in the domains where the number of labels per example is closer to one. Moreover, the models that take the hierarchy into account tend to over-fit less than the models that do not include such information (this is especially true for the HMC trees). Finally, the global methods produce less complex models and are much easier to interpret than the local models offering an overview of the complete output hierarchy.

All in all, the inclusion of hierarchy information improves the performance of the predictive models and the global models are more efficient and easier to interpret than local models.

Acknowledgments. We would like to acknowledge the support of the European Commission through the project MAESTRA - Learning from Massive, Incompletely annotated, and Structured Data (Grant number ICT-2013-612944).

References

1. Kriegel, H.P., Borgwardt, K., Kröger, P., Pryakhin, A., Schubert, M., Zimek, A.: Future trends in data mining. Data Min. Knowl. Disc. **15**, 87–97 (2007)
2. Dietterich, T.G., Domingos, P., Getoor, L., Muggleton, S., Tadepalli, P.: Structured machine learning: the next ten years. Mach. Learn. **73**(1), 3–23 (2008)
3. Silla, C., Freitas, A.: A survey of hierarchical classification across different application domains. Data Min. Knowl. Disc. **22**(1–2), 31–72 (2011)
4. Bakır, G.H., Hofmann, T., Schölkopf, B., Smola, A.J., Taskar, B., Vishwanathan, S.V.N.: Predicting Structured Data. The MIT Press, Cambridge (2007)
5. Blockeel, H.: Top-down induction of first order logical decision trees. Ph.D. thesis, Katholieke Universiteit Leuven, Leuven, Belgium (1998)
6. Kocev, D., Vens, C., Struyf, J., Džeroski, S.: Tree ensembles for predicting structured outputs. Pattern Recogn. **46**(3), 817–833 (2013)
7. Vens, C., Struyf, J., Schietgat, L., Džeroski, S., Blockeel, H.: Decision trees for hierarchical multi-label classification. Mach. Learn. **73**(2), 185–214 (2008)
8. Demšar, D., Džeroski, S., Larsen, T., Struyf, J., Axelsen, J., Bruns-Pedersen, M., Krogh, P.H.: Using multi-objective classification to model communities of soil. Ecol. Modell. **191**(1), 131–143 (2006)
9. Džeroski, S., Demšar, D., Grbović, J.: Predicting chemical parameters of river water quality from bioindicator data. Appl. Intell. **13**(1), 7–17 (2000)
10. Dimitrovski, I., Kocev, D., Loskovska, S., Džeroski, S.: Hierchical annotation of medical images. In: Proceedings of the 11th International Multiconference - Information Society IS 2008, IJS, Ljubljana, pp. 174–181 (2008)
11. Klimt, B., Yang, Y.: The enron corpus: a new dataset for email classification research. In: Boulicaut, J.-F., Esposito, F., Giannotti, F., Pedreschi, D. (eds.) ECML 2004. LNCS (LNAI), vol. 3201, pp. 217–226. Springer, Heidelberg (2004)
12. Lewis, D.D., Yang, Y., Rose, T.G., Li, F.: RCV1: a new benchmark collection for text categorization research. J. Mach. Learn. Res. **5**, 361–397 (2004)
13. Blockeel, H., Struyf, J.: Efficient algorithms for decision tree cross-validation. J. Mach. Learn. Res. **3**, 621–650 (2002)
14. Breiman, L., Friedman, J., Olshen, R., Stone, C.J.: Classification and Regression Trees. Chapman & Hall/CRC, New York (1984)
15. Slavkov, I., Gjorgjioski, V., Struyf, J., Džeroski, S.: Finding explained groups of time-course gene expression profiles with predictive clustering trees. Mol. BioSyst. **6**(4), 729–740 (2010)
16. Clare, A.: Machine learning and data mining for yeast functional genomics. Ph.D. thesis, University of Wales Aberystwyth, Wales, UK (2003)
17. Džeroski, S.: Machine learning applications in habitat suitability modeling. In: Haupt, S.E., Pasini, A., Marzban, C. (eds.) Artificial Intelligence Methods in the Environmental Sciences, pp. 397–412. Springer, Berlin (2009)

18. Lehmann, T., Schubert, H., Keysers, D., Kohnen, M., Wein, B.: The IRMA code for unique classification of medical images. In: Medical Imaging 2003: PACS and Integrated Medical Information Systems: Design and Evaluation, pp. 440–451 (2003)
19. Demšar, J.: Statistical comparisons of classifiers over multiple data sets. J. Mach. Learn. Res. **7**, 1–30 (2006)

Graphs, Networks
and Relational Data

AGWAN: A Generative Model for Labelled, Weighted Graphs

Michael Davis[1](✉), Weiru Liu[1], Paul Miller[1], Ruth F. Hunter[2], and Frank Kee[2]

[1] Centre for Secure Information Technologies, School of Electronics,
Electrical Engineering and Computer Science, Belfast, UK
[2] Centre for Public Health, School of Medicine, Dentistry and Biomedical Sciences,
Queen's University, Belfast, UK
{mdavis05,w.liu,p.miller,ruth.hunter,f.kee}@qub.ac.uk

Abstract. Real-world graphs or networks tend to exhibit a well-known set of properties, such as heavy-tailed degree distributions, clustering and community formation. Much effort has been directed into creating realistic and tractable models for unlabelled graphs, which has yielded insights into graph structure and evolution. Recently, attention has moved to creating models for labelled graphs: many real-world graphs are labelled with both discrete and numeric attributes. In this paper, we present AGWAN (Attribute Graphs: Weighted and Numeric), a generative model for random graphs with discrete labels and weighted edges. The model is easily generalised to edges labelled with an arbitrary number of numeric attributes. We include algorithms for fitting the parameters of the AGWAN model to real-world graphs and for generating random graphs from the model. Using real-world directed and undirected graphs as input, we compare our approach to state-of-the-art random labelled graph generators and draw conclusions about the contribution of discrete vertex labels and edge weights to graph structure.

Keywords: Network models · Graph generators · Random graphs · Labelled graphs · Weighted graphs · Graph mining

1 Introduction

Network analysis is concerned with finding patterns and anomalies in real-world graphs, such as social networks, computer and communication networks, or biological and ecological processes. Real graphs exhibit a number of interesting structural and evolutionary properties, such as power-law or log-normal degree distribution, small diameter, shrinking diameter, and the Densification Power Law (DPL) [6,19,21].

Besides discovering network properties, researchers are interested in the mechanisms of network formation. Generative graph models provide an abstraction of how graphs form: if the model is accurate, generated graphs will obey the same properties as real graphs. Generated graphs are also useful for simulation experiments, hypothesis testing and making predictions about graph

A. Appice et al. (Eds.): NFMCP 2013, LNAI 8399, pp. 181–200, 2014.
DOI: 10.1007/978-3-319-08407-7_12, © Springer International Publishing Switzerland 2014

evolution or missing graph elements. Most existing models are for unlabelled, unweighted graphs [6,19], but some models take discrete vertex labels into account [13,17,22].

In this paper, we present AGWAN, a generative model for labelled, weighted graphs. Weights are commonly used to represent the number of occurrences of each edge: the number of e-mails sent between individuals in a social network [1]; the number of calls to a subroutine in a software call graph [9]; or the number of people walking between a pair of door sensors in a building access control network [8]. In other applications, the edge weight may represent continuous values: donation amounts in a bipartite graph of donors and political candidates [1]; distance or speed in a transportation network [9]; or elapsed time to walk between the sensors in the building network [8]. In some cases, the weight is a multi-dimensional feature vector [8,9].

Our main motivation for this work is to create a model to better understand the laws governing the relationship between graph structure and numeric labels or weights. Furthermore, we want to be able to create realistic random, labelled, weighted graphs for large-scale simulation experiments for our pattern discovery algorithms [8]. Our experiments in Sect. 5 show the extent to which various graph properties are related to labels and weights, and measure exactly how "realistic" our random graphs are. Graphs generated with AGWAN are shown to have more realistic vertex strength distributions and spectral properties than the comparative methods.

This paper is arranged as follows: Sect. 2 is an overview of generative graph models; Sect. 3 presents AGWAN, our generative model for weighted and numeric labelled graphs. We include a fitting algorithm to learn AGWAN's parameters from a real input graph, and an algorithm to generate random graphs from the model. Section 4 gives an overview of the datasets that we use in the experiments, and outlines the statistical measures and tests that we use to evaluate the generated graphs. The experiments in Sect. 5 demonstrate that the vertex labels and edge weights of a graph can predict the graph structure with high accuracy. Conclusions are in Sect. 6.

2   Related Work

Our understanding of the mathematical properties of graph structure was pioneered by Paul Erdős and Alfréd Rényi [10]. Graph formation is modelled as a Bernoulli process, parameterised by the number of vertices and a wiring probability between each vertex pair. While it has been essential to our understanding of component sizes and expected diameter, the Erdős-Rényi model does not explain other important properties of real-world graphs such as degree distribution, transitivity and clustering [6,21].

Barabási and Albert's Preferential Attachment model [2] uses the "rich get richer" principle to grow graphs from a few vertices up to the desired size. The probability of an edge is proportional to the number of edges already connected to a vertex. This generates graphs with power-law degree distributions. A number of variants of Preferential Attachment have been proposed [6,21]. Still, Preferential Attachment models lack some desired properties, such as community structure.

The RMat algorithm [7] solves the community structure problem with its recursive matrix approach. RMat graphs consist of 2^n vertices and E edges, with four probabilities a, b, c, d to determine in which quadrant of the adjacency matrix each edge falls. These parameters allow the specification of power-law or log-normal degree distributions; if $a = b = c = d$, the result will be an Erdős-Rényi graph.

Kronecker Graphs [19] fulfil all the properties mentioned above, as well as the DPL and shrinking diameter effect. The model starts with an initiator matrix. Kronecker multiplication is recursively applied to yield the final adjacency matrix of the desired size. This work synthesises the previous work in random graphs in a very elegant way and proves that RMat graphs are a special case of Stochastic Kronecker graphs.

The models above tend to have a small number of parameters and are analytically tractable, with simple and elegant proofs of the desired properties. However, graph labels are not taken into consideration. Stochastic models are another class of generative algorithm which may not be amenable to analytical proofs, but can be fit to real-world labelled graphs and used to learn the properties of those graphs. Models in this category include the Stochastic Block Model [22] and Latent Space approaches [13].

The Multiplicative Attribute Graph (MAG) model [17] draws on both of the above strands of research. MAG is parameterised by the number of vertices, a set of prior probabilities for vertex label values and a set of *affinity matrices* specifying the probability of an edge conditioned on the vertex labels. The affinity matrices can be learned from real graphs using Maximum Likelihood Estimation [16]. Reference [17] proves that Kronecker Graphs are a special case of MAG graphs, and that suitably-parameterized MAG graphs fulfil all the desired properties: log-normal or power-law degree distribution, small diameter, the existence of a unique giant component and the DPL. The MAG model considers discrete vertex labels only. We believe that our method, described in the next section, is the first generative model to include numeric labels or weights.

3   AGWAN: A Generative Model for Labelled, Weighted Graphs

In this section, we present our generative model, AGWAN (Attribute Graph: Weighted and Numeric). The model is illustrated in Fig. 1 for the Enron graph described in Sect. 4.

Consider a graph $G = (V, E)$ with discrete vertex label values drawn from a set L. In Fig. 1, $u, v \in V$ are vertices and $w_{uv}, w_{vu} \in \mathbb{R}$ are edge weights. Edges $e \in E$ are specified as a 3-tuple $\langle u, v, w_{uv} \rangle$. In the discussion which follows,

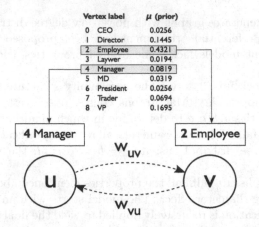

Fig. 1. AGWAN parameters. Vertex labels are selected according to prior probability μ. Edge weight w_{uv} is selected from mixture model Ω^{42} and w_{vu} is selected from mixture model Ω^{24}.

we restrict ourselves to a single label on each vertex; we outline how this can be extended to multiple labels in Sect. 3.3.

We must choose a suitable probability distribution to model the edge weights accurately and efficiently. The Gaussian distribution is popular as it has an analytically tractable Probability Density Function (PDF). However, the edge weights $W^{ij} = \{w_{ij}\}$ follow an arbitrary probability distribution which is not necessarily Gaussian. By using a weighted mixture of Gaussian components, we can get a reasonable approximation to any general probability distribution [3]. The resulting Gaussian Mixture Model (GMM) is quite flexible and is used extensively in statistical pattern recognition [15].

A parametric GMM can be used where we know the number of components in advance. In our case, the number of components—and therefore the number of parameters in the model—changes according to the data. We avoid the problem of knowing the "correct" number of components by using a non-parametric model. We assume that W^{ij} consists of an infinite number of components and use variational inference to determine the optimal number for our model [4].

The AGWAN model is parameterised by μ, a set of prior probabilities over L; and Θ, a set of edge weight mixture parameters: $\Theta = \{\Omega^{ij} | i, j \in L\}$. For directed graphs, $|\Theta| = |L|^2$ and we need to generate both w_{uv} and w_{vu} (see Fig. 1). For undirected graphs, $\Omega^{ij} = \Omega^{ji}$, so $|\Theta| = O(|L|^2/2)$ and $w_{vu} = w_{uv}$.

For each combination of vertex attributes $\langle i, j \rangle$, the corresponding mixture model Ω^{ij} parameterises the distribution of edge weights (with an edge weight of 0 indicating no edge). Ω^{ij} is a GMM with M Gaussian components:

$$\Omega^{ij} = \sum_{m=0}^{M-1} \omega_m^{ij} \cdot \eta(\mu_m^{ij}, (\sigma^2)_m^{ij}) \tag{1}$$

where ω_m^{ij} is the weight of each component and $\eta(\mu_m^{ij}, (\sigma^2)_m^{ij})$ is the Gaussian PDF with mean μ_m^{ij} and variance $(\sigma^2)_m^{ij}$. The mixture weights form a probability distribution over the components: $\sum_{m=0}^{M-1} \omega_m^{ij} = 1$. We can specify Ω^{ij} such that the first mixture component encodes the probability of no edge: $\omega_0^{ij} = 1 - P(e_{ij})$, where $P(e_{ij})$ is the probability of an edge between pairs of vertices with labels $\langle i, j \rangle$. The model degenerates to an unweighted graph if there are two components, $\eta_0(0, 0)$ and $\eta_1(1, 0)$. Furthermore, if the weights ω_m^{ij} are the same for all $\langle i, j \rangle$, the model degenerates to an Erdős-Rényi graph.

As the Gaussian distribution has unbounded support, GMMs can be used to model any set of continuous values. However, if the edge weight is a countable quantity representing the number of occurrences of the edge, then W^{ij} is bounded by $[0, \infty)$. Although this case can be modelled as a GMM, it requires a large number of mixture components to describe the data close to the boundary [20]. We consider alternatives to the GMM for the semi-bounded and bounded cases in Sect. 6.

3.1   Graph Generation

Algorithm 1 describes how to generate a random graph using $\mathrm{AGWAN}(N, L, \mu, \Theta)$. The number of vertices in the generated graph is specified by N. After assigning discrete label values to each vertex (lines 2–3, cf. Fig. 1), the algorithm checks each vertex pair $\langle u, v \rangle$ for the occurrence of an edge (lines 4–7). If $m = 0$, $\langle u, v \rangle$ is not an edge (line 7). If there is an edge, we assign its weight from mixture component m (lines 8–9). The generated graph is returned as $G = (V, E)$.

Algorithm 1. AGWAN Graph Generation

Require: N (no. of vertices), L (set of discrete label values), μ (prior distribution
  over L), $\Theta = \{\Omega^{ij}\}$ (set of mixture models)
1: Create vertex set V of cardinality N, edge set $E = \emptyset$
2: **for all** $u \in V$ **do**
3:    Assign discrete label $l_u \in L$ from prior μ
4: **for all** $u, v \in V : u \neq v$ **do**
5:    $i = l_u, j = l_v$
6:    Select Gaussian m uniformly at random from Ω^{ij}
7:    **if** $m \neq 0$ **then**
8:       Assign edge weight w_{uv} uniformly at random from $\eta(\mu_m^{ij}, (\sigma^2)_m^{ij})$
9:       Create edge $e = \langle u, v, w_{uv} \rangle, E = E \cup \{e\}$
  return $G = (V, E)$

3.2   Parameter Fitting

To create realistic random graphs, we need to learn the parameters μ, Θ from a real-world input graph G. Let W^{ij} be the set of edge weights between pairs of vertices with labels $\langle i, j \rangle$. During parameter fitting, we want to create a model Ω^{ij}

for each W^{ij} in G. Each GMM Ω^{ij} has a finite number of mixture components M. If M is known, Ω^{ij} can be estimated using Expectation Maximisation [12]. However, not only is M unknown, but we expect that it will be different for each Ω^{ij} within a given graph model [8].

We solve this problem by modelling Ω^{ij} as a non-parametric mixture model with an unbounded number of mixture components: a Dirichlet Process Gaussian Mixture Model (DPGMM) [4]. "Non-parametric" does not mean that the model has no parameters; rather, the number of parameters is allowed to grow as more data are observed. In essence, the DPGMM is a probability distribution over the probability distributions of the model.

The Dirichlet Process (DP) over edge weights W^{ij} is a stochastic process $DP(\alpha, H_0)$, where α is a positive scaling parameter and H_0 is a finite measure on W^{ij}; that is, a mapping of the subsets of W^{ij} to the set of non-negative real numbers. If we draw a sample from $DP(\alpha, H_0)$, the result is a random distribution over values drawn from H_0. This distribution H is discrete, represented as an infinite sum of atomic measures. If H_0 is continuous, then the infinite set of probabilities corresponding to the frequency of each possible value that H can return are distributed according to a *stick-breaking process*. The stick-breaking representation of H is given as:

$$\omega_m^{ij}(\mathbf{x}) \prod_{n=1}^{m-1} (1 - \omega_n^{ij}) \qquad\qquad H = \sum_{n=1}^{\infty} \omega_n^{ij}(\mathbf{x}) \delta_{\eta_m^*} \qquad (2)$$

where $\{\eta_1^*, \eta_2^*, \ldots\}$ are the atoms representing the mixture components. We learn the mixture parameters using the variational inference algorithm for generating Dirichlet Process Mixtures described in [4]. The weights of each component are generated one-at-a-time by the stick-breaking process, which tends to return the components with the largest weights first. In our experiments, 3–5 mixture components was sufficient to account for over 99 % of the data. Mixtures with weights summing to less than 0.01 are dropped from the model, and the remaining weights $\{\omega_m^{ij}\}$ are normalised.

Algorithm 2. AGWAN Parameter Fitting

Require: Input graph $G = (V, E)$
1: $L = \{$discrete vertex label values$\}$, $d = |L|$
2: Calculate vertex label priors, apply Laplace smoothing $\forall l \in L : P(l) = \frac{count(l) + \alpha}{N + \alpha d}$
3: $\mu = $ the normalised probability distribution over L such that $\sum_{i=1}^{d} P(l_i) = 1$
4: $\forall i, j \in L : W^{ij} = \emptyset$
5: **for all** $u, v \in V : u \neq v$ **do**
6:     $i = l_u, j = l_v$
7:     $W^{ij} = W^{ij} \cup \{w_{uv}\}$ ▷ If $\langle u, v \rangle$ is not an edge, then w_{uv} has value zero
8: **for all** $i, j \in L$ **do**
9:     estimate Ω^{ij} from W^{ij} using variational inference
10: $\Theta = \{\Omega^{ij}\}$
        return μ, Θ

Algorithm 2 is the algorithm for AGWAN parameter fitting. First, we estimate the vertex priors (lines 1–3). Next, we sample the edge weights for each possible combination of vertex label values, with no edge counting as a weight of zero (lines 4–7). Finally, we estimate the GMMs Ω^{ij} from the appropriate set of samples W^{ij} using the stick-breaking process described above.

3.3   Extending AGWAN to Multiple Attributes

We have presented AGWAN for a single discrete vertex label and a single numeric edge label (the weight). Many graphs have multiple labels on vertices and edges. AGWAN can be extended to multiple numeric edge labels by generalising the concept of edge weight to k dimensions. In this case, the mean of each mixture component becomes a k-dimensional vector and the variance $(\sigma_m^{ij})^2$ is replaced with the $k \times k$ covariance matrix Σ_m^{ij}. The variational algorithm can be accelerated for higher-dimensional data using a kd-tree [18] and has been demonstrated to work efficiently on datasets of hundreds of dimensions.

A more difficult question is how to extend the model to multiple discrete vertex labels. With even a small number of labels, modelling the full joint probability across all possible combinations of label values becomes a complex combinatorial problem with hundreds or thousands of parameters. The MAG model reduces this complexity by assuming that vertex labels are independent, so edge probabilities can be computed as the product of the probabilities from each label [17]. For latent attributes, MAGFIT enforces independence by regularising the variational parameters using mutual information [16]. However, the MAG model has not solved this problem for real attributes, where independence cannot be assumed. Furthermore, multiplying the probabilities sets an upper limit (proportional to $\log N$) on the number of attributes which can be used in the model. In our experiments (Sect. 5), MAG typically produced the best results with one or two latent variables.

An alternative to multiplying independent probabilities is to calculate the GMM for each edge as the weighted summation of the GMM for each individual attribute. It is likely that some attributes have a large influence on graph structure while others affect it little or not at all. The contribution of each attribute could be estimated using a conditional probability distribution as an approximation to the joint probability, for example using Markov Random Fields (MRF) or Factor Graphs. This problem remains a topic for further research.

4   Experiments

We evaluate our approach by comparing AGWAN with the state-of-the-art in labelled graph generation, represented by the MAG model [16,17]. AGWAN and MAG parameters are learned from real-world graphs. We generate random graphs from each model and calculate a series of statistics on each graph. These statistics are used to compare how closely the model maps to the input graph.

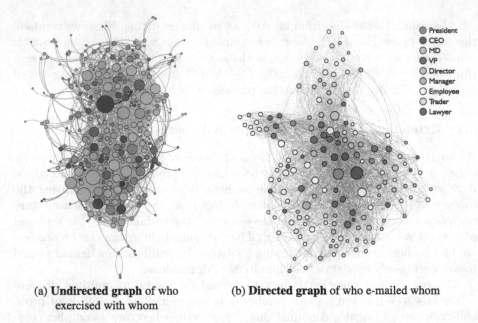

(a) **Undirected graph** of who   (b) **Directed graph** of who e-mailed whom
exercised with whom

Fig. 2. Input graph datasets, from (a) a health study and (b) the Enron e-mail corpus

Our input datasets are a graph of "who exercised with whom" from a behavioural health study [14] (Fig. 2a, $|V| = 279, |E| = 1308$) and the "who communicates with whom" graph of the Enron e-mail corpus [1] (Fig. 2b, $|V| = 159, |E| = 2667$). Vertices in the health study graph are labelled with 28 attributes representing demographic information and health markers obtained from questionnaire data. Edges are undirected and weighted with the number of mutual coincidences between actors during the study. Vertices in the Enron graph are labelled with the job role of the employee. As e-mail communications are not symmetric, edges are directed and weighted with the number of e-mails exchanged between sender and recipient.

We evaluated AGWAN against the following models:

Erdős-Rényi Random Graph (ER): The ER model $G(n,p)$ has two parameters. We set the number of vertices n and the edge probability p to match the input graphs as closely as possible. We do not expect a very close fit, but the ER model provides a useful baseline.

MAG with Real Attributes (MAG-R1): The MAG model with one real attribute is similar to AGWAN with one real attribute, with the difference that the set of GMMs $\Theta = \{\Omega^{ij}\}$ is replaced with a set of binary edge probabilities, $\Theta = \{p^{ij}\}$.

MAG with Latent Attributes (MAG-Lx): The MAG model also allows for modelling the graph structure using latent attributes. The discrete labels provided in the input graph are ignored; instead MAGFIT [16] learns the values

of a set of latent attributes to describe the graph structure. To investigate the
relative contributions of vertex labels and edge weights to graph structure, we
compared MAG models with $x = 1 \ldots 9$ latent binary attributes against AGWAN
models with synthetic attributes taking $2^0 \ldots 2^9$ values.

As ER and MAG do not generate weighted graphs, we set the weight of the
edges in the generated graphs to the mean edge weight from the input graphs.
This ensures that statistics such as average vertex strength are not skewed by
unweighted edges.

To evaluate the closeness of fit of each model, we use the following statistics:

Vertex Strength: For an unweighted graph, one of the most important measures is the degree distribution (the number of in-edges and out-edges of each
vertex). Real-world graphs tend to have heavy-tailed power-law or log-normal
degree distributions [6,21]. For a weighted graph, we generalise the concept of
vertex degree to vertex strength [11]:

$$s_u = \sum_{v \neq u} w_{uv} \tag{3}$$

For the undirected graphs, we plot the Complementary Cumulative Distribution
Function (CCDF) of the total strength of each vertex. For the directed graphs,
we plot the CCDFs for in-strength and out-strength.

Spectral Properties: We use Singular Value Decomposition (SVD) to calculate
the singular values and singular vectors of the graph's adjacency matrix, which
act as a signature of the graph structure. In an unweighted graph, the adjacency
matrix contains binary values, for "edge" or "no edge". In a weighted graph,
the adjacency matrix contains the edge weights (with 0 indicating no edge). For
SVD $U\Sigma V$, we plot Cumulative Distribution Functions (CDFs) of the singular
values Σ and the components of the left singular vector U corresponding to the
highest singular value.

Clustering Coefficients: the clustering coefficient C is an important measure
of community structure. It measures the density of triangles in the graph, or the
probability that two neighbours of a vertex are themselves neighbours [21]. We
extend the notion of clustering coefficients to weighted, directed graphs using
the equation in [11]:

$$C_u = \frac{[\mathbf{W}_u^{[\frac{1}{3}]} + (\mathbf{W}_u^T)^{[\frac{1}{3}]}]_{uu}^3}{2[d_u^{tot}(d_u^{tot} - 1) - 2d_u^{\leftrightarrow}]} \tag{4}$$

where C_u is the weighted clustering coefficient for vertex u, $\mathbf{W}_u$ is the weighted
adjacency matrix for u and its neighbours, $\mathbf{W}^T$ is the transpose of $\mathbf{W}$, d_u^{tot} is
the total degree of a vertex (the sum of its in- and out-degrees) and $d_u^{\leftrightarrow}$ is the
number of bilateral edges in u (the number of neighbours of u which have both
an in-edge and an out-edge between themselves and u).

Triad Participation: Closely related to the clustering coefficient is the concept
of triangle or triad participation. The number of triangles that a vertex is connected to is a measure of transitivity [21]. For the directed graphs, the triangles

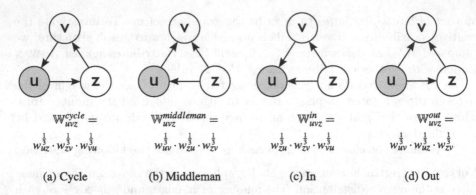

(a) Cycle   (b) Middleman   (c) In   (d) Out

Fig. 3. Triad patterns in a directed graph

have a different interpretation depending on the edge directions. There are four types of triangle pattern [11], as shown in Fig. 3. To generalise the concept of triad participation to weighted, directed graphs, we consider each of the four triangle types separately, and sum the total strength of the edges in each triad:

$$t^y_u = \sum_{v,z \in \mathbf{W}_u \setminus u} \mathbb{W}^y_{uvz} \tag{5}$$

where $y = \{cycle, middleman, in, out\}$ is the triangle type and $\mathbb{W}^y_{uvz}$ is calculated as shown in Fig. 3 for each triangle type y.

To give a more objective measure of the closeness of fit between the generated graphs and the input graph, we use a Kolmogorov-Smirnov (KS) test and the L2 (Euclidean) distance between the CDFs for each statistic. As the CDFs are for heavy-tailed distributions, we use the logarithmic variants of these measures [16]. The KS and L2 statistics are calculated as:

$$KS(D_1, D_2) = max_x |\log D_1(x) - \log D_2(x)| \tag{6}$$

$$L2(D_1, D_2) = \sqrt{\frac{1}{\log b - \log a} \sum_{x=a}^{b} (\log D_1(x) - \log D_2(x))^2} \tag{7}$$

where $[a, b]$ is the interval for the support of distributions D_1 and D_2.

The model that generates graphs with the lowest KS and L2 values for each of the statistics discussed above has the closest fit to the real-world graph.

5   Results

For each model, we generated 10 random graphs and calculated statistics for each. The plots of the averaged CDFs of the 10 graphs for each model are shown in Figs. 4, 5, 6, 7, 8, 9, 10 and 11. Tables 1, 2, 3, 4, 5, 6, 7 and 8 for the closeness of fit of each CDF (KS and L2 statistics) are in the appendix.

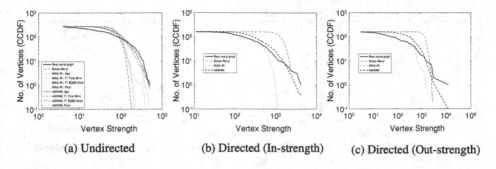

(a) Undirected   (b) Directed (In-strength)   (c) Directed (Out-strength)

Fig. 4. Vertex strength distribution—real attributes

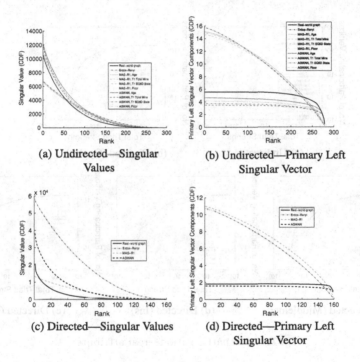

(a) Undirected—Singular   (b) Undirected—Primary Left
          Values   Singular Vector

(c) Directed—Singular Values   (d) Directed—Primary Left
                                      Singular Vector

Fig. 5. Spectral properties—real attributes

5.1 Real Attributes

For the undirected graph (Health Study, Fig. 2a), we show results for four vertex attributes: age; total minutes spent exercising; EQ5D State (a quality-of-life metric determined by questionnaire); and Floor (the building and floor number where the person works; people who work on the same floor were highly likely to exercise together). For the directed graph (Enron, Fig. 2b), we have one vertex attribute, the person's job role.

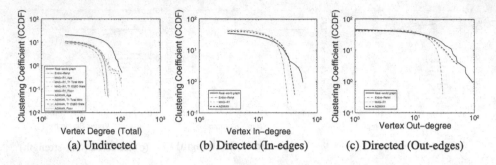

Fig. 6. Clustering coefficients—real attributes

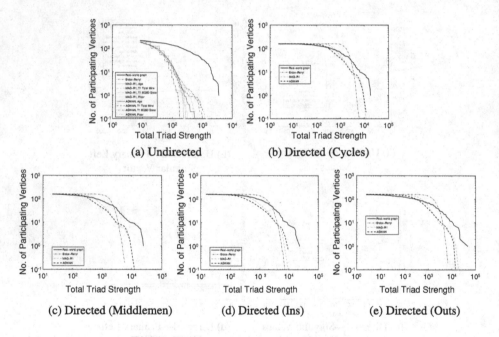

Fig. 7. Triad participation—real attributes

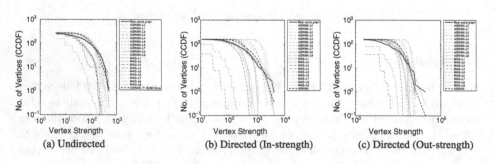

Fig. 8. Vertex strength distribution—synthetic attributes

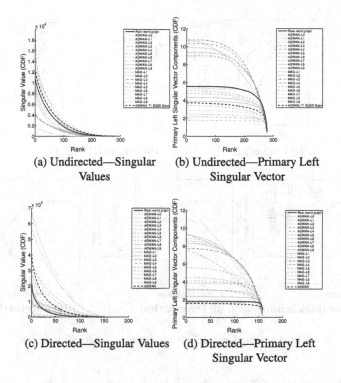

(a) Undirected—Singular Values

(b) Undirected—Primary Left Singular Vector

(c) Directed—Singular Values

(d) Directed—Primary Left Singular Vector

Fig. 9. Spectral properties—synthetic attributes

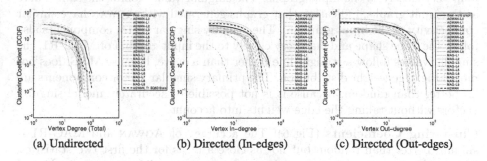

(a) Undirected

(b) Directed (In-edges)

(c) Directed (Out-edges)

Fig. 10. Clustering coefficients—synthetic attributes

Vertex Strength (Fig. 4): The graphs generated from AGWAN have vertex strength distributions which map very closely to the input graphs. The graphs generated from MAG-R1 are better than random (ER), but the vertex strength distribution is compressed into the middle part of the range, with too few high- and low-strength vertices. This indicates that vertex strength depends on both the label distribution and the edge weight distribution; AGWAN models both of these, whereas MAG models only the former.

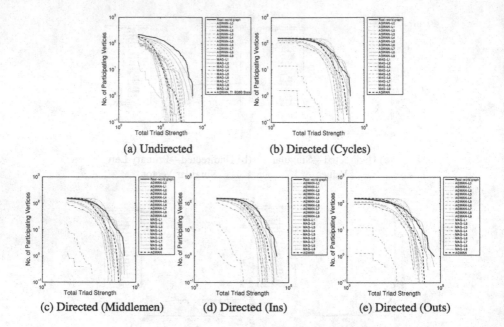

(a) Undirected   (b) Directed (Cycles)

(c) Directed (Middlemen)   (d) Directed (Ins)   (e) Directed (Outs)

Fig. 11. Triad participation—synthetic attributes

Spectral Properties (Fig. 5): The spectral properties of the AGWAN graphs map very closely to the input graphs. The singular values follow the same curve as the input graphs, indicating that graphs generated with AGWAN have similar connectivity to the input graph [6]. The primary singular vector components also follow the same shape and map very closely to the input graph. For MAG-R1, the singular values follow a straight line rather than a curve, because MAG does not model the edge weight distribution. The primary singular vector components are no better than random, because it is not possible to accurately model singular vectors without taking the edge weights into account.

Clustering Coefficients (Fig. 6): The accuracy of AGWAN and MAG-R1 is similar; better than random but not as close a fit as for the first two statistics. The results for vertex strength and spectral properties did not strongly depend on which attribute was chosen, but here it makes a difference: Total Mins and EQ5D State give better results than Age and Floor. This implies that some attributes can predict community formation better than others. As the results for both approaches are similar, we conclude that the processes that give rise to clustering are independent of the edge weight distribution.

Triad Participation (Fig. 7): As triad participation is closely related to clustering, it is no surprise that the results are comparable: the accuracy of AGWAN and MAG-R1 is similar; better than random, but not as close as for vertex strength and spectral properties. Triad participation appears to be dependent to some extent on vertex label values but independent of the edge weight distribution.

One of the findings in [16] was that clustering arises from multiple processes (homophily and core/periphery). "Simplified MAG" (where all attributes are the same) could not model the clustering property, implying that it is not possible to accurately reproduce clustering when the model has only one attribute. We propose to extend our model to more than one attribute as outlined in Sect. 3.3 to investigate whether this produces a more accurate model of clustering and triad participation.

5.2   Synthetic Attributes

An alternate interpretation of the MAG model ignores the true attribute values from the input graph and represents attributes as latent variables, which are learned using a variational inference EM approach [16]. To compare AGWAN with this approach, we replaced the real labels in the input graph with a synthetic vertex attribute taking $2^0 \ldots 2^9$ values allocated uniformly at random, then learned the edge weight distributions using variational inference as normal. We have plotted AGWAN with one real attribute alongside for comparison.

Vertex Strength (Fig. 8): AGWAN with synthetic attributes has similar accuracy to AGWAN-R1. Varying the number of synthetic attributes has a small effect on the accuracy. MAG with latent attributes has similar accuracy to MAG-R1. Varying the number of synthetic attributes causes a large variation in the accuracy. We conclude that vertex strength is dependent on both edge weight and vertex label distribution, but the edge weights play a more important role.

Spectral Properties (Fig. 9): For AGWAN, the spectral properties follow the same curves as the input graphs. For singular values, varying the number of synthetic attributes causes a small variation in the closeness of fit. For singular vectors, the accuracy is highly dependent on the number of synthetic attributes. For MAG, the singular values are almost a straight line, as the edge weight distribution is not taken into account. The singular vectors in general do not match very closely. It is possible to get a good fit using many latent attributes, but this compromises the other statistics which fit better with few latent attributes. We conclude that spectral properties are dependent on both edge weight and vertex label distribution.

Clustering Coefficients (Fig. 10): Both approaches are significantly more accurate using synthetic attributes than they were with real attributes. This implies that while real labels are influenced by the (unobserved) process which gives rise to clustering, synthetic labels with more degrees of freedom can model it more accurately. As before, clustering appears to be independent of the edge weight distribution.

Triad Participation (Fig. 11): As with clustering, synthetic vertex labels can model the process that gives rise to triad participation, while edge weights have little or no influence.

In general, MAG achieves the best results when there are one or two vertex attributes, whereas AGWAN performs best when there are 7 or 8 attributes. MAG assumes that each attribute is independent, so there is a limit on the number of attributes that can be included in the model (proportional to $\log N$). Above this limit, the performance of the model degrades. With AGWAN, there is no independence assumption, so the attributes model the full joint probability. As the number of attribute values (2^x) approaches N, there is a danger of overfitting and the model performance degrades.

6   Conclusions

We presented AGWAN, a model for random graphs with discrete labels and weighted edges. We included a fitting algorithm to learn a model of graph edge weights from real-world data, and a generative algorithm to generate random labelled, weighted graphs with similar characteristics to the real-world graph.

We measured the closeness of fit of our generated graphs to the input graph over a range of graph statistics, and compared our approach to the state-of-the-art in random graph generative algorithms. Our results demonstrate that AGWAN produces an accurate model of the properties of a weighted real-world graph. For vertex strength distribution and spectral properties, AGWAN is shown to produce a closer fit than MAG.

For clustering and triad participation, we achieved a closer fit using synthetic attributes than using real attributes. This is consistent with the results for MAG for unweighted graphs [17]. Further research is required into the relationship between vertex attributes and triangle formation in graphs; our results indicate that edge weights do not play an important part in these processes. We propose to extend AGWAN to multiple vertex labels to investigate the effect on clustering.

In Sect. 3, we considered the case where edge weights are countable quantities bounded by $[0, \infty)$. As GMMs are unbounded, it may be more appropriate to model the edge weights using a truncated GMM [20] or Beta Mixture Model [5]. We propose to investigate these alternatives in future work.

As discussed in Sect. 3.3, MAG's method of combining multiple vertex attributes is unsatisfactory when applied to real attributes, due to the assumption of independence and the limit on the number of attributes which can be modelled. We have proposed a future line of research based on a weighted summation of the GMM for each edge. The fitting algorithm would need to regularise the individual contributions of each edge to take account of dependencies. The complexity of modelling the full joint distribution could be reduced with an approach based on Markov Random Fields or Factor Graphs.

Appendix: KS and L2 Statistics

Table 1. KS statistic for undirected graph, real attributes (Figs. 4, 5, 6, 7)

	E-R	MAG-R1				AGWAN			
		Age	Total Mins	EQ5D State	Floor	Age	Total Mins	EQ5D State	Floor
Vertex Strength	6.064	5.940	2.957	3.689	5.799	0.799	1.081	**0.635**	1.674
Singular Values	36.193	35.644	35.393	35.612	36.001	34.482	**32.319**	33.720	34.946
Singular Vector	1.323	1.239	0.964	0.984	1.134	**0.248**	0.491	0.450	0.371
Clustering Coefficient	5.224	5.048	2.083	3.343	4.895	5.132	2.493	**2.042**	5.161
Triad Participation	7.012	6.877	5.704	5.704	6.685	6.328	**5.106**	5.829	6.768

Table 2. L2 statistic for undirected graph, real attributes (Figs. 4, 5, 6, 7)

	E-R	MAG-R1				AGWAN			
		Age	Total Mins	EQ5D State	Floor	Age	Total Mins	EQ5D State	Floor
Vertex Strength	9.686	7.281	8.265	9.377	10.039	1.829	2.589	**1.765**	3.294
Singular Values	41.815	41.298	41.052	41.227	41.623	39.629	**38.211**	39.100	40.060
Singular Vector	5.004	4.940	4.614	4.742	4.852	**1.486**	3.257	2.914	2.307
Triad Participation	**16.879**	17.334	18.828	18.861	17.101	19.746	19.434	18.348	20.288

Table 3. KS statistic for directed graph, real attributes (Figs. 4, 5, 6, 7)

	E-R	MAG-R1	AGWAN
In-Vertex Strength	2.469	4.700	**1.455**
Out-Vertex Strength	2.708	2.659	**2.303**
Singular Values	37.235	35.752	**34.894**
Singular Vector	1.915	1.801	**0.282**
Clustering Coefficient (In-Edges)	3.444	**2.208**	2.220
Clustering Coefficient (Out-Edges)	3.728	0.769	**0.702**
Clustering Coefficient	4.347	**1.651**	3.163
Triad Participation (Cycles)	4.787	4.248	**3.555**
Triad Participation (Middlemen)	4.382	**4.500**	**4.500**
Triad Participation (Ins)	4.700	4.500	**2.436**
Triad Participation (Outs)	4.382	**4.094**	4.248

Table 4. L2 statistic for directed graph, real attributes (Figs. 4, 5, 6, 7)

	E-R	MAG-R1	AGWAN
In-Vertex Strength	5.679	4.912	**1.816**
Out-Vertex Strength	5.100	3.534	**2.117**
Singular Values	25.044	19.546	**18.360**
Singular Vector	7.316	7.587	**0.988**
Clustering Coefficient (In-Edges)	3.528	1.607	**1.528**
Clustering Coefficient (Out-Edges)	3.145	1.101	**1.002**
Clustering Coefficient	6.949	**1.438**	2.284
Triad Participation (Cycles)	3.823	**3.000**	3.101
Triad Participation (Middlemen)	5.144	**4.178**	4.207
Triad Participation (Ins)	4.630	4.826	**4.332**
Triad Participation (Outs)	3.727	3.295	**3.203**

Table 5. KS statistic for undirected graph, synthetic attributes (Figs. 8, 9, 10, 11)

| | MAG Latent | | | | | | | | | AGWAN |
	1	2	3	4	5	6	7	8	9	EQ5D State
Vertex Strength	**2.243**	5.106	5.886	5.886	5.670	5.481	4.605	5.561	6.234	0.635
Singular Values	**30.901**	46.771	89.148	93.658	81.082	93.413	125.855	72.059	85.863	33.720
Singular Vector	0.645	0.654	0.821	0.694	0.590	0.561	0.645	0.579	**0.313**	0.450
Clustering Coefficient	**1.283**	4.406	3.863	4.575	4.401	3.470	3.256	4.397	4.773	2.042
Triad Participation	**3.829**	6.292	6.709	6.593	6.016	5.768	4.868	5.914	6.877	5.829

| | AGWAN | | | | | | | | | |
	0	1	2	3	4	5	6	7	8	9
Vertex Strength	3.401	2.197	2.303	1.050	1.758	0.916	0.975	0.875	**0.854**	1.589
Singular Values	35.238	35.194	35.226	35.341	35.542	33.763	32.824	**27.713**	34.052	37.384
Singular Vector	0.675	0.827	0.847	0.950	1.139	0.559	**0.183**	0.221	0.258	0.361
Clustering Coefficient	5.353	5.350	3.561	4.615	4.395	4.054	4.470	3.676	**3.401**	3.440
Triad Participation	6.985	7.090	6.994	5.991	5.872	6.607	6.131	5.561	2.238	**1.204**

Table 6. L2 statistic for undirected graph, synthetic attributes (Figs. 8, 9, 10, 11)

| | MAG Latent | | | | | | | | | AGWAN |
	1	2	3	4	5	6	7	8	9	EQ5D State
Vertex Strength	**7.944**	8.473	9.236	10.783	10.103	8.635	9.120	9.603	21.027	1.765
Singular Values	**55.080**	94.881	106.265	109.813	104.160	109.673	120.108	113.166	173.884	39.100
Singular Vector	3.231	3.324	3.895	3.622	2.894	3.092	2.873	3.079	**0.396**	2.914
Triad Participation	12.047	15.550	15.821	17.494	11.038	11.646	**10.367**	14.507	29.136	18.348

| | AGWAN | | | | | | | | | |
	0	1	2	3	4	5	6	7	8	9
Vertex Strength	6.266	4.537	3.754	2.584	2.160	1.731	1.343	0.873	**0.693**	1.229
Singular Values	40.448	40.394	40.391	40.504	40.873	38.980	37.613	**27.296**	44.148	74.019
Singular Vector	4.477	5.513	5.671	6.316	7.530	3.612	**0.866**	1.237	1.719	2.351
Triad Participation	22.841	20.975	23.682	17.878	17.287	16.174	15.254	10.310	5.753	**2.803**

Table 7. KS statistic for directed graph, synthetic attributes (Figs. 8, 9, 10, 11)

| | MAG Latent | | | | | | | | | AGWAN |
	1	2	3	4	5	6	7	8	9	Employee Type
In-Vertex Strength	4.700	**3.602**	5.991	6.522	6.142	5.704	3.951	5.347	5.193	1.455
Out-Vertex Strength	4.942	4.605	5.768	5.991	6.234	5.075	4.317	3.466	**3.401**	2.303
Singular Values	35.715	35.591	27.492	89.063	148.080	32.392	**1.708**	31.555	37.163	34.894
Singular Vector	1.636	1.630	1.453	**0.190**	0.765	1.586	1.525	1.526	1.552	0.282
Clustering Coefficient (In-Edges)	2.961	**0.897**	4.775	5.294	4.578	4.357	3.302	3.770	4.512	2.220
Clustering Coefficient (Out-Edges)	3.164	**0.513**	5.193	5.877	5.463	4.363	3.142	3.273	2.865	0.702
Clustering Coefficient	3.278	**2.347**	5.251	6.255	5.839	4.387	3.739	4.339	4.000	3.163
Triad Participation (Cycles)	3.912	**2.996**	5.347	5.940	6.867	5.247	4.094	3.843	5.704	3.555
Triad Participation (Middlemen)	4.248	**3.401**	4.942	5.920	6.319	4.339	3.602	3.689	5.858	4.500
Triad Participation (Ins)	**3.912**	3.912	5.670	5.940	7.170	5.704	4.700	5.075	6.153	2.436
Triad Participation (Outs)	**1.476**	2.526	4.571	5.695	6.768	5.075	4.500	4.094	4.745	4.248

| | AGWAN | | | | | | | | | |
	0	1	2	3	4	5	6	7	8	9
In-Vertex Strength	2.418	2.513	2.345	2.590	**1.120**	2.303	1.897	2.015	2.303	0.693
Out-Vertex Strength	2.996	2.234	2.090	4.248	**1.122**	1.150	1.514	1.966	1.386	1.204
Singular Values	37.497	37.866	37.377	36.590	36.159	34.801	33.812	32.696	**26.494**	8.327
Singular Vector	1.887	1.962	1.811	1.665	**0.616**	1.130	0.824	0.908	0.887	0.789
Clustering Coefficient (In-Edges)	3.477	3.567	4.386	4.159	3.704	3.682	2.678	0.662	**0.460**	0.492
Clustering Coefficient (Out-Edges)	4.945	4.316	5.134	4.969	4.948	4.747	**2.563**	3.200	2.605	3.204
Clustering Coefficient	4.580	4.018	4.837	2.691	4.369	3.933	1.501	**1.075**	2.620	0.848
Triad Participation (Cycles)	4.500	4.500	2.659	3.912	3.602	3.283	2.996	3.912	**1.204**	1.548
Triad Participation (Middlemen)	4.787	4.787	4.094	5.247	3.843	3.314	3.807	4.248	**1.609**	1.099
Triad Participation (Ins)	4.700	4.700	4.007	5.298	4.700	3.283	2.862	4.094	**1.609**	1.099
Triad Participation (Outs)	4.942	4.942	3.624	4.094	3.977	3.912	3.114	2.862	**1.696**	0.916

Table 8. L2 statistic for directed graph, synthetic attributes (Figs. 8, 9, 10, 11)

| | MAG Latent | | | | | | | | | AGWAN |
	1	2	3	4	5	6	7	8	9	Employee Type
In-Vertex Strength	5.023	**3.055**	8.856	19.820	15.718	8.678	6.171	8.672	7.066	1.816
Out-Vertex Strength	3.001	3.704	7.805	14.329	10.882	3.740	3.120	**2.668**	3.737	2.117
Singular Values	19.285	18.938	13.768	90.672	160.831	28.601	**6.158**	28.074	38.490	18.360
Singular Vector	7.470	7.530	7.100	**0.388**	4.062	7.453	7.200	7.266	7.339	0.988
Clustering Coefficient (In-Edges)	2.507	**1.786**	6.733	12.533	7.692	5.841	4.184	5.705	4.819	1.528
Clustering Coefficient	2.450	**2.419**	10.611	22.886	13.922	5.851	4.568	5.381	7.653	2.284
Triad Participation (Cycles)	2.060	**1.800**	7.788	15.981	16.270	6.781	6.121	5.378	8.763	3.101
Triad Participation (Middlemen)	2.828	**1.771**	11.094	19.126	18.575	7.016	7.204	6.517	11.150	4.207
Triad Participation (Ins)	3.293	**1.902**	11.473	12.061	16.361	9.756	8.905	9.124	13.740	4.332
Triad Participation (Outs)	**1.459**	1.816	6.646	17.093	14.603	5.950	5.399	4.698	6.315	3.203

| | AGWAN | | | | | | | | | |
	0	1	2	3	4	5	6	7	8	9
In-Vertex Strength	5.638	5.774	5.473	4.355	3.151	2.071	1.367	**1.299**	1.412	0.665
Out-Vertex Strength	5.128	4.807	4.732	4.756	3.060	2.224	1.918	2.034	**1.415**	1.045
Singular Values	25.020	25.815	24.922	22.017	20.767	18.270	16.748	15.010	**12.516**	8.758
Singular Vector	7.814	6.764	7.798	5.949	**2.643**	5.471	4.088	4.421	2.725	1.396
Clustering Coefficient (In-Edges)	3.987	4.972	5.834	4.314	3.846	3.413	2.575	1.524	**0.999**	0.686
Clustering Coefficient	7.065	8.188	9.244	6.606	7.581	6.872	4.951	**4.189**	3.658	2.536
Triad Participation (Cycles)	3.212	3.017	2.407	4.816	3.728	3.856	3.566	3.733	**1.113**	1.014
Triad Participation (Middlemen)	4.670	4.310	3.586	7.121	5.734	5.924	5.288	4.942	**2.382**	0.611
Triad Participation (Ins)	4.391	3.757	3.575	7.742	6.376	6.616	5.902	5.306	**2.464**	0.936
Triad Participation (Outs)	4.887	4.537	3.305	4.615	4.540	4.963	4.359	3.978	**1.947**	0.589

References

1. Akoglu, L., McGlohon, M., Faloutsos, C.: OddBall: spotting anomalies in weighted graphs. In: Zaki, M.J., Yu, J.X., Ravindran, B., Pudi, V. (eds.) PAKDD 2010. LNCS, vol. 6119, pp. 410–421. Springer, Heidelberg (2010)

2. Barabási, A.L., Albert, R.: Emergence of scaling in random networks. Science **286**(5439), 509–512 (1999)

3. Bishop, C.M.: Pattern Recognition and Machine Learning. Information Science and Statistics, 3rd edn. Springer, New York (2011)

4. Blei, D.M., Jordan, M.I.: Variational inference for Dirichlet process mixtures. Bayesian Anal. **1**, 121–144 (2005)

5. Bouguila, N., Ziou, D., Monga, E.: Practical Bayesian estimation of a finite Beta mixture through Gibbs sampling and its applications. Stat. Comput. **16**(2), 215–225 (2006)

6. Chakrabarti, D., Faloutsos, C.: Graph Mining: Laws, Tools, and Case Studies. Synthesis Lectures on Data Mining and Knowledge Discovery. Morgan & Claypool Publishers, San Rafael (2012)

7. Chakrabarti, D., Zhan, Y., Faloutsos, C.: R-MAT: a recursive model for graph mining In: Berry, M.W., Dayal, U., Kamath, C., Skillicorn, D.B. (eds.) SDM. SIAM (2004)

8. Davis, M., Liu, W., Miller, P.: Finding the most descriptive substructures in graphs with discrete and numeric labels. J. Intell. Inf. Syst. **42**(2), 307–332 (2014). http://dx.doi.org/10.1007/s10844-013-0299-7, DBLP. http://dblp.uni-trier.de

9. Eichinger, F., Huber, M., Böhm, K.: On the usefulness of weight-based constraints in frequent subgraph mining. In: Bramer, M., Petridis, M., Hopgood, A. (eds.) SGAI Conference, pp. 65–78. Springer, London (2010)

10. Erdős, P., Rényi, A.: On the evolution of random graphs. Publ. Math. Inst. Hung. Acad. Sci. **5**, 17–61 (1960)

11. Fagiolo, G.: Clustering in complex directed networks. Phys. Rev. E **76**(2), 026107 (2007)
12. Figueiredo, M.A.T., Jain, A.K.: Unsupervised learning of finite mixture models. IEEE Trans. Pattern Anal. Mach. Intell. **24**(3), 381–396 (2002)
13. Hoff, P.D., Raftery, A.E., Handcock, M.S.: Latent space approaches to social network analysis. J. Am. Stat. Assoc. **97**(460), 1090–1098 (2002)
14. Hunter, R.F., Davis, M., Tully, M.A., Kee, F.: The physical activity loyalty card scheme: development and application of a novel system for incentivizing behaviour change. In: Kostkova, P., Szomszor, M., Fowler, D. (eds.) eHealth 2011. LNICST, vol. 91, pp. 170–177. Springer, Heidelberg (2012)
15. Jain, A.K., Duin, R.P.W., Mao, J.: Statistical pattern recognition: a review. IEEE Trans. Pattern Anal. Mach. Intell. **22**(1), 4–37 (2000)
16. Kim, M., Leskovec, J.: Modeling social networks with node attributes using the Multiplicative Attribute Graph model. In: Cozman, F.G., Pfeffer, A. (eds.) UAI, pp. 400–409. AUAI Press, Corvallis (2011)
17. Kim, M., Leskovec, J.: Multiplicative Attribute Graph model of real-world networks. Internet Math. **8**(1–2), 113–160 (2012)
18. Kurihara, K., Welling, M., Vlassis, N.A.: Accelerated variational Dirichlet process mixtures. In: Schölkopf, B., Platt, J.C., Hoffman, T. (eds.) NIPS, pp. 761–768. MIT Press, Cambridge (2006)
19. Leskovec, J., Chakrabarti, D., Kleinberg, J.M., Faloutsos, C., Ghahramani, Z.: Kronecker graphs: an approach to modeling networks. J. Mach. Learn. Res. **11**, 985–1042 (2010)
20. Lindblom, J., Samuelsson, J.: Bounded support Gaussian mixture modeling of speech spectra. IEEE Trans. Speech Audio Process. **11**(1), 88–99 (2003)
21. Newman, M.: Networks: An Introduction. OUP, New York (2010)
22. Wang, Y., Wong, G.: Stochastic block models for directed graphs. J. Am. Stat. Assoc. **82**(397), 8–19 (1987)

Thresholding of Semantic Similarity Networks Using a Spectral Graph-Based Technique

Pietro Hiram Guzzi[✉], Pierangelo Veltri, and Mario Cannataro

Department of Medical and Surgical Sciences,
University Magna Graecia of Catanzaro, 88100 Catanzaro, Italy
{hguzzi,veltri,cannataro}@unicz.it

Abstract. The functional similarity among terms of an ontology is evaluated by using Semantic Similarity Measures (SSM). In computational biology, biological entities such as genes or proteins are usually annotated with terms extracted from Gene Ontology (GO) and the most common application is to find the similarity or dissimilarity among two entities through the application of SSMs to their annotations. More recently, the extensive application of SSMs yielded to the Semantic Similarity Networks (SSNs). SSNs are edge-weighted graphs where the nodes are concepts (e.g. proteins) and each edge has an associated weight that represents the semantic similarity among related pairs of nodes. Community detection algorithms that analyse SSNs, such as protein complexes prediction or motif extraction, may reveal clusters of functionally associated proteins. Because SSNs have a high number of arcs with low weight, likened to noise, the application of classical clustering algorithms on raw networks exhibits low performance. To improve the performance of such algorithms, a possible approach is to simplify the structure of SSNs through a preprocessing step able to delete arcs likened to noise. Thus we propose a novel preprocessing strategy to simplify SSNs based on an hybrid global-local thresholding approach based on spectral graph theory. As proof of concept we demonstrate that community detection algorithms applied to filtered (thresholded) networks, have better performances in terms of biological relevance of the results, with respect to the use of raw unfiltered networks.

Keywords: Semantic similarity measures · Semantic similarity networks

1 Introduction

The accumulation of raw experimental data about genes and proteins has been accompanied by the accumulation of functional information, i.e. knowledge about function. The assembly, organization and analysis of this data has given a considerable impulse to research [9].

Usually biological knowledge is encoded by using annotation terms, i.e. terms describing for instance function or localization of genes and proteins. Terms are organized into ontologies, that offer a formal framework to represent biological

A. Appice et al. (Eds.): NFMCP 2013, LNAI 8399, pp. 201–213, 2014.
DOI: 10.1007/978-3-319-08407-7_13, © Springer International Publishing Switzerland 2014

knowledge [16]. For instance, Gene Ontology (GO) provides a set of descriptions of biological aspects (namely GO Terms), structured into three main taxonomies: Molecular Function (MF), Biological Process (BP), and Cellular Component (CC). Terms are then linked to the related biological concept (e.g. proteins) by a process known as annotation. Then, for each protein a set of related terms (or annotations) is currently available and stored in publicly available databases, such as the Gene Ontology Annotation (GOA) database [7].

The similarity among terms belonging to the same ontology is usually evaluated by using Semantic Similarity Measures (SSM). A SSM takes in input two or more terms of the same ontology and produces as output a numeric value representing their similarity. Since proteins are annotated with set of terms extracted from an ontology, the use of SSMs to evaluate the functional similarity among proteins is becoming a common task. Consequently the use of SSMs to analyze biological data is gaining a broad interest from researchers [16].

More specifically, the existing analysis methods based on the use of SSMs may be categorized in four main classes: (i) the definition of ad-hoc semantic similarity measures tailored to the characteristics of the biomedical domain [17]; (ii) the definition of measures of comparison among genes and proteins; (iii) the introduction of methodologies for the systematic analysis of metabolic networks; and more recently (iv) the analysis of organisms on a system scale using the so called *semantic similarity networks* [28].

A semantic similarity network (SSN) of proteins is an edge-weighted graph $G_{ssu} = (V, E)$, where V is the set of proteins, E is the set of edges, and such that for each pair of proteins having a semantic similarity greater than zero, an edge connecting them is added to the graph. Each edge has an associated weight that represents the semantic similarity among the connected nodes.

The main class of algorithms that analyse SSNs are the so called community detection algorithms, such as protein complexes prediction or motif extraction, that may reveal clusters of functionally associated proteins. As we pointed out in [16], the probability to have a semantic similarity value greater than zero for any given pair of protein is quite high (especially for complex organisms). Consequently, resulting graph is often a complete graph with n nodes and $\frac{n \times (n-1)}{2}$ edges, containing a relevant number of meaningless edges.

Thus, to take into account the meaningless edges (e.g. edges with little weight), preprocessing algorithms can be used to remove meaningless edges improving SSNs analysis performance, e.g. community detection or clustering algorithms.

Usually such preprocessing is named *thresholding* because main algorithms set a similarity threshold and then prune all edges of the SSN with a weight lower than the threshold. The definition of a correct threshold value able to retain only the meaningful relationships is obviously relevant. Indeed, an high treshold value to define significant edge weight may implie loss of relationchips, while low value threshold may introduce a lot of noise.

Many methods for networks thresholding have been defined in several application fields: for instance the use of an arbitrary global threshold [14], the use of only a fraction of the highest relationship [1], as well as statistical-based methods [30]. Methods based on global threshold, prune all edges with weights lower than the threshold, while those based on local thresholds usually compute a different threshold for each node or group of nodes.

Internal characteristics of SSMs (as investigated in [27]) bring to exclude the use of global thresholds. In fact, small regions of relatively low similarities may be due to the characteristics of measures while proteins or genes have high similarity. The use of local threshold may constitute an efficient way, i.e. retaining only top k-edges for each node [20]. Although this consideration, this choice may be influenced by the presence of local noise and in general may cause the presence of biases in different regions.

Starting from these considerations, we developed a novel hybrid threshold-ing method employing both local and global approaches and based on spectral graph theory. We apply a local threshold for each node, i.e. we retain only edges whose weight is higher than the average of all its adjacent. The choice of the threshold is made by considering a global aspect: the evidentiaition of nearly-disconnected components. The evidence of the presence of these components is analyzed by calculating the eigenvalues of the Laplacian matrix [12,26]. The choice of this simplification has a biological counterpart on the structure of bio-logical networks. It has been proved in many works that these biological networks tend to have a modular structure in which hub proteins (i.e. relevant proteins) have many connections [3,22,33]. Hub proteins usually connect small modules (or communities), i.e. small dense regions with few link to other regions [31] in which proteins share a common function [18].

As proof-of-concept, we initially build different SSNs, then we apply different thresholding (varying the threshold level), and finally we mine these networks by using Markov Clustering [13]. We show that clustering algorithms on thresholded networks have in general better performances and that the best ones are reached when networks are nearly disconnected.

The rest of the paper is structured as follows: Sect. 2 discusses main related approaches, Sect. 3 presents the proposed approaches, Sect. 4 presents results through a case study, finally Sect. 5 concludes the paper.

2   Related Work

2.1   Spectral Graph Analysis

Spectral graph theory [10] refers to the study of the properties of a graph by looking at the properties of the eigenvalues and eigenvectors of matrices associ-ated to the graph. In particular, we focus on the Laplacian matrix of a graph as defined in [5,11].

Given an edge-weighted graph G with n nodes, the weighted adjacency matrix A is the $n \times n$ matrix in which the element $a_{i,j}$ is defined as:

$$a_{i,j} = \begin{cases} w_{i,j} \in (0,1] & \text{if i,j are connected;} \\ 0 & \text{if i,j are not connected} \end{cases} \tag{1}$$

For each node v_i the degree vol is defined as the sum of the weights of all the adjacent edges

$$vol_{v_i} = \Sigma_j w_{i,j} \tag{2}$$

Then we may define the Degree Matrix D as follows:

$$d_{i,j} = \begin{cases} vol_{v_i}, & \text{if i=j;} \\ 0, & \text{elsewhere} \end{cases} \tag{3}$$

Finally, the Laplacian Matrix L is defined as $L = D - A$ [23].

The eigenvector correspondent to the smallest nonzero eigenvalue of a Laplacian Matrix often referred to as Fiedler vector [25] is particularly relevant. The algebraic multiplicity of this eigenvalue in case of both un-weighted and weighted graphs corresponds to the number of connected components.

Starting from this consideration, Ding et al. [12] extended the analysis and they observed that nearly-disconnected components may also identified by analyzing the Fiedler vector.

2.2   Semantic Similarity Measures

A semantic similarity measure ($SSMs$) is a formal instrument to quantify the similarity of two or more terms of the same ontology. Measures comparing only two terms are often referred to as pairwise semantic measures, while measures that compare two sets of term yielding a global similarity among sets are referred to as groupwise measures.

Since proteins and genes are associated to a set of terms coming from Gene Ontology, $SSMs$ are often extended to proteins and genes. Similarity of proteins is then translated in the determination of similarity of set of associated terms [28,32]. Many similarity measures have been proposed (see for instance [16] for a complete review) that may be categorized according to different strategies used for evaluating similarity. We here do not discuss deeply $SSMs$ for lack of space, but we introduce some of them.

For instance, the Resnik's similarity measure sim_{res} of two terms T_1 and T_2 of GO is based on the determination of the Information Content (IC) of the their Most Informative Common Ancestor ($MICA$) [29]:

$$sim_{res} = IC(MICA(T_1, T_2)) \tag{4}$$

A drawback of the Resnik's measure is that it considers mainly the common ancestor and it does not take into account the distance among the compared terms and the shared ancestor.

The Lin's measure [21], sim_{Lin}, faces with this problem by considering both terms and yielding to the following formula:

$$sim_{Lin} = \frac{IC(MICA(T_1, T_2)}{IC(T_1) + IC(T_2)} \tag{5}$$

In a similar way the Jiang and Conrath's measure, sim_{JC}, takes into account this distance by calculating the following formula:

$$sim_{JC} = 1 - IC(T_1) + IC(T_2) - 2 * IC(MICA(T_1, T_2)) \qquad (6)$$

2.3   Thresholding of Networks

In other fields many methods for thresholding networks have been defined. These approach may be categorized on the basis of the approaches in global thresholding, i.e. a single value of threshold is applied for all the edges, and local thresholding, in which a different threshold is applied for each node. For instance in [14], the use of an arbitrary global threshold is proposed, while in [1] the use of only a fraction of the highest relationship is used. Differently a statistical-based method is used in [30].

Internal characteristics of SSMs (as investigated in [27]) do not suggest the use of global thresholds. In fact, the relatively low similarities of small regions of the network, may be due to the characteristics of measures used, while proteins or genes have high similarity.

On the other hand, the use of local threshold, e.g. retaining only top k-edges for each node [20], may constitute an efficient solution for preprocessing a SSN. Although this consideration, this choice may be influenced by the presence of local noise and in general may cause the presence of biases in different regions.

2.4   Extraction of Modules in Biological Networks

Markov Clustering (MCL) is a well known algorithm used to find clusters on graphs, robust with respect to noise and graph alterations. Brohee and Van Helden demostrated, in an extensive comparison [6], that MCL outperforms other clustering algorithms, such as MCODE [2], RNSC [19] and Super Paramagnetic Clustering [4]. More recently, MCL has been employed in a network alignment algorithm [24] to identify protein complexes on single PINs.

MCL simulates a stochastic flow on the network that resembles a set of random walks on the graph. MCL consists of two main operations: expand and inflate. The expand step spreads the flow out of a vertex to potentially new vertices, particularly enhancing the flow toward those vertices that are reachable by multiple (and short) paths. The inflation step introduces a modification into the process, enhancing the flows within the clusters and weakening the inter-cluster flows. In this way the initial distribution of flows, relatively uniform, becomes more and more non-uniform, inducing the emergence of a cluster structure, i.e. local regions with high level of flow.

3   The Proposed Approach

We here introduce a method for threshold selection on weighted graphs, based on the spectrum of the associated Laplacian matrix. The pruning algorithm examines each node in the input graph. For each node it stores all the weights of the adjacent edges. Then it determines a local threshold $k = \mu + \alpha \times sd$, where μ is the average of weights, sd is the standard deviation of weights, and α is a variable threshold that is fixed globally. In this way we realize an hybrid approach since the threshold k has a global component α and a local one given by the average and standard deviation of the weights of the adjacent.

If the weight of an edge is greater than k considering the adjacent of both its nodes, then it will be inserted into the novel graph with unitary weight. Otherwise, if the weight of an edge is greater than k considering only one of its adjacent nodes, then it will be inserted into the novel graph with weight $0, 5$. At the end of this process, the Laplacian of the spectrum of the graph is analyzed as described in Ding et al. [12]. If the graph presents nearly disconnected components, then the process stops, alternatively a novel graph with a more stringent threshold k is generated.

3.1 Building Semantic Similarity Networks

The following algorithm explains the building of the semantic similarity network G_{ssu} by iteratively calculating semantic similarity among each pair of proteins. For each step two proteins are chosen and the semantic similarity among them is calculated. Then nodes are added to the graph and an edge is inserted when the semantic similarity is greater than 0.

Algorithm 1. Building Semantic Similarity Networks

Data: Protein Dataset P, Semantic Similarity Measure SS
Result: Semantic Similarity Network $G_{ssu}{=}V_{ssu}, E_{ssu}$
initialization;
forall the p_i **in** P **do**
    read p_i;
    add p_i in V_{ssu} ;
    forall the p_j **in** P, $j \neq i$ **do**
        Let $\sigma{=}$SS(p_i,p_j) ;
        if $\sigma > 0$ **then**
            add the weighted edge (p_i,p_j,σ) to E_{ssu};
        end
    end
end

3.2 Pruning Semantic Similarity Networks

This Section explains the pruning of a semantic similarity network through an example. To better clarify the process, we use an auxiliary graph G_{pr} that is the final output of pruning. The graph is built in an incremental way by considering all the nodes of G_{ssu}. The pruning algorithm examines each node $i \in G_{ssu}$. For each node it stores all the weights of the adjacent edges. Then it determines a local threshold. At the end of this step, the node i and all the adjacent ones are inserted in to G_{pr} (only if they are not yet present).

Then each edge adjacent to i with weight greater with the determined local threshold is inserted into G_{pr}. If the considered edge is not present in G_{pr}, the edge will have weight 0,5, otherwise the weight of the edge is set to 1. We used in this work two simple thresholds, the average and the median of all the weights. Finally all the nodes with $degree = 1$ are deleted from G_{pr}.

The rationale of this process is that edges that are *relevant* considering the neighborhood of both nodes will compare in the pruned graph with unitary weight while edges that are *relevant* considering one node will compare with 0.5 weight. For instance, let us consider the network depicted in Fig. 1 and let us suppose that threshold $k = \mu + \alpha \times sd$ is represented by the average. Without loss of generality we suppose $\alpha = 0$ in this example. Let $AVG(node_i)$ be the average of the weights of edges adjacent to $node_i$ that is used as threshold. Figure 1 depicts the overall process.

The algorithm initially explores $node_0$, since it has degree 1, it is discarded from the analysis. Then it explores the neighbors of $node_1$ and it reaches $node_2$. Then it adds into G_{pr} $node_0$, $node_1$, $node_2$, and $node_3$ and the edge $(node_2, node_3)$ with weight 0,5 - (the average of the weights of the neighbours of $node_2$ is equal to

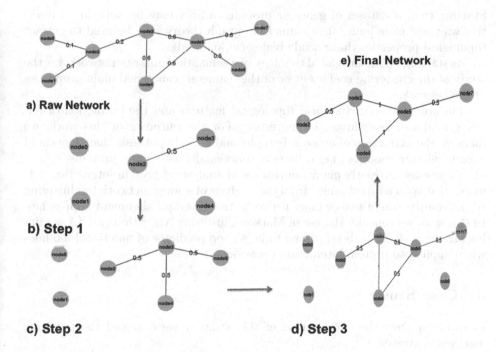

a) Raw Network

b) Step 1

c) Step 2

d) Step 3

e) Final Network

Fig. 1. The thresholding process.

0,13 and other two edges have a lower weight). Figure 1(b) depicts the produced graph at this step.

Then the algorithm considers $node_3$. Then $node_4$, and $node_5$ are inserted into G_{pr}. The $AVG(node_3)$ is equal to 0,46, so only edges $(node_3, node_4)$ and $(node_3, node_5)$ are inserted into G_{pr} with weight 0.5. Figure 1(b) depicts G_{pr} after this step. Then $node_4$ is reached. Since all the adjacent nodes have been inserted into G_{pr}, no nodes are added into this step. The $AVG(node_4)$ is equal to 0,6, so all the edges must be inserted. In particular edge $(node_4, node_3)$ is yet present, so its weight is updated to 1,0. Diversely, $(node_4, node_5)$ is inserted with weight equal to 0,5. Figure 1(d) depicts G_{pr} after this step.

At this point $node_5$ is reached. $node_7$ and $node_8$ are inserted into G_{pr}. The $AVG(node_5)$ is 0,575. Consequently the weight of the edges $(node_5, node_3)$, $(node_5, node_4)$ in G_{pr} are updated to 1, $(node_6, node_7)$ is inserted into G_{pr}. Finally, $node_7$ and $node_8$ are visited but discarded since they have degree equal to 1. In the last step all the nodes with zero degree are eliminated from G_{pr}, producing the resulting graph depicted in Fig. 1(e).

The generation of pruned graph is repeated until the graph has nearly disconnected components. This may be evident by analyzing the spectrum of the associated laplacian for value of threshold.

3.3   Analysis of Semantic Similarity Networks

Starting from a dataset of genes or proteins, a SSN may be built in an iterative way, and once built, algorithms from graph theory may be used to extract topological properties that encode biological knowledge.

As starting point, the global topology of a semantic similarity network, i.e. the study of the clustering coefficient or of the diameter, can reveal main properties of the network.

The study of recurring local topological features and the extraction of relevant modules, i.e. cliques, is interesting. For the purposes of this work, we focus on the extraction of dense subgraphs under the hypothesis that they could encode relevant modules, i.e. subsets of functionally associated proteins.

There exist currently many approaches of analysis of protein interaction networks that span a broad range, from the analysis of a single network by clustering to the comparison of two or more networks trough graph alignment approaches. In this work we consider the use of Markov Clustering Algorithm (MCL) as mining strategy. MCL has been proved to be a good predictor of functional modules when applied to protein interaction networks.

4   Case Study

In order to show the effectiveness of this strategy we designed the following assessment strategy:

1. we downloaded three datasets of proteins (the CYC2008[1] dataset, the MIPS [15] catalog, and the Annotated Yeast High-Throughput Complexes[2]);
2. for each pair of proteins within a dataset we calculated different semantic similarities among them using the FastSemSim[3] tool;
3. we considered 11 semantic similarity measures from those available in Fast-SemSim (Czekanowsky-Dice, Dice, G-Sesame, Jaccard, Kin, NTO, SimGic, SimICND, SimIC, SimUI, TO), see [16]);
4. we used the Biological Process (BP) and Molecular Function (MF) ontologies, therefore we generated 22 SSNs for each input dataset;
5. for each SSN we applied our thresholding algorithm at increasing levels of the threshold value, therefore for each SSN we generated a list of thresholded networks; Since the local threshold computed on each node is $k = \mu + \alpha \times sd$, we varied $\alpha \in [0, 1]$;
6. we applied Markov clustering on the raw and simplified networks showing the improvements of our strategy that is reported in terms of functional enrichment of modules (i.e. the quantification of biological meaning of modules).

As final step we compare our thresholding technique with other global strategies demonstrating the effectiveness of the local thresholding.

4.1   Results

For each generated network we used the markov clustering algorithm (MCL) to extract modules. The effectiveness of the use of MCL for detecting modules in networks has been demonstrated in many works (see for instance [8]). We here have two main objectives: (i) to assess how MCL is able to discover *functionally coherent* modules in different semantic similarity networks; (ii) to show how this clustering process is positively influenced by the proposed network thresholding.

In particular, we show how the process of simplification improves the overall results and how the best results are obtained when networks presents nearly disconnected components. For MCL the inflation parameter is set to 1.2.

We evaluated the obtained results in terms of *functional coherence* of extracted modules. We define *functional coherence FC* of a module M as the average of semantic similarity values of all the pair of nodes (i,j) composing it, as summarized in the following formula where $i, j \in M$, and N is the number of the proteins of the module M.

$$FC = \sum_{i,j} \frac{SSM(i,j)}{N} \tag{7}$$

Starting from this definition, we consider the FC of all the extracted modules, and then we average this value to obtain a single value for each set of modules

---

[1] http://wodaklab.org/cyc2008/
[2] http://wodaklab.org/cyc2008/
[3] http://fastsemsim.sourceforge.net

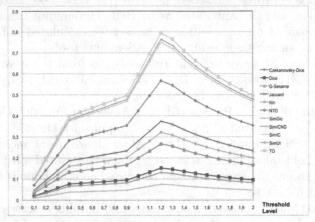

Fig. 2. Comparison of average FC at different threshold levels on the CYC2008 dataset. Each point represents the average functional coherence obtained for a single network.

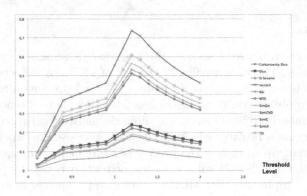

Fig. 3. Comparison of average FC at different threshold levels on the MIPS dataset. Each point represents the average functional coherence obtained for a single network. Dataset.

extracted by a single network. We consider this average value as a representative quality index for the thresholded network.

Figures 2, 3, and 4 summarize these results. In particular, Figs. 2, 3, and 4 show the average Functional Coherence of all the extracted modules at different level of thresholds for all the used similarity measures.

Results confirm an uniform behaviour for all the measures demonstrating the robustness of the proposed thresholding methods when varying the semantic similarity measures. Moreover we point out that the improvement on average functional coherence is not uniform, nor linear, but presents a maximum around $k = 1.2$, that also guarantees the insurgence of nearly disconnected components.

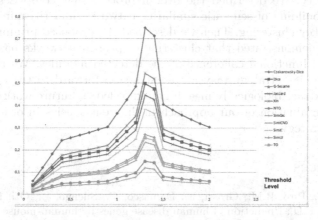

Fig. 4. Comparison of average FC at different threshold levels on the annotated high throughput complexes datasets. Each point represents the average functional coherence obtained for a single network.

A similar behaviour is also shown for all the considered datasets, i.e. the best improvement in FC is obtained when network presents nearly disconnected components. This result is biologically sound since modules correspond to high dense subnetworks that are often connected by few edges. Moreover Figs. 3, and 4 evidence that the extraction of modules on raw networks produces always worse results with respect to thresholded networks.

Finally we want to exclude that results are influenced positively or negatively by the MCL parameters. Consequently we repeated the tests by considering Inflation parameter of MCL in the range [1.0 ÷ 3.0]. Results confirmed that the behaviour of thresholded networks with respect to raw ones is still the same.

5   Conclusion

Semantic Similarity Networks of proteins, that embed into edges weights the functional similarity among proteins, are becoming an important tool for biological research. The analysis of SSNs comprises two main steps: (i) Build a SSN starting from a set of proteins; (ii) mine the network to determine clusters of semantically related proteins.

SSNs are often huge and may contain meaningless edges among proteins; thus they may be pruned using thresholding algorithms. We presented a thresholding strategy to improve the use of SSNs to obtain biological knowledge in a simple way.

Taking into account the drawbacks of existing strategies, such as local and global thresholding, we presented a novel thresholding technique that improves the use of Semantic Similarity Networks as a framework to mine biological knowledge in a semantic/functional space. To face main drawbacks of current strategies, we proposed a hybrid global-local thresholding strategy based on spectral graph theory. We described the overall process that comprises the following steps: (i) building of semantic similarity networks; and (ii) mining of them by using Markov clustering. Then we described the proposed pruning strategy and finally we demonstrated that clustering of pruned networks has better results in terms of functional coherence of the detected modules. We think that such kind of pruning strategy may represent a step forward to the extensive use of SSNs to extract biologically meaningful knowledge. Future work will regard the extensive evaluation of our approach in different species in order to refine the pruning strategy.

References

1. Ala, U., Piro, R.M., Grassi, E., Damasco, C., Silengo, L., Oti, M., Provero, P., Cunto, F.D.: Prediction of human disease genes by human-mouse conserved coexpression analysis. PLoS Comput. Biol. **4**(3), e1000043 (2008)
2. Bader, G.D., Hogue, C.W.V.: An automated method for finding molecular complexes in large protein interaction networks. BMC Bioinform. **27**, 1–27 (2003)
3. Bertolazzi, P., Bock, M.E., Guerra, C.: On the functional and structural characterization of hubs in protein-protein interaction networks. Biotechnol. Adv. **31**(2), 274–286 (2013)
4. Domany, E., Blatt, M., Wiseman, S.: Superparamagnetic clustering of data. Phys. Rev. Lett. **76**(18), 3251–3254 (1996)
5. Bolla, M., Tusnády, G.: Spectra and optimal partitions of weighted graphs. Discrete Math. **128**(1), 1–20 (1994)
6. Brohée, S., van Helden, J.: Evaluation of clustering algorithms for protein-protein interaction networks. BMC Bioinform. **7**, 488 (2006)
7. Camon, E., Magrane, M., Barrell, D., Lee, V., Dimmer, E., Maslen, J., Binns, D., Harte, N., Lopez, R., Apweiler, R.: The gene ontology annotation (goa) database: sharing knowledge in uniprot with gene ontology. Nucl. Acids Res. **32**(suppl-1), D262–D266 (2004)
8. Cannataro, M., Guzzi, P.H., Veltri, P.: Protein-to-protein interactions: technologies, databases, and algorithms. ACM Comput. Surv. **43**, 1:1–1:36 (2010)
9. Cannataro, M., Guzzi, P.H., Sarica, A.: Data mining and life sciences applications on the grid. Wiley Interdisc. Rev.: Data Mining Knowl. Discov. **3**(3), 216–238 (2013)
10. Chung, F.: Spectral Graph Theory. Regional Conference Series in Mathematics, vol. 92. American Mathematical Society, Providence (1994)
11. Cvetković, D., Simić, S.K.: Towards a spectral theory of graphs based on the signless laplacian, ii. Linear Algebra Appl. **432**(9), 2257–2272 (2010)
12. Ding, C., He, X., Zha, H.: A spectral method to separate disconnected and nearly-disconnected web graph components. In: Proceedings of the Seventh ACM International Conference on Knowledge Discovery and Data Mining, San Francisco, 26–29 August 2001
13. Enright, S., Van Dongen, A.J., Ouzounis, C.A.: An efficient algorithm for large-scale detection of protein families. Nucleic Acids Res. **30**(7), 1575–1584 (2002)
14. Freeman, T.C., Goldovsky, L., Brosch, M., van Dongen, S., Maziere, P., Grocock, R.J., Freilich, S., Thornton, J., Enright, A.J.: Construction, visualization, and clustering of transcription networks from microarray expression data. PLoS Comput. Biol. **3**(10), e206 (2007)

15. Guldener, U., Munsterkotter, M., Oesterheld, M., Pagel, P., Ruepp, A., Mewes, H.W., Stumpflen, V.: Mpact: the mips protein interaction resource on yeast. Nucleic Acids Res. **34**, D436–D441 (2006)
16. Guzzi, P.H., Mina, M., Guerra, C., Cannataro, M.: Semantic similarity analysis of protein data: assessment with biological features and issues. Briefings Bioinform. **13**(5), 569–585 (2012)
17. Harispe, S., Sanchez, D., Ranwez, S., Janaqi, S., Montmain, J.: A framework for unifying ontology-based semantic similarity measures: a study in the biomedical domain. J. Biomed. Inform. **48**, 38–53 (2014)
18. Ji, J., Zhang, A., Liu, C., Quan, X., Liu, Z.: Survey: functional module detection from protein-protein interaction networks. IEEE Trans. Knowl. Data Eng. **99**(PrePrints), 1 (2013)
19. King, A.D., Przulj, N., Jurisica, I.: Bioinformatics (Oxford, England)
20. Lee, H.K., Hsu, A.K., Sajdak, J., Qin, J., Pavlidis, P.: Coexpression analysis of human genes across many microarray data sets. Genome Res. **14**, 1085–1094 (2004)
21. Lin, D.: An Information-Theoretic Definition of Similarity. Morgan Kaufmann, San Francisco (1998)
22. Ma, X., Gao, L.: Biological network analysis: insights into structure and functions. Briefings Funct. Genomics **11**(6), 434–442 (2012)
23. Merris, R.: Laplacian matrices of graphs: a survey. Linear Algebra Appl. **197**, 143–176 (1994)
24. Mina, M., Guzzi, P.H.: Alignmcl: comparative analysis of protein interaction networks through markov clustering. In: BIBM Workshops, pp. 174–181. IEEE (2012)
25. Mohar, B.: The laplacian spectrum of graphs. Graph Theor. Comb. Appl. **2**, 871–898 (1991)
26. Ng, A.Y., Jordan, M.I., Weiss, Y., et al.: On spectral clustering: analysis and an algorithm. Adv. Neural Inf. Process. Syst. **2**, 849–856 (2002)
27. Guzzi, P., Mina, M.: Investigating bias in semantic similarity measures for analysis of protein interactions. In: Proceedings of 1st International Workshop on Pattern Recognition in Proteomics, Structural Biology and Bioinformatics (PR PS BB 2011), pp. 71–80, 13 September 2011 (2012)
28. Pesquita, C., Faria, D., O Falcão, A., Lord, P., Couto, F.M.: Semantic similarity in biomedical ontologies. PLoS Comput. Biol. **5**(7), e1000443 (2009)
29. Resnik, P.: Using information content to evaluate semantic similarity in a taxonomy. In: IJCAI, pp. 448–453 (1995)
30. Rito, T., Wang, Z., Deane, C.M., Reinert, G.: How threshold behaviour affects the use of subgraphs for network comparison. Bioinformatics **26**(18), i611–i617 (2010)
31. Su, G., Kuchinsky, A., Morris, J.H., States, D.J., Meng, F.: Glay: community structure analysis of biological networks. Bioinformatics **26**(24), 3135–3137 (2010)
32. Wang, H., Zheng, H., Azuaje, F.: Ontology- and graph-based similarity assessment in biological networks. Bioinformatics **26**(20), 2643–2644 (2010)
33. Zhu, X., Gerstein, M., Snyder, M.: Getting connected: analysis and principles of biological networks. Genes Dev. **21**(9), 1010–1024 (2007)

A Relational Unsupervised Approach to Author Identification

Fabio Leuzzi[1], Stefano Ferilli[1,2(✉)], and Fulvio Rotella[1]

[1] Dipartimento di Informatica, Università di Bari, Bari, Italy
[2] Centro Interdipartimentale per la Logica e sue Applicazioni,
Università di Bari, Bari, Italy
{fabio.leuzzi,stefano.ferilli,fulvio.rotella}@uniba.it

Abstract. In the last decades speaking and writing habits have changed. Many works faced the author identification task by exploiting frequency-based approaches, numeric techniques or writing style analysis. Following the last approach we propose a technique for author identification based on First-Order Logic. Specifically, we translate the complex data represented by natural language text to complex (relational) patterns that represent the writing style of an author. Then, we model an author as the result of clustering the relational descriptions associated to the sentences. The underlying idea is that such a model can express the typical way in which an author composes the sentences in his writings. So, if we can map such writing habits from the unknown-author model to the known-author model, we can conclude that the author is the same. Preliminary results are promising and the approach seems viable in real contexts since it does not need a training phase and performs well also with short texts.

1 Introduction

Speaking and writing habits have changed in the last decades, and many works have investigated the author identification task by exploiting frequency-based approaches, numeric techniques and writing style analysis. The spreading of documents across the Internet made the writing activity faster and easier compared to past years. Thus, author identification became a primary issue, due to the increasing number of plagiarism cases. In order to face such problems, several approaches have been attempted in the Machine Learning field [1,4,13,20].

The authorship attribution task is well-understood (given a document, determine who wrote it) although amenable to many variations (given a document, determine a profile of the author; given a pair of documents, determine whether they were written by the same author; given a document, determine which parts of it were written by a specific person), and its motivation is clear. In applied areas such as law and journalism knowing the author's identity may save lives.

The most common approach for testing candidate algorithms is to cast the problem as a text classification task: given known sample documents from a small, finite set of candidate authors, assess if any of those authors wrote a

A. Appice et al. (Eds.): NFMCP 2013, LNAI 8399, pp. 214–228, 2014.
DOI: 10.1007/978-3-319-08407-7_14, © Springer International Publishing Switzerland 2014

questioned document of unknown authorship. A more lifelike approach is: given a set of documents by a single author and a questioned document, determine whether the questioned document was written by that particular author or not. This is more interesting for professional forensic linguistics because it is a primary need in that environment.

This setting motivated us to face the following task: given a small set (no more than 10, possibly just one) of "known" documents written by a single person and a "questioned" document, determine whether the latter was written by the same person who wrote the former.

Performing a deep understanding of the author to seize his style is not trivial, due to the intrinsic ambiguity of natural language and to the huge amount of common sense and linguistic/conceptual background knowledge needed to switch from a purely syntactic representation to the underlying semantics. Traditional approaches are not able to seize the whole complex network of relationships, often hidden, between events, objects or a combination of them. Conversely relational approaches treat natural language texts as complex data from which mining complex patterns.

So after extracting and making explicit the typed syntactical dependencies of each sentence, we formally express them in a First-Order Logic representation. In this way the unstructured texts in natural language are expressed by complex (relational) patterns on which automatic techniques can be applied. Exploiting such patterns, the author's style can be modeled in order to classify a new document as written by the same author or not.

For the sake of clarity, from now on we refer to the known author used for training as the *base*, and to the unknown author that must be classified as *target*.

This work is organized as follows: the next section describes related works; Sect. 3 outlines the proposed approach, that is evaluated subsequently. Lastly, we conclude with some considerations and future works.

2 Related Work

There is a huge amount of research conducted on Author Identification in the last 10 years. With the spread of anonymous documents in Internet, authorship attribution becomes important. Researches focus on different properties of texts, the so-called *style markers*, to quantify the writing style under different labels and criteria. Five main types of features can be found: lexical, character, syntactic, semantic and application specific. The lexical and character features consider a text as a mere sequence of word-tokens or characters, respectively. An example of the first category is [2] in which new lexical features are defined for use in stylistic text classification, based on taxonomies of various semantic functions of certain choice words or phrases. While this work reaches interesting results, it is based on the definition of arbitrary criteria (such as the 675 lexical features and taxonomies) and requires language-dependent expertise. Another example is the approach in [19] which is based on comparing the similarity between the given documents and a number of external (impostor) documents, so that documents can be classified as having been written by the same author, if they are shown to be more similar to each other than to the impostors, in a number of trials exploiting different lexical feature sets.

In [23] the authors build a suffix tree representing all possible character n-grams of variable length and then extract groups of character n-grams as features. An important issue of such approaches based on character feature is the choice of n, because a larger n captures more information but increases the dimensionality of the representation. On the other hand, a small n might not be adequate to learn an appropriate model.

Syntactic features are based on the idea that authors tend to unconsciously use similar syntactic patterns. Therefore they exploit information such as PoS-tags, sentence and phrase structures. These approaches carry on two major drawbacks: the former is the need of robust and accurate NLP tools to perform syntactic analysis of texts and the latter is the huge amount of extracted features they require (e.g., in [21] there are about 900 K features). For instance, the work in [6] defines a set of coherence features together with some stylometric features. Since such features are unavailable for non-English languages, they exploit corresponding translations produced by Google Translate service. It is easy to note that not only this work suffers by the drawbacks of the NLP tools, but it also introduces noise in the representation due to the automatic translation.

Semantic approaches rely on semantic dependencies obtained by external resources, such as taxonomies or thesauri. In [14] the authors exploit WordNet [5] to detect "semantic" information between words. Although the use of an external taxonomic or ontological resource can be very useful for these purposes, such resources are not always available and often do not exist at all for very specific domains.

Finally, there are non-general-purpose approaches, that define application-specific measures to better represent the style in a given text domain. Such measures are based on the use of greetings and farewells in the messages, types of signatures, use of indentation, paragraph length, and so on [12].

While the various approaches faced the problem from different perspectives, a common feature to all of them is their using a flat (vectorial) representation of the documents/phrases. Even the two before the last approach, although starting from syntactic trees or word/concept graphs, subsequently create new flat features, losing in this way the relations embedded in the original texts. For example, [22] builds graphs based on POS sequences and then extracts sub-graph patterns. This graph-based representation attempts to capture the sequence among the sentence words, as well as the sequence among their PoS tags, with the aim of feeding a graph mining tool which extracts some relevant relational features. But they lose all relational features when a feature vector for each document is built upon them as input for a Support Vector Machine classifier.

A different approach that preserves the phrase structure is presented in [16]. In this work a probabilistic context-free grammar (PCFG) is built for each author and then each test document is assigned to the author whose PCFG produced the highest likelihood for such a document. While this approach takes into account the syntactic tree of the sentences, it needs many documents per author to learn the right probabilities. Thus it is not applicable in settings in which a small set of documents of only one author is available. Moreover we believe that the exploitation of only parse trees is not enough to characterize the author's style, conversely the syntactical relationships would be better enriched with grammatical ones.

Differently from all of these approaches, our proposal aims at preserving the informative richness of textual data by extracting and exploiting complex patterns from such complex data.

3   Proposed Approach

Natural Language Text is a complex kind of data encoding implicitly the author's style. We propose to translate textual data into a relational description in order to make explicit the complex patterns representing the author's style. The relational descriptions are clustered using the similarity measure presented in [7], where the threshold to be used as a stopping criterion is automatically recognized. We apply this technique to build both base and target models. Then, the classification results from the comparison of these two models. The underlying idea is that the target model describes a set of ways in which the author composes the sentences. If we can bring such writing habits back to the base model, then we can conclude that the author is the same.

3.1   The Representation Formalism

Natural language texts are processed by ConNeKTion [10] (acronym for 'CONcept NEtwork for Knowledge representaTION'), a framework for conceptual graph learning and exploitation. This framework aims at partially simulating some human abilities in the text understanding and concept formation activity, such as: extracting the concepts expressed in given texts and assessing their relevance [8]; obtaining a practical description of the concepts underlying the terms, which in turn would allow to generalize concepts having similar description [17]; applying some kind of reasoning 'by association', that looks for possible indirect connections between two identified concepts [11]; identifying relevant keywords that are present in the text and helping the user in retrieving useful information [18].

In this work we exploit ConNeKTion in order to obtain a relational representation of the syntactic features of the sentences. In particular exploiting the *Stanford Parser* and *Stanford Dependencies* tools [3,9] we obtain phrase structure trees and a set of grammatical relations (typed dependencies) for each sentence. These dependencies are expressed as binary relations between pairs of words, the former of which represents the governor of the grammatical relation, and the latter its dependent. Words in the input text are normalized using lemmatization instead of stemming, which allows to distinguish their grammatical role and is more comfortable to read by humans. ConNeKTion also embeds JavaRAP, an implementation of the classic Anaphora Resolution Procedure [15]. Indeed, the subjects/objects of the sentences in long texts are often expressed as pronouns, referred to the latest occurrence of the actual subject/object. After applying all these pre-processing steps, we translate each sentence into a relational pattern. In particular, each sentence is translated into a Horn Clause of the form:

$$sentence(IdSentence) : -description(IdSentence).$$

where $description(Idsentence)$ is a combination of atoms built on the following predicates, that express the relations between the words in the sentence:

- $phrase(Tag, IdSentence, Pos)$ represents a constituent whose Tag is the type of phrase (e.g. NP, VP, S,...) and Pos is the term position in the phrase;
- $term(IdSentence, Pos, Lemma, PosTag)$ defines a single term whose position in the sentence is Pos, its lemma is $Lemma$ and its part-of-speech (e.g. N,V,P,...) is $PosTag$;
- $sd(IdSentence, Type, PosGov, PosDep)$ represents the grammatical relation $Type$ (e.g. dobj, subj,...) between the governor word in position $PosGov$ and the dependent word in position $PosDep$.

This allows us to represent all the relationships between the terms, their grammatical relations and the phrases to which they belong.

3.2   The Similarity Measure

The similarity strategy exploited here was presented in [7]. It takes values in $]0, 4[$ and is computed by repeated applications of the following formula to different parameters extracted from the relational descriptions:

$$sf(i', i'') = sf(n, l, m) = \alpha\frac{l+1}{l+n+2} + (1-\alpha)\frac{l+1}{l+m+2}$$

where:

- i' and i'' are the two items under comparison;
- n represents the information carried by i' but not by i'';
- l is the common information between i' and i'';
- m is the information carried by i'' but not by i';
- α is a weight that determines the importance of i' with respect to i'' (0.5 means equal importance).

More precisely, the overall similarity measure carries out a layered evaluation that, starting from simpler components, proceeds towards higher-level ones repeatedly applying the above similarity formula. At each level, it exploits the information coming from lower levels and extends it with new features. At the

---

Algorithm 1. Relational pairwise clustering.
Interface: *pairwiseClustering(M, T)*.

Input: M is the similarity matrix; T is the threshold for similarity function.
Output: set of clusters.

> $pairs \leftarrow empty$
> $averages \leftarrow empty$
> **for all** $O_i \mid i \in O$ **do**
> $newCluster \leftarrow O_i$
> $clusters.add(newCluster)$
> **end for**
> $merged \leftarrow true$
> **while** $merged = true$ **do**
> $merged \leftarrow false$
>   **for all** $pair(C_k, C_z) \mid C \in clusters \wedge k, z \in [0, clusters.size[$ **do**
>   **if** $completeLink(C_k, C_z, T)$ **then**
> $pairs.add(C_k, C_z)$
> $averages.add(getScoreAverage(C_k, C_z))$
>   **end if**
>   **end for**
>   **if** $pairs.size() > 0$ **then**
> $pair \leftarrow getBestPair(pairs, averages)$
> $merge(pair)$
> $merged = true$
>   **end if**
> **end while**

---

completeLink(matrix, cluster₁, cluster₂, threshold) → TRUE if complete link assumption for the passed clusters holds, FALSE otherwise.
getBestPair(pairs, averages) → returns the pair having the maximum average.

---

basic level terms (i.e., constants or variables in a Datalog setting) are considered, that represent objects in the world and whose similarity is based on their properties (expressed by unary predicates) and roles (expressed by their position as arguments in n-ary predicates). The next level involves atoms built on n-ary predicates: the similarity of two atoms is based on their "star" (the multiset of predicates corresponding to atoms directly linked to them in the clause body, that expresses their similarity 'in breadth') and on the average similarity of their arguments. Since each of the four components ranges into $]0, 1[$, their sum ranges into $]0, 4[$. Then, the similarity of sequences of atoms is based on the length of their compatible initial subsequence and on the average similarity of the atoms appearing in such a subsequence. Finally, the similarity of clauses is computed according to their least general generalization, considering how many literals and terms they have in common and on their corresponding lower-level similarities.

Algorithm 2. Best model identification.
Interface: $getBestModel(M, T_{lower}, T_{higher})$

Input: M is the similarity matrix; T_{lower} is the starting threshold, T_{higher} is the maximum threshold that can be attempted.
Output: $bestModel$ that is the model having the best threshold.

$t \leftarrow T_{lower}$
$models \leftarrow \emptyset$
$thresholds \leftarrow \emptyset$
while $t < T_{higher}$ **do**
  $clusters \leftarrow pairwiseClustering(M, t)$
  $models.add(clusters)$
  $thresholds.add(t)$
  $t \leftarrow t + 0.005$
end while
$maxHop \leftarrow 0$
$bestModel \leftarrow null$
for all $m_i \mid m_i \in models, i > 0$ **do**
  $hop \leftarrow (m_i.size * 100)/m_{i-1}.size$
  if $maxHop < hop$ **then**
    $maxHop \leftarrow hop$
    $bestModel \leftarrow m_{i-1}$
  end if
end for
return $bestModel$

3.3   Building Models

After obtaining a relational description for each sentence as described in Sect. 3.1, we applied the similarity measure described in Sect. 3.2 to pairs of sentences. In particular, for each training-test pair we computed an upper triangular similarity matrix between each pair sentences. As can be seen in Fig. 1 the global matrix can be partitioned into three parts, the top-left submatrix (filled with diagonal lines) contains the similarity scores between each pair of sentences of known documents (base). The bottom-right one (filled with solid gray) includes the similarities between pairs of sentences belonging to the unknown document (target). The top-right submatrix reports the similarity scores across known and unknown documents.

Then, we performed an agglomerative clustering to both base and target submatrices according to Algorithm 1. Initially each description makes up a different singleton cluster; then the procedure works by iteratively finding the next pair of clusters to be merged according to a *complete link* strategy. Complete link states that *the distance of the farthest items of the involved clusters must be less than a given threshold*. Since more than one pair can satisfy such a requirement, the procedure needs the ranking of the pairs, because the ordering of the pairs affects the final model. Then for each iteration only the pair with the highest average similarity is merged. In this work we refer to a *model* as a

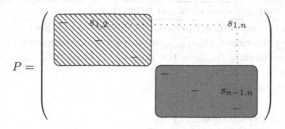

Fig. 1. Global similarity matrix. Each $s_{i,j}$ represents the similarity between sentences i and j calculated as explained in Sect. 3.2

possible grouping of similar descriptions, as obtained by running the clustering algorithm with a given threshold.

In this perspective, there is the need of establishing the threshold by which pairwise clustering is carried out for each *model*. Our approach is based on the idea that as long as the threshold increases, also the number of clusters grows, and thus the merging becomes more and more difficult. We consider as a cut point the largest gap between the number of clusters obtained with a threshold and the next one obtained by performing clustering with a greater threshold. It is easy to note that a given difference value obtained with many clusters is less significant than the one obtained with a smaller number of clusters.

Taking into account such considerations we have defined the following function that encodes such intuitive assumptions. Given a sequence of models $< m_1, ..., m_n >$ obtained by repeating the clustering procedure with the cut-threshold progressively incremented by step 0.005, and function $c(m_i)$ that computes the number of clusters in the i-th model, we can define:

$$g(i) = \frac{c(m_{i+1})}{c(m_i)} \quad \text{and} \quad \theta = \arg \max_i g(i)$$

where $0 \leq i < n$ and θ is the desired threshold associated to the model m_i yielding the greatest distance from the model m_{i+1} (see Algorithm 2). Since our similarity measure ranges in $]0, 4[$, the threshold varies within such range as well.

Once the appropriate thresholds are chosen, we have defined base and target models, each of which having its own threshold, thus we can perform the classification. As can be seen in Algorithm 3, this phase of the procedure works on clusters having more than one item. For each cluster in the target model, if it can be merged with at least one cluster in the base model (under the complete link assumption), the author is the same, otherwise it is not. Such merging check exploits the similarities in the top-right submatrix and the maximum threshold between base and target model, making harder a full alignment between such models. This choice encourages accurate classifications.

Algorithm 3. Complete classification procedure.

Input: O_{known} is the set of descriptions (represented as in Sect. 3.1) obtained from the known-author documents; $O_{unknown}$ is the set of descriptions obtained from the unknown-author document; T_{lower} is the starting threshold, T_{higher} is the maximum threshold that can be attempted.
Output: Classification outcome.

$M_{known} \leftarrow getSimilarities(O_{known})$
$model_{known} \leftarrow getBestModel(M_{known}, T_{lower}, T_{higher})$
$t_{known} \leftarrow getBestThreshold(M_{known}, T_{lower}, T_{higher})$
$M_{unknown} \leftarrow getSimilarities(O_{unknown})$
$model_{unknown} \leftarrow getBestModel(M_{unknown}, T_{lower}, T_{higher})$
$t_{unknown} \leftarrow getBestThreshold(M_{unknown}, T_{lower}, T_{higher})$
$t \leftarrow max(t_{known}, t_{unknown})$
$O \leftarrow O_{known}$
$O.add(O_{unknown})$
$M \leftarrow getSimilarities(O)$
for all $(C_k, C_u) \mid C_k \in model_{known} \wedge C_u \in model_{unknown}$ **do**
  if $(|C_k| \leq 3 \wedge instances(C_k) \leq instances(C_u) * 0.2)$
  $\vee (|C_u| \leq 3 \wedge instances(C_u) \leq instances(C_k) * 0.2)$ **then**
    return *null*
  else
    if $!completeLink(M, C_k, C_u, t)$ **then**
      return *false*
    end if
  end if
end for
return *true*

completeLink(matrix,cluster₁,cluster₂,threshold) → TRUE if complete link assumption for the passed clusters holds, FALSE otherwise.
getSimilarities(list) → returns the similarity matrix between all pairs of objects in 'list'.
instances(model) → returns the number of instances composing the clusters in 'model'.

3.4   The Gray Zone

In preliminary evaluations performed on the training set (Table 1), we have noted that in problem 'EN23' summing all the sentences of the known documents we get half of the sentences belonging to the corresponding unknown one. This strange situation has brought us to hypothesize that, although our system can build a reliable model using few texts, it cannot deal with too poor text (just like a human). Due to such a particular behavior we have defined as *gray zone* a portion of cases for which the approach must not provide a classification since it could be unreliable.

According to our hypothesis, we have left out the problems having models composed just by one, two or three clusters, since they were too poor. For each model that does not suffer of poorness, if the number of the instances of a model

Table 1. Training-set details and outcomes.

ID	Known docs			Unknown doc		Outcomes		
	$\#_{docs}$	$\#_{clauses}$	μ_{length}	$\#_{clauses}$	μ_{length}	Expected	Class	Score
EN04	4	261	121.06	62	136.60	Y	Y	1.0
EN07	4	260	121.48	44	195.47	N	Y	1.0
EN11	2	109	185.87	39	160.41	Y	Y	1.0
EN13	3	109	156.99	65	134.65	N	N	0.6
EN18	5	274	154.25	53	165.49	Y	Y	1.0
EN19	3	139	164.35	56	210.05	Y	N	0.37
EN21	2	109	210.89	24	269.21	N	N	0.67
EN23	2	51	217.29	97	277.29	Y	N	0.92
EN24	5	242	147.06	89	169.08	N	N	0.69
EN30	2	95	189.87	33	322.09	N	N	0.8

Table 2. Dataset composition.

Set	English	Greek	Spanish
Training	10	20	5
Test 1	20	20	10
Test 2	29	30	25
Total	59	70	40

is less than 20 % of the other one, the approach does not try a classification, since the obtained models could be considered unreliable (see Algorithm 3).

4 Evaluation

We evaluated our procedure using the dataset provided in the 9th evaluation lab on uncovering plagiarism, authorship, and social software misuse (PAN) held as part of the CLEF 2013 conference.

The dataset composition is as shown in Table 2: Training is the training dataset, Test 1 is an early-bird evaluation dataset that is a subset of the complete evaluation dataset Test 2 (it should have included 30 instances, unfortunately we found only 29 instances in the archive). Since our approach does not require a training phase, we were able to consider the training set as part of the dataset. In this evaluation we have considered the English problems only, since the current version of ConNeKTion is based on the Stanford NLP tools, that cannot deal natively with the other two languages. However, our approach can be easily extended to the other languages, as long as suitable NLP tools for them are available.

In Table 3 we have reported some statistics about the datasets, such as the minimum, the average and the maximum value for each perspective, that are useful to understand the amount of information with which we deal in order to face this task. In particular there is the number of documents, the total number

Table 3. Dataset details.

Set	Known docs								
	#docs			#clauses			μ_{length}		
	min	μ	max	min	μ	max	min	μ	max
Training	2	3.20	5	51	178.87	274	121.06	166.82	216.37
Test 1	3	4.45	9	29	146.79	329	107.35	218.24	322.82
Test 2	2	4.27	14	29	145.59	367	100.81	209.55	319.59
Total	2	3.96	14	29	157.08	367	100.81	198.2	322.82

Set	Unknown doc					
	#clauses			μ_{length}		
	min	μ	max	min	μ	max
Training	24	56.10	96	134.65	194.22	322.09
Test 1	9	117.31	238	117.01	228.11	358.41
Test 2	9	56.00	301	110.29	213.85	351.18
Total	9	76.47	301	110.29	212.06	358.41

Table 4. Outcomes overview that sums up true positives (T.P.), true negatives (T.N.), false positives (F.P.), false negatives (F.N.) and not classified (N.C.).

Type	Set	T.P. + T.N.	F.P.	F.N.	N.C.
Boolean evaluation	Training	0.7	0.3	0.0	0.0
	Test 1	0.7	0.15	0.15	0.0
	Test 2	0.45	0.31	0.24	0.0
	Total	0.58	0.25	0.17	0.0
Smoothed evaluation	Training	0.7	0.1	0.0	0.2
	Test 1	0.65	0.1	0.05	0.2
	Test 2	0.41	0.14	0.14	0.31
	Total	0.55	0.12	0.08	0.25

of clauses built from such documents and their average length for both known and unknown documents.

Thus we performed an evaluation aimed at investigating how good the approach is with and without using the gray zone. In Table 4, the procedure without the use of the gray zone is referred to as *boolean evaluation*. Conversely, the exploitation of the gray zone is referred to as *smoothed evaluation*. Considering each sub-dataset along with the related performance, we can see the difference between the misclassifications (i.e. F.P. + F.N.) with and without the use of the gray zone as a *gain* (e.g. in Test 1 we have $0.3 - 0.15 = 0.15$), whereas the difference between the correct classifications (i.e. T.P. + T.N.) as a *loss* (e.g. in Test 1 we have $0.7 - 0.65 = 0.05$). For each sub-dataset the gain is much more than the loss. Obviously, this situation is verified for the entire dataset.

Table 5 reports the performance using the standard measures Precision, Recall and F-measure. Let us to consider each sub-dataset. The difference between the Precision scores with and without the use of the gray zone can be seen as a *gain*

Table 5. Evaluation of the *grey zone* application.

Type	Set	Precision	Recall	F-measure
	Training	0.7	0.7	0.7
Boolean evaluation	Test 1	0.7	0.7	0.7
	Test 2	0.45	0.45	0.45
	Total	0.58	0.58	0.58
	Training	0.87	0.7	0.77
Smoothed evaluation	Test 1	0.81	0.65	0.72
	Test 2	0.6	0.41	0.49
	Total	0.73	0.55	0.62

(since reducing the amount of cases in which a classification is given, we keep only the most reliable cases, cutting out several misclassifications), whereas the difference between the Recall scores can be seen as a *loss* (since reducing the amount of cases in which a classification is given, also some relevant cases having a borderline classification are lost, although they are correct). Unlike the previous perspective of gain, here both gain and loss are referred to the correct classifications. In particular, the gain represents the decreasing misclassifications with respect to the cases in which our approach gives a response, whereas the loss represents the correct classifications over the entire dataset. Among the three sub-datasets, the gain is much more than the loss. Such good performances given by the application of the gray zone affects the F-measure since it combines Precision and Recall. Hence the score of the F-measure obtained in the second approach is a further evidence that the use of the gray zone affects positively the performance of our approach.

5   Conclusions

This work proposed a technique for author identification based on First-Order Logic. It is motivated by the assumption that making explicit the typed syntactical dependencies in the text one may obtain significant features on which basing the predictions. Thus, this approach translates the complex data represented by natural language text to complex (relational) patterns that allow to model the writing style of an author. Then, these models can be exploited to classify a novel document as written by the author or not. Our approach consists in translating the sentences into relational descriptions, then clustering these descriptions (using an automatically computed threshold to stop the clustering procedure). The resulting clusters represent our model of an author. So, after building the models of the base (known) author and the target (unknown) one, the comparison of these models suggests a classification (i.e., whether the target author is the same as the base one or not). The underlying idea is that the model describes a set of ways in which an author composes the sentences in its writings. If we can bring back such writing habits from the target model to the base model, we can conclude that the author is the same. There could be some cases in which the amount of text (i.e., the amount of information from which capturing the writing style) is not enough. In order to identify such cases we defined the *gray zone* that aims at capturing the indeterminacy.

It must be underlined a small number of documents is sufficient, using this approach, to build an author's model. This is important because, in real life, only a few documents are available for the base author, on which basing a classification. We wanted to stress specifically this aspect in our experiments, using the dataset released for the PAN 2013 challenge. Preliminary results are promising. Our approach seems viable in real contexts since it does not need a training phase and performs well also with short texts.

The current work in progress concerns the refinement of the identification of the gray zone, in order to keep out as much indeterminacy as possible. As a future work, we plan to study the quality of the clusters, pursuing an intensional understanding thereof. In particular, we want to study whether generalizing the clustered clauses we can obtain a theory expressing the typical sentence construction that the author exploits in his texts. Such theory would be the intentional model of the author, which would allow to carry on the investigation in the learning field.

Acknowledgments. We wish to express our sincere thanks to Paolo Gissi, for many useful discussions and for the inspiring concept of *gray zone*. This work was partially funded by Italian FAR project DM19410 MBLab "Laboratorio di Bioinformatica per la Biodiversità Molecolare" and Italian PON 2007-2013 project PON02_00563_3489339 "Puglia@Service".

References

1. Argamon, S., Saric, M., Stein, S.S.: Style mining of electronic messages for multiple authorship discrimination: first results. In: Getoor, L., Senator, T.E., Domingos, P., Faloutsos, C. (eds.) Proceedings of the Ninth ACM SIGKDD International Conference on Knowledge Discovery and Data Mining, pp. 475–480. ACM (2003)
2. Argamon, S., Whitelaw, C., Chase, P., Hota, S.R., Garg, N., Levitan, S.: Stylistic text classification using functional lexical features: research articles. J. Am. Soc. Inf. Sci. Technol. **58**(6), 802–822 (2007)
3. De Marneffe, M.C., Maccartney, B., Manning, C.D.: Generating typed dependency parses from phrase structure parses. In: Proceedings of International Conference on Language Resources and Evaluation (LREC), pp. 449–454 (2006)
4. Diederich, J., Kindermann, J., Leopold, E., Paass, G.: Authorship attribution with support vector machines. Appl. Intell. **19**(1–2), 109–123 (2003)
5. Fellbaum, C. (ed.): WordNet: An Electronic Lexical Database. MIT Press, Cambridge (1998)
6. Feng, V.W., Hirst, G.: Authorship verication with entity coherence and other rich linguistic features notebook for PAN at CLEF 2013. In: Forner, P., Navigli, R., Tufis, D. (ed.) CLEF 2013 Labs and Workshops - Online Working Notes, Padua, Italy, September 2013. PROMISE (2013)
7. Ferilli, S., Basile, T.M.A., Di Mauro, N., Esposito, F.: Plugging numeric similarity in first-order logic horn clauses comparison. In: Pirrone, R., Sorbello, F. (eds.) AI*IA 2011. LNCS, vol. 6934, pp. 33–44. Springer, Heidelberg (2011)

8. Ferilli, S., Leuzzi, F., Rotella, F.: Cooperating techniques for extracting conceptual taxonomies from text. In: Proceedings of The Workshop on Mining Complex Patterns at AI*IA XIIth Conference (2011)
9. Klein, D., Manning, C.D.: Fast exact inference with a factored model for natural language parsing. In: Becker, S., Thrun, S., Obermayer, K. (eds.) Advances in Neural Information Processing Systems, vol. 15. MIT Press, Cambridge (2003)
10. Leuzzi, F., Ferilli, S., Rotella, F.: ConNeKTion: a tool for handling conceptual graphs automatically extracted from text. In: Catarci, T., Ferro, N., Poggi, A. (eds.) IRCDL 2013. CCIS, vol. 385, pp. 93–104. Springer, Heidelberg (2014)
11. Leuzzi, F., Ferilli, S., Rotella, F.: Improving robustness and flexibility of concept taxonomy learning from text. In: Appice, A., Ceci, M., Loglisci, C., Manco, G., Masciari, E., Ras, Z.W. (eds.) NFMCP 2012. LNCS, vol. 7765, pp. 170–184. Springer, Heidelberg (2013)
12. Li, J., Zheng, R., Chen, H.: From fingerprint to writeprint. Commun. ACM **49**(4), 76–82 (2006)
13. Lowe, D., Matthews, R.: Shakespeare vs. fletcher: a stylometric analysis by radial basis functions. Comput. Humanit. **29**(6), 449–461 (1995)
14. Mccarthy, P.M., Lewis, G.A., Dufty, D.F., Mcnamara, D.S.: Analyzing writing styles with coh-metrix. In: Sutcliffe, G., Goebel, R. (eds.) Proceedings of the Florida Artificial Intelligence Research Society International Conference (FLAIRS), pp. 764–769. AAAI Press (2006)
15. Qiu, L., Kan, M.-Y., Chua, T.-S.: A public reference implementation of the RAP anaphora resolution algorithm. In: Proceedings of the Fourth International Conference on Language Resources and Evaluation, LREC 2004, 26–28 May 2004, Lisbon, Portugal, pp. 291–294. European Language Resources Association (2004)
16. Raghavan, S., Kovashka, A., Mooney, R.: Authorship attribution using probabilistic context-free grammars. In: Proceedings of the ACL 2010 Conference Short Papers, ACLShort '10, pp. 38–42, Stroudsburg, PA, USA, Association for Computational Linguistics (2010)
17. Rotella, F., Ferilli, S., Leuzzi, F.: An approach to automated learning of conceptual graphs from text. In: Ali, M., Bosse, T., Hindriks, K.V., Hoogendoorn, M., Jonker, C.M., Treur, J. (eds.) IEA/AIE 2013. LNCS, vol. 7906, pp. 341–350. Springer, Heidelberg (2013)
18. Rotella, F., Ferilli, S., Leuzzi, F.: A domain based approach to information retrieval in digital libraries. In: Agosti, M., Esposito, F., Ferilli, S., Ferro, N. (eds.) IRCDL 2012. CCIS, vol. 354, pp. 129–140. Springer, Heidelberg (2013)
19. Seidman, S.: Authorship verification using the impostors method notebook for pan at clef 2013. In: Forner, P., Navigli, R., Tufis, D. (eds.) CLEF 2013 Labs and Workshops - Online Working Notes, Padua, Italy, September 2013. PROMISE (2013)
20. Tweedie, F.J., Singh, S., Holmes, D.I.: Neural network applications in stylometry: the federalist papers. Comput. Humanit. **30**(1), 1–10 (1996)
21. van Halteren, H.: Linguistic profiling for author recognition and verification. In: Proceedings of the 42nd Annual Meeting on Association for Computational Linguistics, ACL '04, Stroudsburg, PA, USA. Association for Computational Linguistics (2004)

22. Vilarino, D., Pinto, D., Gomez, H., Leo, S., Castillo, E.: Lexical-syntactic and graph-based features for authorship verification - notebook for pan at clef 2013. In: Forner, P., Navigli, R., Tufis, D. (eds.) CLEF 2013 Labs and Workshops - Online Working Notes, Padua, Italy, September 2013. PROMISE (2013)
23. Zheng, R., Li, J., Chen, H., Huang, Z.: A framework for authorship identification of online messages: writing-style features and classification techniques. J. Am. Soc. Inf. Sci. Technol. **57**(3), 378–393 (2006)

Machine Learning and Music Data

From Personalized to Hierarchically Structured Classifiers for Retrieving Music by Mood

Amanda Cohen Mostafavi[1]([✉]), Zbigniew W. Raś[1,2],
and Alicja A. Wieczorkowska[3]

[1] Department of Computer Science, University of North Carolina,
Charlotte, NC 28223, USA
[2] Institute of Computer Science, Warsaw University of Technology,
00-665 Warsaw, Poland
[3] Polish-Japanese Institute of Information Technology, 02-008 Warsaw, Poland
acohen24@uncc.edu

Abstract. With the increased amount of music that is available to the average user, either online or through their own collection, there is a need to develop new ways to organize and retrieve music. We propose a system by which we develop a set of personalized emotion classifiers, one for each emotion in a set of 16 and a set unique to each user. We train a set of emotion classifiers using feature data extracted from audio which has been tagged with a set of emotions by volunteers. We then develop SVM, kNN, Random Forest, and C4.5 tree based classifiers for each emotion and determine the best classification algorithm. We then compare our personalized emotion classifiers to a set of non-personalized classifiers. Finally, we present a method for efficiently developing personalized classifiers based on hierarchical clustering.

Keywords: Music information retrieval · Classification · Clustering

1 Introduction

With the average size of a person's digital music collection expanding into the hundreds and thousands, there is a need for creative and efficient ways to search for and index songs. This problem shows up in several sub-areas of Music Information Retrieval (MIR) such as genre classification, automatic artist Identification, and instrument detection. Here we focus on indexing music by emotion, as in how the song makes the listener feel. This way the user could select songs that make him/her happy, sad, excited, depressed, or angry depending on what mood the listener is in (or wishes to be in). However, the way a song makes someone feel, or the emotions he associates with the music, varies from person to person for a variety of reasons ranging from personality and taste to upbringing and the music the listener was exposed to growing up. This means that any sort of effective emotion indexing system must be personal and/or adaptive to the user. This is so far a mostly unexplored area of MIR research, as many researchers that

A. Appice et al. (Eds.): NFMCP 2013, LNAI 8399, pp. 231–245, 2014.
DOI: 10.1007/978-3-319-08407-7_15, © Springer International Publishing Switzerland 2014

attempt to personalize their music emotion recognition systems do so from the perspective of finding how likely the song is to be tagged with certain emotions rather than finding a way to create a system that can be personalized.

We present a system through which we can build and train personalized user classifiers, which are unique for individual users. We built these classifiers based on user data accumulated through an online survey and music data collected via a feature extraction toolkit called MIRToolbox [1]. We then use four classification algorithms to determine the best algorithm for this data: support vector machines (SVM), k-nearest neighbors (kNN), random forest, and C4.5 trees. Based on the best algorithm, we build a broad non-personalized classifier to compare the personalized classifiers to. Finally, we present a more efficient method for building personalized classifiers with comparable accuracy and consistency. This method uses agglomerative clustering to group users based on background and mood states, then builds classifiers for each of these individual groups.

2 Related Work

There is some discussion as to the possible usefulness of creating a personalized music recommender system. On the one hand, [2] demonstrated that emotion in music is not so subjective that it cannot be modeled; on the other hand, the results from researchers who attempt to build personalized music emotion recommendation systems are very promising, suggesting personalization is at least a way to improve emotion classification accuracy. Yang et al. in [3] was one of the earliest to study the relationship between music emotion recognition and personality. The authors looked at users demographic information, musical experience, and user scores on the Big Five personality test to determine possible relationships and build their system. Classifiers were built based on support vector regression, and test regressors trained on general data and personalized data. The results were that the personalized regressors outperformed the general regressors in terms of improving accuracy, first spotlighting the problem of trying to create personalized recommendation systems for music and mood based on general groups. However, there has been continued work on collaborative filtering, as well as hybridizing personalized and group based preferences. Lu and Tseng in [4] proposed a system that combined emotion-based, content-based, and collaborative-based recommendation and achieved an overall accuracy of 90 %. In [5], the authors first proposed the idea of using clustering to predict emotions for a group of users. The results were good, but some improvement was needed. The users were clustered into only two groups based on their answers to a set of questions, and the prediction was based on MIDI files rather than real audio. In this work, we propose creating personalized classifiers first (trained on real audio data), clustering users, creating representative classifiers for each cluster, and then allowing the classifiers to be altered based on user behavior.

Each of the possible classification algorithms has been used commonly in previous MIR research, with varying results. kNN had been evaluated previously in [6,7] for genre classification. Mckay and Fujinaga achieved a 90 %–98 % classification accuracy by combining kNN and Neural Network classifiers and applying them to MIDI files using a 2-level genre hierarchical system. On the other hand, [7] only achieved a 61 % accuracy at the highest using real audio and k of 3. Random Forest was used principally in [8] for instrument classification in noisy audio. Sounds were created with one primary instrument and artificial noise of varying levels added in incrementally. The authors found that the percentage error was overall much lower than previous work done with SVM classifiers on the same sounds up until the noise level in the audio reached 50 %. They also observed that Random Forest could indicate the importance of certain attributes in the classification based on the structure of the resulting trees and the attributes used in the splitting. SVM classification is one of the more common algorithms used in MIR for a variety of tasks, such as [9] for mood classification, [10] for artist identification (compared with kNN and Gaussian Mixture Models), and [11] for mood tracking. It has also been evaluated beside other classifiers in [7] for genre identification. These evaluations have shown the SVM classifier to be remarkably accurate, particularly in predicting mood. Regarding C4.5 decision trees, the authors in [12] in a comparison of the J48 implementation of C4.5 to Bayesian network, logical regression, and logically weighted learning classification models for musical instrument classification found that J48 was almost universally the most accurate classifier (regardless of the features used to train the classifier). The classification of musical instrument families (specifically string or woodwind) using J48 ranged in accuracy from 90–92 %, and the classification of actual instruments ranged from 60–75 % for woodwinds and 60–67 % for strings.

3   Data Composition and Collection

3.1   Music Data

Music data was collected from 100 audio clips 25–30 s in length culled from one of the author's personal music collection. These clips were split into 12–15 segments (depending on the length of the original clip) of roughly 0.8 seconds in order to allow for changes in annotation as the clip progresses, resulting in a total of 1440 clips. These clips originated from several film and video game sound tracks in order to achieve a similar effect to the dataset composed in [13] (namely a set composed of songs that are less known and more emotionally evocative). As such the music was mainly instrumental with few if any intelligible vocals. The MIRToolbox [1] collection was then used to extract musical features. MIRToolbox is a set of functions developed for use in MATLAB which uses, among others, MATLAB's Signal Processing toolbox. It reads .wav files at a sample rate of 44100 Hz. The following features were extracted using this toolbox.

– **Rhythmic Features** (fluctuation peak, fluctuation centroid, frame-based tempo estimation, autocorrelation, attack time, attack slope): Rhythmic features refer to the set of audio features that describe a song's rhythm and tempo, or how fast the song is, although features such as attack time and attack slope are better indicators of the rhythmic style of the audio rather than pure tempo estimation. Fluctuation based features are based on calculations to a fluctuation summary (calculated from the estimated spectrum with a Bark-band redistribution), while the rest of the rhythmic features are based on the calculation of an onset detection curve (which shows the rhythmic pulses in the song in the form of amplitude peaks for each frame).

- **Timbral Features** (spectral centroid, spectral spread, coefficient of spectral skewness, kurtosis, spectral flux, spectral flatness, irregularity, Mel-Frequency Cepstral Coefficients (MFCC) features, zero crossings, brightness): Timbral features describe a piece's sound quality, or the sonic texture of a piece of audio. The timbre of a song can change based on instrument composition as well as play style. Most of these features are derived from analysis of the audio spectrum, a decomposition of an audio signal. MFCC features are based on analysis of audio frequencies (based on the Mel scale, which replicates how the human ear processes sound). Brightness and zero crossings are calculated based on the audio signal alone.
- **Tonal Features** (pitch, chromagram peak and centroid, key clarity, mode, Harmonic Change Detection Function (HCDF)): Tonal features describe the tonal aspects of a song such as key, dissonance, and pitch. They are based primarily on a pitch chromagram, which shows the distribution of energy across pitches based on the calculation of dominant frequencies in the audio.

3.2 User Data

We have created a questionnaire so that individuals can go through multiple times and annotate different sets of music based on their moods on a given day. 68 users completed the questionnaire between 1 and 8 times, resulting in almost 400 unique user sessions.

Questionnaire Structure. The Questionnaire is split into 5 sections:

- Demographic Information (where the user is from, age, gender, ethnicity),
- General Interests (favorite books, movies, hobbies),
- Musical Tastes (what music the user generally likes, what he listens to in various moods),
- Mood Information (a list of questions based on the Profile of Mood States),
- Music Annotation (where the user annotates a selection of musical pieces based on mood).

The demographic information section is meant to compose a general picture of the user (see Fig. 1). The questions included ask for ethnicity (based on the NSF definitions), age, what level of education the user has achieved, what field they work or study in, where the user was born, and where the user currently lives. Also included is whether the user has ever lived in a country other than where he/she was born or where he/she currently lives for more than three years. This question is included because living in another country for that long would expose the user to music from that country.

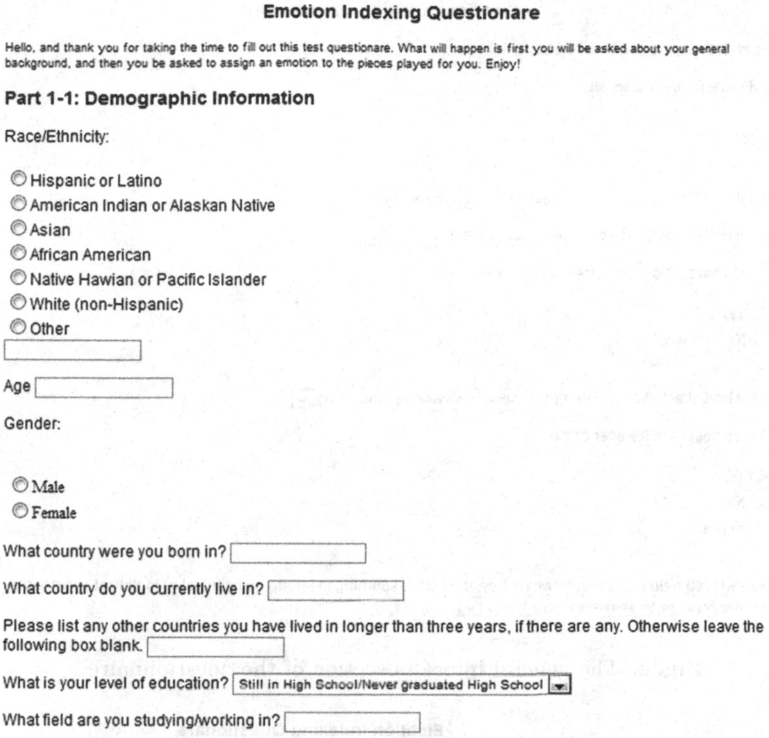

Emotion Indexing Questionare

Hello, and thank you for taking the time to fill out this test questionare. What will happen is first you will be asked about your general background, and then you be asked to assign an emotion to the pieces played for you. Enjoy!

Part 1-1: Demographic Information

Race/Ethnicity:

- Hispanic or Latino
- American Indian or Alaskan Native
- Asian
- African American
- Native Hawian or Pacific Islander
- White (non-Hispanic)
- Other

Age

Gender:

- Male
- Female

What country were you born in?

What country do you currently live in?

Please list any other countries you have lived in longer than three years, if there are any. Otherwise leave the following box blank.

What is your level of education? | Still in High School/Never graduated High School |

What field are you studying/working in?

Fig. 1. The demographic information section of the questionnaire

The general interests section gathers information on the user's interests outside of music (see Fig. 2). It asks for the user's favorite genre of books, movies, and what kind of hobbies he/she enjoys. It also asks whether the user enjoyed math in school, whether he/she has a pet or would want one, whether he/she believes in an afterlife, and how he/she would handle an aged parent. These questions are all meant to build a more general picture of the user.

The musical taste section is meant to get a better picture of how the user relates to music (Fig. 3). It asks how many years of formal musical training the user has had, his/her level of proficiency in reading/playing music if any, and what genre of music the user listens to when they are happy, sad, angry, or calm.

The mood information section is a shortened version of the Profile of Mood States [14]. The Profile of Mood States asks users to rate how strongly he/she has been feeling a set of emotions over a period of time from the following list of possible responses:

– Not at all; A little; Moderately; Quite a bit; Extremely.

The possible emotions asked about in the mood information session are:

– Tense; Shaky; Uneasy; Sad; Unworthy; Discouraged; Angry; Grouchy; Annoyed; Lively; Active; Energetic; Efficient.

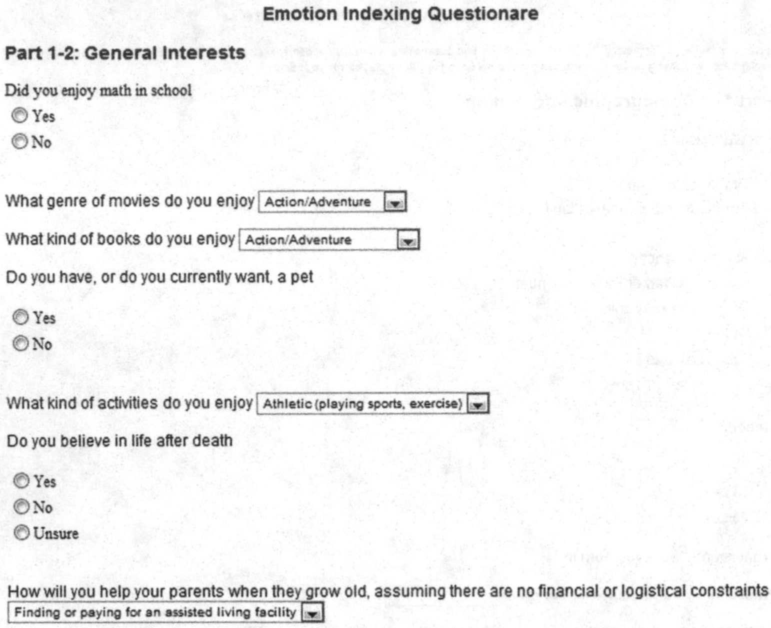

Emotion Indexing Questionare

Part 1-2: General Interests

Did you enjoy math in school
- ⊙ Yes
- ⊙ No

What genre of movies do you enjoy [Action/Adventure ▾]

What kind of books do you enjoy [Action/Adventure ▾]

Do you have, or do you currently want, a pet

- ⊙ Yes
- ⊙ No

What kind of activities do you enjoy [Athletic (playing sports, exercise) ▾]

Do you believe in life after death

- ⊙ Yes
- ⊙ No
- ⊙ Unsure

How will you help your parents when they grow old, assuming there are no financial or logistical constraints
[Finding or paying for an assisted living facility ▾]

Fig. 2. The general interests section of the questionnaire

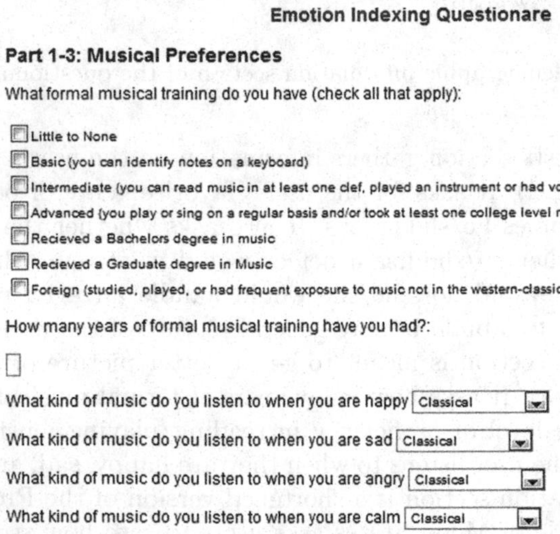

Emotion Indexing Questionare

Part 1-3: Musical Preferences

What formal musical training do you have (check all that apply):

- ☐ Little to None
- ☐ Basic (you can identify notes on a keyboard)
- ☐ Intermediate (you can read music in at least one clef, played an instrument or had vocal training)
- ☐ Advanced (you play or sing on a regular basis and/or took at least one college level music course)
- ☐ Recieved a Bachelors degree in music
- ☐ Recieved a Graduate degree in Music
- ☐ Foreign (studied, played, or had frequent exposure to music not in the western-classical style)

How many years of formal musical training have you had?:

☐

What kind of music do you listen to when you are happy [Classical ▾]

What kind of music do you listen to when you are sad [Classical ▾]

What kind of music do you listen to when you are angry [Classical ▾]

What kind of music do you listen to when you are calm [Classical ▾]

Fig. 3. The musical taste/background information section of the questionnaire

A sample of these questions can be seen in Fig. 4. This is the section that is filled out every time the user returns to annotate music, since their mood would affect how they annotate music on a given day. These answers are later converted into a mood vector for each session, which describes the user's mood state at the time of the session.

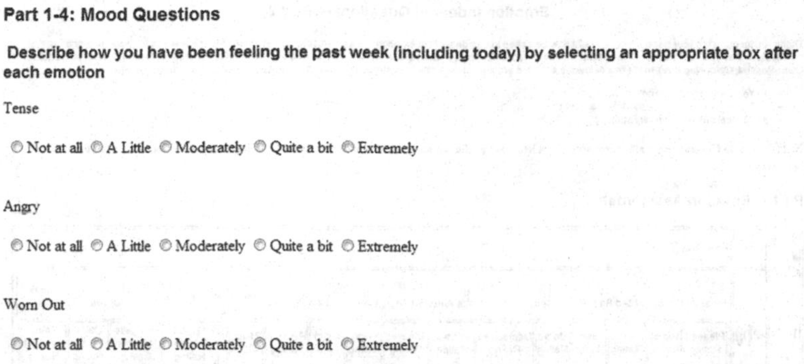

Part 1-4: Mood Questions

Describe how you have been feeling the past week (including today) by selecting an appropriate box after each emotion

Tense

○ Not at all ○ A Little ○ Moderately ○ Quite a bit ○ Extremely

Angry

○ Not at all ○ A Little ○ Moderately ○ Quite a bit ○ Extremely

Worn Out

○ Not at all ○ A Little ○ Moderately ○ Quite a bit ○ Extremely

Fig. 4. Part of the mood state information section of the questionnaire. This section is filled out every time the user reenters the questionnaire (the user starts on this page once he/she has filled out the rest of the questionnaire once)

Mood Vector Creation. For each session the user is asked to select an answer describing how much he/she has been feeling a selection of emotions. Once this is finished, his/her answers are then converted to a numerical mood vector as follows: each answer is given a score based on the response, with 0 representing "Not at all", 1 – "A little", 2 – "moderately", 3 – "Quite a bit", and 4 – "Extremely". From here, sets of mood scores corresponding to different emotions are added together into a set of scores:

$$TA = Tense + Shaky + Uneasy \tag{1}$$

Where TA stands for Tension/Anxiety,

$$DD = Sad + Unworthy + Discouraged \tag{2}$$

Where DD stands for Depression/Dejection,

$$AH = Angry + Grouchy + Annoyed \tag{3}$$

Where AH stands for Anger/Hostility,

$$VA = Lively + Active + Energetic \tag{4}$$

Where VA stands for Vigor/Activity,

$$FI = WornOut + Fatigued + Exhausted \tag{5}$$

Where FI stands for Fatigue/Inertia,

$$CB = (Confused + Muddled) - Efficient \tag{6}$$

Where CB stands for Confusion/Bewilderment.
These scores are recorded, along with a total score calculated as follows:

$$Total = (TA + DD + AH + FI + CB) - VA \tag{7}$$

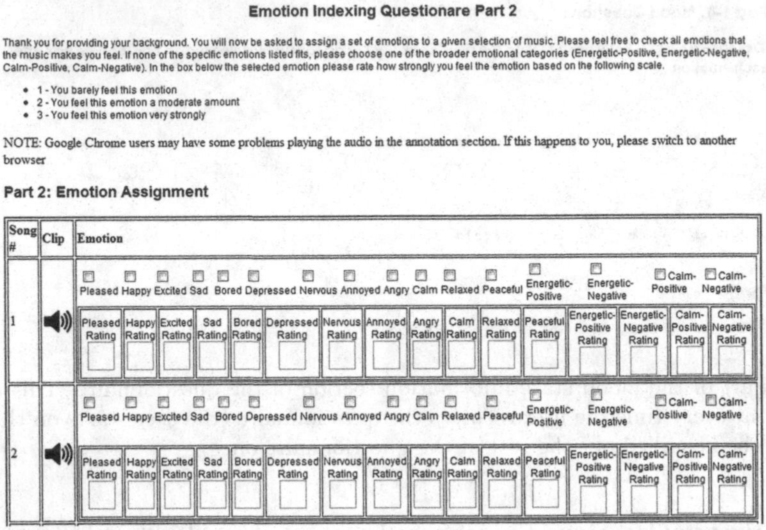

Emotion Indexing Questionare Part 2

Thank you for providing your background. You will now be asked to assign a set of emotions to a given selection of music. Please feel free to check all emotions that the music makes you feel. If none of the specific emotions listed fits, please choose one of the broader emotional categories (Energetic-Positive, Energetic-Negative, Calm-Positive, Calm-Negative). In the box below the selected emotion please rate how strongly you feel the emotion based on the following scale.

- 1 - You barely feel this emotion
- 2 - You feel this emotion a moderate amount
- 3 - You feel this emotion very strongly

NOTE: Google Chrome users may have some problems playing the audio in the annotation section. If this happens to you, please switch to another browser

Part 2: Emotion Assignment

Song #	Clip	Emotion
1		Pleased, Happy, Excited, Sad, Bored, Depressed, Nervous, Annoyed, Angry, Calm, Relaxed, Peaceful, Energetic-Positive, Energetic-Negative, Calm-Positive, Calm-Negative Pleased Rating, Happy Rating, Excited Rating, Sad Rating, Bored Rating, Depressed Rating, Nervous Rating, Annoyed Rating, Angry Rating, Calm Rating, Relaxed Rating, Peaceful Rating, Energetic-Positive Rating, Energetic-Negative Rating, Calm-Positive Rating, Calm-Negative Rating
2		Pleased, Happy, Excited, Sad, Bored, Depressed, Nervous, Annoyed, Angry, Calm, Relaxed, Peaceful, Energetic-Positive, Energetic-Negative, Calm-Positive, Calm-Negative Pleased Rating, Happy Rating, Excited Rating, Sad Rating, Bored Rating, Depressed Rating, Nervous Rating, Annoyed Rating, Angry Rating, Calm Rating, Relaxed Rating, Peaceful Rating, Energetic-Positive Rating, Energetic-Negative Rating, Calm-Positive Rating, Calm-Negative Rating

Fig. 5. The music emotion annotation section, also filled out every time the user goes through the questionnaire. The user clicks on a speaker to hear a music clip, then checks an emotion and supplies a rating 1–3

The mood vector is then defined for user u and session s as

$$m(u, s) = (TA(u, s), DD(u, s), AH(u, s), FI(u, s), CB(u, s), VA(u, s), Total(u, s)) \tag{8}$$

Finally, the music annotation section is where users go to annotate a selection of clips. 40 clips are selected randomly from the set of 1440 clips mentioned in Sect. 3.1. The user is then asked to check the checkbox for the emotion he/she feels in the music, along with a rating from 1–3 signifying how strongly the user feels that emotion (1 being very little, 3 being very strongly). The user has a choice of 16 possible emotions to pick (to be specific, 12 emotions and 4 generalizations), based on a 2-D hierarchical emotional plane (see Fig. 6 for the emotion plane and Fig. 5 for a view of the questionnaire annotation section).

When the user goes through the questionnaire any time after the first time, he only has to fill out the mood profile and the annotations again. Each of these separate sections (along with the rest of the corresponding information) is treated as a separate user, so each individual session has classifiers trained for each emotion, resulting in 16 emotion classifiers for each user session.

Emotion Model. This model was first presented in [5], and implements a hierarchy on the 2-dimensional emotion model, while also implementing discrete elements. The 12 possible emotions are derived from various areas of the 2-dimensional arousal-valence plane (based on Thayer's 2-dimensional model of arousal and valence [15]). However, there are also generalizations for each area of

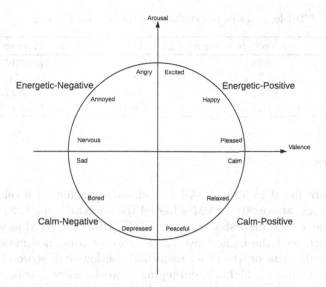

Fig. 6. A diagram of the emotional model

the plane (excited-positive, excited-negative, calm-positive, and calm-negative) that the users can select as well. This compensates for songs that might be more ambiguous to the user; if a user generally knows that a song is high-energy and positive feeling but the words excited, happy, or pleased do not adequately describe it, they can select the generalization of energetic-positive.

3.3 Classifier Development

Personalized classifiers were trained and tested using the classification algorithms listed previously (C4.5, SVM, Random Forest, kNN). The user annotation data was first converted so that each annotation for each song was represented as a vector of 16 numbers with each number representing the emotion labeling. The numbers ranged from 0 to 3, with 0 representing an emotion that was not selected by the user and the remaining numbers being the strength the user entered with the annotation. These vectors for all the users were then linked with the feature data extracted from the corresponding music clips. From this resulting table all the annotations and music data linked with individual user IDs were separated and used to train and test personalized classifiers for each emotion. This resulted in each user having at most 16 personalized classifiers (depending on whether the user used a given emotion during the course of annotating), where for each classifier the class attribute was one of the 16 possible emotions. The classifiers were evaluated via Weka [16] using 10-fold cross validation. For the C4.5 classifier we used the J48 implementation in Weka and for kNN we used Weka's IBk. Analysis of the results indicates which classifier is most effective for personalized classification and, therefore, the most effective cluster-driven classifier.

Table 1. Table of classifier accuracies and F-scores

Classifier	Average accuracy (%)	Average F-score	Average Kappa
SVM	82.35	0.90	0.137658
IBk	85.7	0.87	0.153468
J48	86.62	0.89	0.076869
Random forest	84.25	0.90	0.133027

4 Results

The results are listed in Table 1. All four classifiers achieved a relatively high average accuracy, above 80 %. SVM achieved the lowest accuracy, 82.35 %, while J48 trees achieved the highest accuracy, 86.62 %. However, SVM as well as Random Forest achieved the highest average F-score (a combined measure of precision and recall). IBk on the other hand had the lowest F-score of 0.92. SVM was expected to have a higher accuracy as it works so well with music data, but our previous success with J48 means the high accuracy and F-score are not surprising.

The Kappa statistic reveals further insights into the effectiveness of each classifier. This statistic measures the agreement between a true class and the prediction, and the closer to 1 the statistic is the more agreement (1 represents complete agreement). None of the classifiers reaches higher than 0.1, although again IBk has the highest average Kappa (J48, again, the lowest). This suggests that while J48 is overall very accurate it is more inconsistent in terms of this particular set of data, while SVM is moderately accurate and very consistent.

As it proved to be the most accurate classifier, we have chosen J48 as the algorithm to use to build the non-personalized classifiers for comparison. We again built 16 emotion classifiers, this time using all the user annotations to train and test rather than individual user annotations. The results compared to the personalized J48 classifiers are shown in Table 2.

The average accuracy does not change too much between personalized and non-personalized classifiers (only 0.7 % point). However, this was mainly due to the fact that several of the emotions were not used to the same extent as others when tagging (for example, the generalized emotions), and in that case all the classifier did was predict '0' (for emotions that were not selected). This raised the accuracy for those classifiers, but it is not nearly as indicative as to the quality of the classifier as the F-Score and Kappa, which showed a great

Table 2. Comparison of classifier accuracies and F-scores between personalized and non-personalized classifiers

Classifier	Average accuracy (%)	Average F-score	Average Kappa
Personalized J48	86.62	0.89	0.076869
Non-personalized J48	87.29	0.82	0.00015

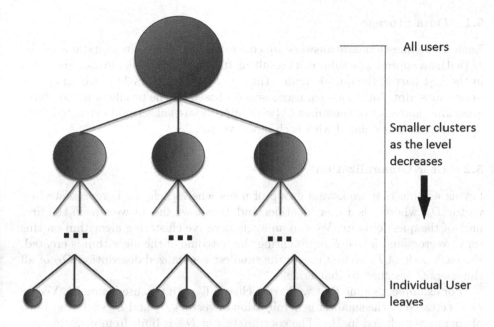

All users

Smaller clusters
as the level
decreases

Individual User
leaves

Fig. 7. A visual of the proposed classifier hierarchy

deal of improvement in the personalized classifier. The average F-score for the
non-personalized classifiers is 0.07 less than the average F-score for personalized
classifiers, and the average Kappa for the non-personalized classifiers is far less
than the personalized classifiers. These both signify a significant loss in classifier
consistency once the classifiers are no longer personalized.

5   Hierarchical Cluster Driven Classifiers

We have also developed a more efficient method of developing personalized classifiers. Using agglomerative clustering, we have developed hierarchical cluster-
based classifiers (see Fig. 7). These clusters are built based on the user data
gathered through the questionnaire (as described in Sect. 3.2).

We can now build a tree structure of classifiers where the leaf nodes of the tree
are labeled by vectors representing individual users and the root of any subtree
is labeled by a smallest generalization vector covering vector labels associated
with all leaves of that subtree. Each node of the tree has its own set of 16
emotion classifiers, one for each possible emotion, based on the annotation data
from the group of users assigned to that node. The lower the node on the tree,
the more specialized its classifiers are. Now, if there is a need, we can assign a
new user to a correct node of the tree structure which is the lowest one labeled
by generalization vector containing the vector label of that user. Then, we can
apply the classifiers associated with that node to annotate the music.

5.1   Data Storage

Each user's questionnaire answers are converted into three sets of data: a vector D (with an appropriate subscript) resulting from the questions the user answered in the first part of the questionnaire, the set of mood vectors M (with an appropriate subscript) built for each user's session based on the profile of mood states questions, and a set of classifiers C (with appropriate subscripts) extracted from decision tables associated with each mood vector.

5.2   User Generalization

Let us assume that we have a group of users where each one is represented by vector D_a, where a is a user identifier and D_a shows the answers from the first part of the questionnaire. We run an agglomerative clustering algorithm on this set of vectors and a tree T representing the outcome of the algorithm is created. For each node of T, we first create the smallest generalized description D_C of all the users C assigned to that node.

For example, assume that we have a cluster C with two users a and b. Vector D_C is created as the smallest generalization of vectors D_a and D_b such that both of them are included in D_C. The coordinate i of D_C is built from coordinates i of D_a and D_b as:

$$D_{Ci} = \{k : min(D_{ai}, D_{bi}) \leq k \leq max(D_{ai}, D_{bi})\} \tag{9}$$

The mood vectors and decision tables assigned to the nodes of T which are not leaves are generalized on a more conditional basis, using the distance between mood vectors. For example, let us assume that users $a1, a2$ end up in the same cluster C and their sessions are represented by mood vectors m[a1,8], m[a2,5], m[a1,2]. If the distance between any of these vectors is less than λ (a given threshold), then they are added together by following the same strategy we used for vectors D representing the first part of the questionnaire. It should be noted that mood vectors representing sessions of different users can be added together, whereas vectors representing the same user may remain separated. When the new mood vectors for a node of T are built, then the new decision table for each of these new mood vectors is built by taking the union of all decision tables associated with mood vectors covered by this new mood vector (Fig. 8).

5.3   New User Placement

When a new user, x, fills out the first part of the questionnaire his/her representative vector D_x is created. Then it is checked to see if this vector is equal to one of the representative vectors representing leaves of the tree structure. If this is not the case, then the representative sets of vectors belonging to the parents of these leaves are checked to see if one of them contains D_x. This is accomplished by comparing the individual column values of each column, D_{xi}, to each column range D_{Ci}. If every D_{xi} fits in the range of each D_{Ci}, then the user is

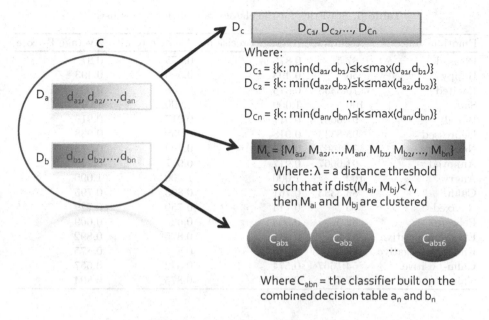

Fig. 8. A visualization of the data stored for each cluster

assigned to that cluster. If D_x does not fit within any node on a given level, then it is compared to the parent of the node that D_x most closely matches (based on the number of column ranges in D_C that D_x does fit in). For example, if on the bottom level D_x cannot be assigned to any cluster, but all i in D_{xi} fit in the ranges for D_{Ci} except for one, this would make it the closest matched cluster, and then D_x would be compared to the parent of D_C. The goal is to assign the user to the cluster on the lowest level it can fit in order to utilize the most specialized classifiers built (since lower level classifiers are built on data from a more homogenized group of users, and therefore a more unified group of annotations). These classifiers will therefore be more accurate to the new user, solving the "cold-start" issue inherent in collaborative recommender systems.

5.4   Test Case: New User

To demonstrate the effectiveness of this system, we analyze one new user using this system for the first time. This user has filled out the same questionnaire questions, however they only have one mood vector which (along with their other questionnaire answers) is used to assign this user to a cluster. This user also annotated a different set of songs, which had the same feature data extracted from them as the initial training set.

Once this user is assigned to a node in our tree structure, they are given the set of emotion classifiers made for the cluster in that part of the tree. The user's new annotations were then used to evaluate the classifiers' effectiveness. The resulting statistics are shown in Table 3.

Table 3. Statistics for the classifiers built for a new user

Emotion	Accuracy	Average precision	Average recall	Average F-score
Pleased	87.5	0.875	0.875	0.87
Happy	50	0.493	0.500	0.493
Excited	70.8333	0.675	0.708	0.691
Sad	100	1.000	1.000	1.000
Bored	91.6667	0.917	0.917	0.917
Depressed	95.8333	0.918	0.958	0.938
Nervous	79.1667	0.756	0.792	0.773
Annoyed	91.6667	0.917	0.917	0.917
Angry	100	1.000	1.000	1.000
Calm	83.3333	0.761	0.833	0.795
Relaxed	75	0.714	0.750	0.729
Peaceful	62.5	0.594	0.625	0.609
Energetic-positive	87.5	0.911	0.875	0.892
Energetic-negative	87.5	0.875	0.875	0.875
Calm-positive	54.1667	0.574	0.542	0.557
Calm-negative	87.5	0.915	0.875	0.894

The accuracy and F-scores can stay relatively high for each classifier, although it is not consistent. This could be explained by the user not annotating songs with certain emotions, which would make the accuracies very difficult to judge. This could indicate also that certain emotions are easier to classify for songs than others. Observe that the lower accuracies (aside from Happy) are for emotion classifiers from the calm-positive quadrant of the arousal-valence plane (Calm, Relaxed, Peaceful), and Calm-Positive is the least accurate quadrant classifier (Energetic-Positive, Energetic-Negative, Calm-Positive, Calm-Negative). This could imply that emotions with a high valence and low arousal are particularly difficult to detect in music, but this would require further investigation.

6   Conclusion

We have presented a system through which we build personalized music emotion classifiers based on user data accumulated through an online survey. Using this data, we have built classifiers that are about as accurate as standard, non-personalized ones but far more consistent. In a real world situation, this would mean that music that accurately reflects the user's mood (or desired mood) would be recommended far more often than not. Future work would involve using these classifiers in a full music player, and improving classifiers by retraining them through usage and by ensuring the clusters are disjoint, using Michalski's STAR method applied in $AQ15$ to build disjoint D-vector representations [17].

Acknowledgments. This project was partially supported by the Research Center of PJIIT, supported by the Polish Ministry of Science and Higher Education.

References

1. Lartillot, O., Toiviainen, P., Eerola, T.: MIRtoolbox. University of Jyväskylä (2008)
2. Laurier, C., Herrera, P.: Automatic detection of emotion in music: interaction with emotionally sensitive machines. In: Vallverdu, D., Casacuberta, D. (eds.) Handbook of Research on Synthetic Emotions and Sociable Robotics: New Applications in Affective Computing and Artificial Intelligence, pp. 9–32. IGI Global, Hershey (2009)
3. Yang, Y.H., Su, Y.F., Lin, Y.C., Chen, H.H.: Music emotion recognition: the role of individuality. In: Proceedings of International Workshop on Human-Centered Multimedia 2007 (HCM'07), Augsburg, Germany, September 2007. ACM (2007)
4. Lu, C.C., Tseng, V.S.: A novel method for personalized music recommendation. Expert Syst. Appl. **36**(6), 10035–10044 (2009)
5. Grekow, J., Raś, Z.W.: Detecting emotions in classical music from MIDI files. In: Rauch, J., Raś, Z.W., Berka, P., Elomaa, T. (eds.) ISMIS 2009. LNCS, vol. 5722, pp. 261–270. Springer, Heidelberg (2009)
6. Mckay, C., Fujinaga, I.: Automatic genre classification using large high-level musical feature sets. In: ISMIR 2004, pp. 525–530 (2004)
7. Silla Jr, C.N., Koerich, A.L., Kaestner, C.A.A.: A machine learning approach to automatic music genre classification. J. Braz. Comp. Soc. **14**(3), 7–18 (2008)
8. Kursa, M., Rudnicki, W., Wieczorkowska, A., Kubera, E., Kubik-Komar, A.: Musical instruments in random forest. In: Rauch, J., Raś, Z.W., Berka, P., Elomaa, T. (eds.) ISMIS 2009. LNCS, vol. 5722, pp. 281–290. Springer, Heidelberg (2009)
9. Laurier, C., Meyers, O., Marxer, R., Bogdanov, D., Serrà, J., Gómez, E., Herrera, P., Wack, N.: Music classification using high-level models. In: ISMIR 2009 (2010)
10. Mandel, M.I., Ellis, D.P.W.: Song-level features and support vector machines for music classification. In: Reiss, J.D., Wiggins, G.A. (eds.) ISMIR 2005, London, U.K., vol. 6, pp. 594–599. (2005)
11. Panda, R., Paiva, R.P.: Using support vector machines for automatic mood tracking in audio music. In: 130th Audio Engineering Society Convention (2011)
12. Zhang, X., Ras, Z.: Differentiated harmonic feature analysis on music information retrieval for instrument recognition. In: IEEE GrC 2006, Atlanta, Georgia, pp. 578–581 (2006)
13. Eerola, T., Vuoskoski, J.K.: A comparison of the discrete and dimensional models of emotion in music. Psychol. Music **39**(1), 18–49 (2011)
14. McNair, D.M., Lorr, M., Droppleman, L.F.: Profile of Mood States (POMS) (1971)
15. Thayer, R.E.: The Biopsychology of Mood and Arousal. Oxford University Press, New York (1989)
16. Hall, M., Frank, E., Holmes, G., Pfahringer, B., Reutemann, P., Witten, I.H.: The WEKA data mining software. ACM SIGKDD Explor. Newslett. **11**(1), 10 (2009)
17. Michalski, R., Mozetic, T., Hong, J., Lavarac, N.: The multi-purpose incremental learning system AQ15 and its testing application to three medical domains. In: AAAI-86 Proceedings, pp. 1041–1045. AAAI (1986)

Mining Audio Data for Multiple Instrument Recognition in Classical Music

Elżbieta Kubera[1](✉) and Alicja A. Wieczorkowska[2]

[1] University of Life Sciences in Lublin, Akademicka 13, 20-950 Lublin, Poland
elzbieta.kubera@up.lublin.pl
[2] Polish-Japanese Institute of Information Technology,
Koszykowa 86, 02-008 Warsaw, Poland
alicja@poljap.edu.pl

Abstract. This paper addresses the problem of identification of multiple musical instruments in polyphonic recordings of classical music. A set of binary random forests was used as a classifier, and each random forest was trained to recognize the target class of sounds. Training data were prepared in two versions, one based on single sounds and their mixes, and the other containing also sound frames taken from classical music recordings. The experiments on identification of multiple instrument sounds in recordings are presented, and their results are discussed in this paper.

Keywords: Music information retrieval · Sound recognition · Random forests

1 Introduction

Music information retrieval (MIR) became a topic of broad interest for researchers several years ago, see e.g. [29,32], and one of the most challenging tasks within this area is to automatically extract meta-information from audio waveform [23]. Audio data stored as sound amplitude values changing over time represent very complex data, where multiple sounds of a number of instruments are represented by a single value (i.e. amplitude value of a complex sound) in each time instant in the case of monophonic recordings, or by a single value in each recording channel. Extraction of information about timbre of particular sounds is difficult; still, it has been addressed in audio research last years, with various accuracy [12]. On the contrary, the identification of music titles through query-by-example, including excerpts replayed on mobile devices, has been quite successfully solved [31,34], as well as finding pieces of music through query-by-humming [25]. However, identification of instruments in audio excerpts is still a challenge, sometimes addressed through multi-pitch tracking [11], often supported with external provision of pitch data, or limited to the sound identification of a predominant instrument [3,8].

In this paper, we deal with the identification of multiple sounds of multiple instruments in the recordings of classical music. No pitch tracking is required, and the classification is performed on the data as is. There are no pre-assumptions

A. Appice et al. (Eds.): NFMCP 2013, LNAI 8399, pp. 246–260, 2014.
DOI: 10.1007/978-3-319-08407-7_16, © Springer International Publishing Switzerland 2014

regarding the number of instruments playing together (i.e. the polyphony level), and the recordings can contain any instrument sound, including instruments for which our classifiers are not trained. This is because we use a set of binary classifiers, where each classifier is trained to identify a target sound class. If none of the classifiers recognizes its target class, the analyzed audio sample represents unknown instrument(s) or silence. Classification is performed for mono or stereo input data, and a mix (average) of the channels is taken as input for stereo data.

1.1   Identification of Instruments in Audio Signal

Automatic identification of musical instruments has been performed so far by many researchers, and usually on different sets of instruments, number of classes, sound parametrization and the resulting feature vector, number of sounds used, and classifiers used in the research. Identification of a single instrument in a single isolated sound is the easiest case, and virtually all available classification tools have been applied for this purpose, including k-nearest neighbors, neural networks, support vector machines (SVM), rough set based classifiers, decision trees etc. Quality of the recognition depends heavily on the number of sounds and instruments/classes applied; it can even reach 100 % for a few classes, or be as low as 40 % if there are 30 or more classes; for detailed review see [12].

Identification of musical instruments in polyphonic recordings is much more challenging, since sounds blend with each other, and overlap in time and frequency domains. Still, it is indispensable in order to progress in the endeavor to achieve the ultimate goal: automatic music transcription, which requires extraction of all notes from the audio material and assigning them into particular instruments in the score [2,26]. The presented overview of research in this domain shows approaches taken and results obtained. Since the research is performed on various sets of data, it is not feasible to compare all results directly, or to compare our results with those obtained by other researchers, but it sketches a general outlook of the state of the art in this area.

In the research performed so far, the recognition of instruments in polyphonic environment has been addressed in various ways. Usually initial assumptions are made: on the number of instruments in the polyphony, on pitch data as input, on the instrument set in the analyzed recordings, or on identifying predominant sound, see e.g. [8,11]. Since the final goal of such research is score extraction, such assumptions are understandable, and in some cases the research addresses sound separation into single sounds. These external data are often manually provided, not extracted from audio recordings. Training and testing is often performed on artificial mixes of single sounds. Training sets often include single sounds, but the recognition improves when mixes are added to the training set [22]. Training and testing on real recordings is also performed, but it requires tedious labeling to get ground truth data. This is why mixes are commonly applied in such a research.

The polyphony in the research performed so far varies, and in the simplest case is limited to two sounds being played together, i.e. to duets [6,14,21,35]. Since overlapping partials in audio spectrum hamper sound identification, these partials are sometimes omitted in the recognition process. For instance, in [6] missing feature approach was applied, where the timefrequency regions that are dominated by interfering sounds are considered missing or unreliable. These components are marked in a binary mask, and then excluded from the classification process. Gaussian Mixture Models yielded about 60 % accuracy for duets from 5-instrument set when using this approach. In [35], a different approach was taken. Note spectra are represented as sums of instrument-dependent log-power spectra, and chord spectra are represented as sums of note power spectra. Hidden Markov Models have been applied to model note duration. Training performed on solo excerpts allows learning sound of time-varying power, vibrato, and so on. In [5], the approach applied was inspired by non-negative matrix factorization, with an explicit sparsity control. The experiments were performed on violin and piano, with tests on mixes, which is often the case in such research, as this way ground truth data are easily obtained.

Direct spectrum/template matching is sometimes performed in instrument sound identification, without feature extraction [14,15]. In [14], this approach was applied to 2-instrumental mixes for 26-instrument set. In [15], an adaptive method for template matching was used for 3 instruments (flute, violin and piano), with polyphony up to 3 sounds played together. When musical context was integrated into their system, the recognition rate reached 88 %.

A variety of classification methods has been applied to the instrument identification task in polyphonic recordings. SVM, decision trees, and k-nearest neighbor classifiers were used in [22]. This research aimed at recognizing the dominant instrument. The outcomes vary with the level of target obfuscation, and obtained correctness was around 50–80 %. In [17], linear discriminant analysis (LDA) was used. Their approach consists in finding partials which were least influenced by overlapping, and when musical context was also taken into account, they obtained 84.1 % recognition rate for duets, 77.6 % for trios, and 72.3 % for quartets. Linear discrimination was also applied in [1]. Their strategy consists in exploring the spectral disjointness among instruments by identifying isolated partials. For each such a partial, a pairwise comparison approach (for each pair of instruments) was taken to determine which instrument most likely has generated that partial; unisons were excluded from the research because of fully overlapping partials. For 25 instruments (and therefore 300 pairs) a high recall of 86–100 % was obtained, with 60 % average precision. In [7], SVMs were applied in a hierarchical classification scheme, extracted through hierarchical clustering. The obtained taxonomy of musical ensembles (for jazz in their case) yielded the average accuracy of 53 %, for the polyphony up to four instruments. Clustering was also applied in [24] for polyphony up to 4 notes for 6 instruments. Their approach is based on spectral clustering, and their goal is to achieve sound separation. Spectral basis decomposition is performed in their research using Principal Component Analysis (PCA). This approach yielded 46 % recall when evaluated on 4-note mixtures, with a precision of 56 %.

The task of instrument identification is often combined with sound source separation, as it is necessary anyway for the automatic music transcription, see e.g. [11]. Sound separation of instruments can be addressed using statistical tools when the number of channels is at least equal to the number of instruments, but this usually is not the case (e.g. CD recordings are in stereo format). In [16], semi-automatic music transcription is addressed through shift-variant non-negative matrix deconvolution (svNMD) based on constant-Q spectrogram using multiple spectral templates per instrument and pitch. K-means clustering was applied for learning, and best accuracy was obtained when each pitch was represented by more than one spectral template. Mixtures of 2 to 5 instruments were investigated; the more instruments, the lower the accuracy, below 40 % in real case scenario for 5 instruments.

In our research, we would like to perform instrument recognition (without sound separation) with no pre-assumptions, and also without initial data segmentation. We have already performed similar research, for jazz recordings [18], but it required tedious segmentation and labeling of small frames of the recordings in order to obtain ground-truth data. In order to facilitate research, we decided to perform annotation for 0.5-s excerpts; MIDI files and scores were used as guidance. Classification was performed using a set of random forests, since such a classifier proved quite successful in our previous research [18], and it is resistant to overtraining; also, the accuracy of random forests outperformed SVM classifier (considered state of the art and comparing well with the others) by an order of magnitude when compared on the same data [20].

2   Random Forests

A random forest (RF) is a classifier based on a tree ensemble; such classifiers are gaining increasing popularity last years [30]. RF is constructed using procedure minimizing bias and correlations between individual trees. Each tree is built using a different N-element bootstrap sample of the N-element training set. Since the elements of the N-element sample in bootstrapping are obtained through drawing with replacement from the original N-element set, roughly $1/3$ of the training data are not used in the bootstrap sample for any given tree. Assuming that objects are described by a vector of K attributes (features), k attributes out of all K attributes are randomly selected ($k \ll K$, often $k = \sqrt{K}$) at each stage of tree building, i.e. for each node of any particular tree in RF. The best split on these k attributes is used to split the data in the node.

The best split of the data in the node is determined as minimizing the Gini impurity criterion, which is the measure of how often an element would be incorrectly labeled if labeled randomly, according to the distribution of labels in the subset. Each tree is grown to the largest extent possible, without pruning. By repeating this randomized procedure M times a collection of M trees is obtained, constituting a random forest. Classification of each object is made by simple voting of all trees [4].

The classifier used in our research consists of a set of binary random forests. Each RF is trained to identify the target sound class, representing an instrument, whether this particular timbre is present in the sound frame under investigation, or not. If the percentage of votes of the trees in the RF is 50 % or more, then the answer of the classifier is considered to be positive, otherwise it is considered to be negative.

3 Audio Data

Our experiments focused on musical instrument sounds of chordophones (stringed instruments) and aerophones (wind instruments), typical for classical music. These instruments produce sounds of definite pitch, but information about pitch was not used nor retrieved in our research. This way we avoid provision of external data on pitch or calculating pitch in polyphonic data, as this can introduce additional errors.

In our experiments, we chose wind and stringed instruments, played in various ways, i.e. with various articulation, including bowing vibrato and pizzicato. Percussive instruments, i.e. idiophones and membranophones (basically drums) were excluded from the described research. If none of the classifiers gives positive answer, then we can conclude that an unknown instrument or instruments are playing in the investigated sound frame, or it represents silence.

The following sound classes were investigated in the reported research:

- flute (fl),
- oboe (ob),
- bassoon (bn),
- clarinet (cl),
- French horn (fh),
- violin (vn),
- viola (va),
- cello (ce),
- double bass (db), and
- piano (pn).

All sounds were recorded at 44.1 kHz sampling rate with 16-bit resolution, or converted to this format. If the RMS level for a sound segment (frame) was below 300, this segment is treated as silence; silence segments are used as negative training examples in our experiments. The silence threshold was empirically set in our previous experiments.

3.1 Training Data

Training of the classifiers was performed in two versions, in both cases for the ten instruments mentioned above, on 40-ms sound frames, as this is the length of the analyzing frame applied in the parameterization procedure. The training frames were taken from the audio data with no overlap between the frames used, to have as diversified data as possible.

The first training $(T1)$ was based on single sounds of musical instruments, taken from RWC [10], MUMS [28], and IOWA [33] sets of single sounds of musical

Table 1. Number of pieces in RWC Classical Music Database with the selected instruments playing together

Instrument	clarinet	cello	dbass	flute	Fhorn	piano	bassoon	viola	violin	oboe
clarinet	0	8	7	5	6	1	6	8	8	5
cello	8	0	13	9	9	4	8	17	19	8
dbass	7	13	0	9	9	2	8	13	13	8
flute	5	9	9	1	7	1	7	9	9	6
Fhorn	6	9	9	7	3	2	8	9	10	8
piano	1	4	2	1	2	5	1	2	5	0
bassoon	6	8	8	7	8	1	0	8	8	7
viola	8	17	13	9	9	2	8	0	17	8
violin	8	19	13	9	10	5	8	17	17	8
oboe	5	8	8	6	8	0	7	8	8	2

instruments, and also mixes of up to three instrument sounds were added to this training set. In this training, two sets of data were created, of the size of 20,000 and 40,000 frames, to observe whether increasing the training data set influences the obtained results. The training set for $T1$ consisted of 5,000 (or 10,000) frames of single sounds constituting positive examples for a target instrument, 5,000 (or 10,000) frames of single sounds constituting negative examples, 5,000 (or 10,000) mixes constituting positive examples, and 5,000 (or 10,000) mixes constituting negative examples. In positive examples, a target instruments is playing, and in negative examples is not.

Mixes constitute a single chord or unison, and a set of instruments is always typical for classical music. Up to 3 instrument sounds are taken for this mix, and the probability of instruments playing together in the mix reflects the probability of these instruments playing together in the RWC Classical Music Database. Table 1 shows how often the instruments investigated play together in the RWC Classical set.

The second training ($T2$) consisted of single sounds and mixes from RWC, MUMS, and IOWA sets, and also sounds taken from recordings, with no initial segmentation to separate single sounds; these were both solo and polyphonic recordings. The recordings were taken from RWC Classical Music Database [9], CDs and .mp3 files converted to .au format. The recordings used include:

- pieces from RWC Classical Music Database: No. 03 – 10, 12, 13, 17, 19, 21, 26 – 28, 36, 41 – 43.
- additional pieces with mainly solo (and also polyphonic) recordings: N. Rimsky-Korsakov - Scheherezade II. The Kalendar Prince; D. Shostakovich - Symphony no. 9 op. 70, Largo; J. Hummel - Concerto for Bassoon and Orchestra in F-major, Allegro Moderato; J.S. Bach - Suite for cello no. 2, Prelude; J.S. Bach - Partita in A-minor for solo flute, Sarabande; J.S. Bach - Suite BWV 1011; L. Berio - Psy for double bass; J. Sperger - Jagdmusik (Hornduette), Adagio (Morgensegen), Allegro Moderato, Menuetto, Allegro; Mozart concerto for Flute in

G-major (KV 313), for Oboe in C-major (KV 314), for Bassoon in B-flat Major (KV 191), String Quartet no. 19 in C-major (KV 465) - Movement 1;
- .mp3 files: viola Suite No. 1 in G-major BWV 1007, J.S. Bach, for cello solo transcribed for viola (Prelude and Allemande, [19]); Suite no 1 in G-major, BWV 1007, J.S. Bach, for cello transcribed for double bass; Clarinet Concerto in A-major, KV622, W.A. Mozart - Allegro and Adagio; Horn Concerto in E-flat Major, KV 495, W.A. Mozart - Allegro moderato; Concerto for Flute, Oboe, Bassoon, Strings and continuo in F-Major (Tempesta di mare), RV570, A. Vivaldi; Trio For Piano, Clarinet and Viola in E-flat Major (Kegelstatt), KV498 W.A. Mozart - Menuetto; Duett-Concertino for Clarinet, Bassoon, and Strings, R. Strauss - Allegro Moderato; Serenade No. 10 for 12 Winds and Contrabass in B-flat Major (Gran partita), KV 361, W.A. Mozart - Adagio; Trio for piano, violin, cello No. 1 op. 49, F. Mendelssohn - Movement 1.

When solo and polyphonic segments are taken as training data, such a training set represents more realistic sounds, which can be encountered in all classical music recordings, including their compressed file version. This training was also performed for 20,000 and 40,000 frames data sets. The training sets consisted of 2,500 (or 5,000) single sounds, mixes, solos and polyphonic recordings as both positive and negative examples.

3.2   Testing Data

Testing was performed on RWC Classical Music Database recordings [9], and for presentation purposes the first minute of each investigated piece was used. The following pieces were used:

- No. 1, F.J. Haydn, Symphony no.94 in G major, Hob.I-94 'The Surprise'. 1st mvmt., with the following instruments playing in the first minute of the recording: flute, oboe, bassoon, French horn, violin, viola, cello, double bass;
- No. 2, W.A. Mozart, Symphony no.40 in G minor, K.550. 1st mvmt.; instruments playing in the first minute: flute, oboe, bassoon, French horn, violin, viola, cello, double bass;
- No. 16, W.A. Mozart, Clarinet Quintet in A major, K.581. 1st mvmt.; instruments playing in the first minute: clarinet, violin, viola, cello;
- No. 18, J. Brahms, Horn Trio in Eb major, op.40. 2nd mvmt.; instruments playing in the first minute: piano, French horn, violin;
- No. 44, N. Rimsky-Korsakov, The Flight of the Bumble Bee; flute and piano.

These pieces represent various polyphony and pose diverse difficulties for the classifier, including short sounds, and multiple instruments playing at the same time. No training data were used in our tests. Testing was performed on 40-ms frames (with 10-ms hop size, i.e. with overlap, to marginalize errors on note edges), but labeling the ground truth data for such a frame is a tedious and difficult task, so labeling was done on 0.5 s segments. For each instrument, the outputs of the classifier over all frames within the 0.5-s segment were averaged, and if the result exceeded the 50 % threshold, the output of our binary RF classifier was considered to be "yes", meaning that the target instrument is playing in this 0.5-s segment. This way we also adjust the granularity of the instrument identification to sound segments of more perceivable length.

4 Feature Set

Audio data are usually parameterized before classification is applied, since raw data representing amplitude changes vs. time undergo dramatic changes in a fraction of second, and the amount of the data is overwhelming. The identification of musical instruments in audio data depends on the sound parametrization applied, and there is no one feature set used worldwide; each research group utilizes a different feature set. Still, some features are commonly used, and our feature set is also based on these features. We decided to utilize a feature set which proved successful in our previous, similar research [18]. Our parametrization is performed for 40-ms frames of audio data. No data segmentation or pitch extraction are needed, thus multi-pitch extraction is avoided, and no labeling particular sounds in polyphonic recording with the appropriate pitches is needed. The feature vector consists of basic features, describing properties of an audio frame of 40 ms, and additionally difference features, calculated as the difference between the feature calculated for a 30 ms sub-frame starting from the beginning of the frame, and a 30 ms sub-frame starting with 10 ms offset. Fourier transform was used to calculate spectral features, with Hamming window. Most of the features applied represent MPEG-7 low-level audio descriptors, often used in audio research [13]. Identification of instruments is performed frame by frame, for consequent frames, with 10 ms hop size. Final classification result is calculated as an average of classifier output over 0.5-s segment of the recording, in order to avoid tedious labeling of ground-truth data over shorter frames.

The feature vector we applied consists of the following 91 parameters [18]:

- *Audio Spectrum Flatness*, $flat_1, \ldots, flat_{25}$ — a multidimensional parameter describing the flatness property of the power spectrum within a frequency bin for selected bins; 25 out of 32 frequency bands were used;
- *Audio Spectrum Centroid* — the power weighted average of the frequency bins in the power spectrum; coefficients are scaled to an octave scale anchored at 1 kHz [13];
- *Audio Spectrum Spread* — RMS (root mean square) value of the deviation of the log frequency power spectrum wrt. *Audio Spectrum Centroid* [13];
- *Energy* — energy (in log scale) of the spectrum of the parametrized sound;
- *MFCC* — a vector of 13 mel frequency cepstral coefficients. The cepstrum was calculated as the logarithm of the magnitude of the spectral coefficients, and then transformed to the mel scale, to better reflect properties of the human perception of frequency. 24 mel filters were applied, and the obtained results were transformed to 12 coefficients. The 13th coefficient is the 0-order coefficient of MFCC, corresponding to the logarithm of the energy [27];
- *Zero Crossing Rate*; a zero-crossing is a point where the sign of the time-domain representation of the sound wave changes;

254 E. Kubera and A.A. Wieczorkowska

- *Roll Off* — the frequency below which an experimentally chosen percentage equal to 85 % of the accumulated magnitudes of the spectrum is concentrated; parameter originating from speech recognition, where it is applied to distinguish between voiced and unvoiced speech;
- *NonMPEG7 - Audio Spectrum Centroid* — a linear scale version of *Audio Spectrum Centroid*;
- *NonMPEG7 - Audio Spectrum Spread* — a linear scale version of *Audio Spectrum Spread*;
- changes (measured as differences) of the above features for a 30 ms sub-frame of the given 40 ms frame (starting from the beginning of this frame) and the next 30 ms sub-frame (starting with 10 ms shift), calculated for all the features shown above;
- *Flux* — the sum of squared differences between the magnitudes of the DFT points calculated for the starting and ending 30 ms sub-frames within the main 40 ms frame; this feature by definition describes changes of magnitude spectrum, thus it is not calculated in a static version.

Audio data were in mono or stereo format; mixes of the left and right channel (i.e. the average value of samples in both channels) were taken if the audio signal was in stereo format.

5 Experiments and Results

The experiments aimed at investigating how many instruments can be identified correctly in real polyphonic recordings, and whether adding real recordings representing solos and polyphonic recordings of the same style of music, i.e. classical music (rather than isolated single sounds and their mixes), can improve the performance of the classifier. The main problem with such classification is the recall, which is usually quite low, i.e. instruments in recordings are missed by classifiers. Another problem is how to assess the results, since many instruments can be playing in the same segment. As mentioned before, the classification was performed using RFs, since they proved quite successful in our previous research [18], and outperformed SVM (considered state of the art and comparing well with the others). Since we deal with binary classifiers, possible errors include false negatives (missed target instrument) and false positives (false indication of the target instrument, not playing in a given segment). The details of classification results for both training versions, $T1$ and $T2$ are shown in Tables 2 and 3 for 20,000 frames training, and in Tables 4 and 5 for 40,000 frames training. Precision, recall, f-measure and accuracy were calculated as follows, on the basis of true positives (TP), true negatives (TN), false positives (FP) and false negatives (FN):

- precision pr was calculated as $pr = TP/(TP + FP)$,
- recall rec was calculated as $rec = TP/(TP + FN)$,
- f-measure f_{meas} was calculated as $f_{meas} = 2 \cdot pr \cdot rec/(pr + rec)$,
- accuracy acc was calculated as $acc = (TP + TN)/(TP + TN + FP + FN)$.

Table 2. Results of the recognition of musical instruments in the selected RWC Classical recordings, for training on single sounds and mixes ($T1$), for 20,000 training frames

Result	bn	ob	cl	fl	fh	pn	ce	va	vn	db	Average
TP	38	31	10	88	19	96	178	110	250	107	
FP	64	9	9	2	94	100	162	77	04	083	
FN	78	70	32	78	143	70	43	122	124	45	
TN	420	490	549	432	344	334	217	291	222	365	
Precision	37 %	78 %	53 %	98 %	17 %	49 %	52 %	59 %	98 %	56 %	60 %
Recall	33 %	31 %	24 %	53 %	12 %	58 %	81 %	47 %	67 %	70 %	48 %
F-measure	35 %	44 %	33 %	69 %	14 %	53 %	63 %	53 %	80 %	63 %	51 %
Accuracy	76 %	87 %	93 %	87 %	61 %	72 %	66 %	67 %	79 %	79 %	77 %

Table 3. Results of the recognition of musical instruments in the selected RWC Classical recordings, for training on single sounds and mixes, and also sounds from real recordings, representing solos and polyphonic segments ($T2$), for 20,000 training frames

Result	bn	ob	cl	fl	fh	pn	ce	va	vn	db	Average
TP	61	34	23	108	40	146	186	188	292	151	
FP	98	31	282	49	140	187	95	140	9	173	
FN	55	67	19	58	122	20	35	44	82	1	
TN	386	468	276	385	298	247	284	228	217	275	
Precision	38 %	52 %	8 %	69 %	22 %	44 %	66 %	57 %	97 %	47 %	50 %
Recall	53 %	34 %	55 %	65 %	25 %	88 %	84 %	81 %	78 %	99 %	66 %
F-measure	44 %	41 %	13 %	67 %	23 %	59 %	74 %	67 %	87 %	63 %	54 %
Accuracy	75 %	84 %	50 %	82 %	56 %	66 %	78 %	69 %	85 %	71 %	72 %

Table 4. Results of the recognition of musical instruments in the selected RWC Classical recordings, for training on single sounds and mixes ($T1$), for 40,000 training frames

Result	bn	ob	cl	fl	fh	pn	ce	va	vn	db	Average
TP	30	30	11	82	17	104	184	93	261	103	
FP	53	6	6	1	82	93	143	60	2	78	
FN	86	71	31	84	145	62	37	139	113	49	
TN	431	493	552	433	356	341	236	308	224	370	
Precision	36 %	83 %	65 %	99 %	17 %	53 %	56 %	61 %	99 %	57 %	63 %
Recall	26 %	30 %	26 %	49 %	10 %	63 %	83 %	40 %	70 %	68 %	47 %
F-measure	30 %	44 %	37 %	66 %	13 %	57 %	67 %	48 %	82 %	62 %	51 %
Accuracy	77 %	87 %	94 %	86 %	62 %	74 %	70 %	67 %	81 %	79 %	78 %

If the denominator in the formula for calculating precision or recall is equal to zero, then these measures are undefined.

As we can see, the classifier built for the training on single sounds and mixes ($T1$) gives quite low recall, but using real unsegmented recordings for training ($T2$) improves the recall significantly, no matter the size of the training set. Also,

Table 5. Results of the recognition of musical instruments in the selected RWC Classical recordings, for training on single sounds and mixes, and also sounds from real recordings, representing solos and polyphonic segments (*T*2), for 40,000 training frames

Result	bn	ob	cl	fl	fh	pn	ce	va	vn	db	Average
TP	65	32	20	115	39	144	191	190	294	151	
FP	87	28	260	39	144	164	99	136	7	168	
FN	51	69	22	51	123	22	30	42	80	1	
TN	397	471	298	395	294	270	280	232	219	280	
Precision	43%	53%	7%	75%	21%	47%	66%	58%	98%	47%	52%
Recall	56%	32%	48%	69%	24%	87%	86%	82%	79%	99%	66%
F-measure	49%	40%	12%	72%	23%	61%	75%	68%	87%	64%	55%
Accuracy	77%	84%	53%	85%	56%	69%	79%	70%	86%	72%	73%

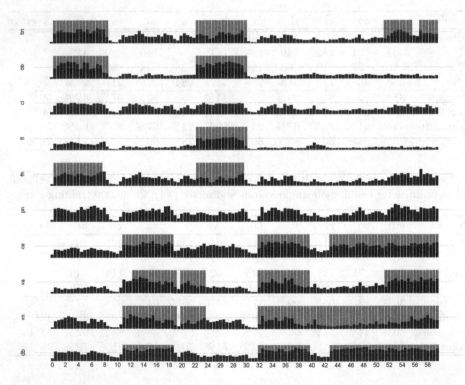

Fig. 1. Outcome of each random forest for the RWC Classical Music No. 1, for each 0.5-s segment of the first minute of the recording, for the training *T*1 on 40,000 training frames. If the result for a forest (trained to recognize a target instrument) is 0.5 or more, then this classifier indicates that the target instrument is playing in this segment. Ground-truth data are marked in grey.

we can observe that adding more samples to the training set (i.e. increasing the size from 20,000 to 40,000 samples) does not change results much. The accuracy compares favorably with [6], since they obtained about 60 % for duets from 5-instrument set, and with [16], as in their experiments the accuracy dropped below 40 % in real case scenario for 5 instruments, whereas we obtained above 70 % for 10 instruments and sometimes even higher polyphony.

For some instruments, adding real recordings into the training set improves the results - for example, bassoon, French horn and cello are better recognized, both precision and recall are improved. In some cases, precision decreases when real recordings are added to training data, for example in the case of oboe, clarinet, flute, and double bass. In the case of clarinet, the precision drops dramatically. However, the results indicated by the RF are just below the 0.5 threshold before adding real recordings to the training data, and after that the results shift just above this threshold (at least this is the case for the first minute of RWC Classical Music No. 1, see Figs. 1 and 2). Therefore, we think the results can be improved by playing with this threshold. If the $T1$-trained classifier shows

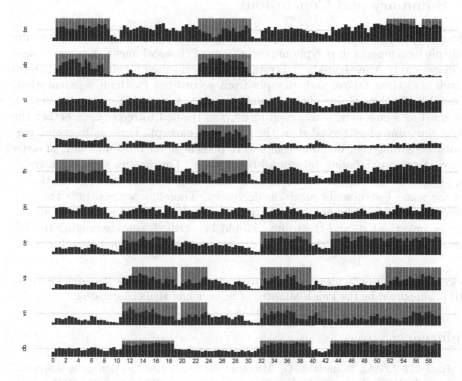

Fig. 2. Outcome of each random forest for the RWC Classical Music No. 1, for each 0.5-s segment of the first minute of the recording, for the training $T2$ on 40,000 training frames. If the result for a forest (trained to recognize a target instrument) is 0.5 or more, then this classifier indicates that the target instrument is playing in this segment. Ground-truth data are marked in grey.

positive outcome, the indication is just above 0.5. The outcomes for the $T2$-trained classifier are much higher, although there are errors, especially for string instruments. However, this piece is difficult for recognition as an orchestral piece of high polyphony, so errors in this case were rather unavoidable. Also, the sound level of particular instruments can be much lower than the other, predominant instruments, and in such a case (especially in high polyphony) identification of all instruments playing together is indeed very difficult.

For all instruments, the recall improves after adding real recordings to training and overall precision decreases. This can be seen as the usual trade-off between precision and recall, but since F-measure improves, we can conclude that adding real recordings to training slightly improves instrument identification, but the choice of the training data can depend on the task in hand – whether precision or recall is more important. We should also keep in mind that working with real recordings is a tedious task, because it requires labeling, prone to errors. This is why mixes of single sounds are so readily used in this research.

6   Summary and Conclusions

The research presented in this paper aimed at the difficult task of recognizing multiple instruments in polyphonic recordings of classical music. Ten instrumental classes were investigated. The training was performed for single instrumental sounds and their mixes, and excerpts from recordings (without segmentation) were added in the training of the second classifier. A set of binary random forests was used as a classifier, where each forest was trained to recognize whether the target instrument is recorded in the analyzed excerpt. Forty-millisecond segments were analyzed, but the results were presented for 0.5-s segment, in order to avoid tedious labeling of ground-truth data. The results show that training performed on unsegmented real recordings improves the recall dramatically, but for some instruments precision decreases. Therefore we conclude that the training data should be adjusted to the target in hand, RF technique should be further investigated, and the results should be verified, since averaging the RF output for all the frames through 0.5-s segment may deteriorate overall results.

Acknowledgments. This project was partially supported by the Research Center of PJIIT, supported by the Polish Ministry of Science and Higher Education.

References

1. Barbedo, J.G.A., Tzanetakis, G.: Musical instrument classification using individual partials. IEEE Trans. Audio Speech Lang. Process. **19**(1), 111–122 (2011)
2. Benetos, E., Dixon, S., Giannoulis, D., Kirchhoff, H., Klapuri, A.: Automatic music transcription: breaking the glass ceiling. In: 13th International Society for Music Information Retrieval Conference (ISMIR), pp. 379–384 (2012)

3. Bosch, J.J., Janer, J., Fuhrmann, F., Herrera, P.: A comparison of sound segrega-
   tion techniques for predominant instrument recognition in musical audio signals.
   In: 13th International Society for Music Information Retrieval Conference (ISMIR),
   pp. 559–564 (2012)
4. Breiman, L.: Random forests. Mach. Learn. **45**, 5–32 (2001)
5. Cont, A., Dubnov, S., Wessel, D.: Realtime multiple-pitch and multiple-instrument
   recognition for music signals using sparse non-negativity constraints. In: Proceedings of the 10th International Conference on Digital Audio Effects (DAFx-07), pp.
   85–92 (2007)
6. Eggink, J., Brown, G.J.: Application of missing feature theory to the recognition
   of musical instruments in polyphonic audio. In: 4th International Conference on
   Music Information Retrieval ISMIR (2003)
7. Essid, S., Richard, G., David, B.: Instrument recognition in polyphonic music based
   on automatic taxonomies. IEEE Trans. Audio Speech Lang. Process. **14**(1), 68–80
   (2006)
8. Fuhrmann, F.: Automatic musical instrument recognition from polyphonic music
   audio signals. Ph.D. Thesis, Universitat Pompeu Fabra (2012)
9. Goto, M., Hashiguchi, H., Nishimura, T., Oka, R.: RWC music database: popular, classical, and jazz music databases. In: Proceedings of the 3rd International
   Conference on Music Information Retrieval, pp. 287–288 (2002)
10. Goto, M., Hashiguchi, H., Nishimura, T., Oka, R.: RWC music database: music
    genre database and musical instrument sound database. In: 4th International Con-
    ference on Music Information Retrieval ISMIR, pp. 229–230 (2003)
11. Heittola, T., Klapuri, A., Virtanen, A.: Musical instrument recognition in poly-
    phonic audio using source-filter model for sound separation. In: Proceedings of
    the 10th International Society for Music Information Retrieval Conference (ISMIR
    2009) (2009)
12. Herrera-Boyer, P., Klapuri, A., Davy, M.: Automatic classification of pitched musi-
    cal instrument sounds. In: Klapuri, A., Davy, M. (eds.) Signal Processing Methods
    for Music Transcription. Springer Science+Business Media LLC, New York (2006)
13. ISO: MPEG-7 Overview. http://www.chiariglione.org/mpeg/
14. Jiang, W., Wieczorkowska, A., Raś, Z.W.: Music instrument estimation in poly-
    phonic sound based on short-term spectrum match. In: Hassanien, A.-E., Abraham,
    A., Herrera, F. (eds.) Foundations of Computational Intelligence Volume 2. SCI,
    vol. 202, pp. 259–273. Springer, Heidelberg (2009)
15. Kashino, K., Murase, H.: A sound source identification system for ensemble music
    based on template adaptation and music stream extraction. Speech Commun. **27**,
    337–349 (1999)
16. Kirchhoff, H., Dixon, S., Klapuri, A.: Multi-template shift-variant non-negative
    matrix deconvolution for semi-automatic music transcription. In: 13th Interna-
    tional Society for Music Information Retrieval Conference (ISMIR), pp. 415–420
    (2012)
17. Kitahara, T., Goto, M., Komatani, K., Ogata, T., Okuno, H.G.: Instrument iden-
    tification in polyphonic music: feature weighting to minimize influence of sound
    overlaps. EURASIP J. Appl. Signal Process. **2007**, 1–15 (2007)
18. Kubera, E., Kursa, M.B., Rudnicki, W.R., Rudnicki, R., Wieczorkowska, A.A.: All
    that jazz in the random forest. In: Kryszkiewicz, M., Rybinski, H., Skowron, A.,
    Raś, Z.W. (eds.) ISMIS 2011. LNCS (LNAI), vol. 6804, pp. 543–553. Springer,
    Heidelberg (2011)
19. Kuperman, M.: Suite N 1 in G-Dur BWV 1007. http://www.viola-bach.info/

20. Kursa, M., Rudnicki, W., Wieczorkowska, A., Kubera, E., Kubik-Komar, A.: Musical instruments in random forest. In: Rauch, J., Raś, Z.W., Berka, P., Elomaa, T. (eds.) ISMIS 2009. LNCS (LNAI), vol. 5722, pp. 281–290. Springer, Heidelberg (2009)
21. Leveau, P., Vincent, E., Richard, G., Daudet, L.: Instrument-specific harmonic atoms for mid-level music representation. IEEE Trans. Audio Speech Lang. Process. 16(1), 116–128 (2008)
22. Little, D., Pardo, B.: Learning musical instruments from mixtures of audio with weak labels. In: 9th International Conference on Music Information Retrieval ISMIR (2008)
23. Martin, K.D.: Toward automatic sound source recognition: identifying musical instruments. Presented at the 1998 NATO Advanced Study Institute on Computational Hearing, Il Ciocco, Italy (1998)
24. Martins, L.G., Burred, J.J., Tzanetakis, G., Lagrange, M.: Polyphonic instrument recognition using spectral clustering. In: 8th International Conference on Music Information Retrieval ISMIR (2007)
25. MIDOMI: Search for Music Using Your Voice by Singing or Humming. http://www.midomi.com/
26. Müller, M., Ellis, D., Klapuri, A., Richard, G.: Signal processing for music analysis. IEEE JSTSP 5(6), 1088–1110 (2011)
27. Niewiadomy, D., Pelikant, A.: Implementation of MFCC vector generation in classification context. J. Appl. Comput. Sci. 16(2), 55–65 (2008)
28. Opolko, F., Wapnick, J.: MUMS – McGill University Master Samples. CD's (1987)
29. Raś, Z.W., Wieczorkowska, A.A. (eds.): Advances in Music Information Retrieval. SCI, vol. 274. Springer, Heidelberg (2010)
30. Richards, G., Wang, W.: What influences the accuracy of decision tree ensembles? J. Intell. Inf. Syst. 39, 627–650 (2012)
31. Shazam Entertainment Ltd., http://www.shazam.com/
32. Shen, J., Shepherd, J., Cui, B., Liu, L. (eds.): Intelligent Music Information Systems: Tools and Methodologies. Information Science Reference, Hershey (2008)
33. The University of IOWA Electronic Music Studios: Musical Instrument Samples. http://theremin.music.uiowa.edu/MIS.html
34. TrackID – Sony Smartphones. http://www.sonymobile.com/global-en/support/faq/xperia-x8/internet-connections-applications/trackid-ps104/
35. Vincent, E., Rodet, X.: Music transcription with ISA and HMM. In: Puntonet, C.G., Prieto, A.G. (eds.) ICA 2004. LNCS, vol. 3195, pp. 1197–1204. Springer, Heidelberg (2004)

Author Index

Printed in the United States
By Bookmasters